Wheat - Recent Advances

Edited by Mahmood-ur-Rahman Ansari

Published in London, United Kingdom

Wheat - Recent Advances
http://dx.doi.org/10.5772/intechopen.98157
Edited by Mahmood-ur-Rahman Ansari

Contributors
Timothy J. Tse, Farley Chicilo, Daniel J. J. Wiens, Martin J.T. Reaney, Ihsan Elahi, Hira Nawaz, Uzair Saeed, Abdul Wadood, Amjad Abbas, Sohail Jabbar, Mukaddes Kayim, Abdulkreem Alsalmo, Muhammad Javid Iqbal, Kalsoom Fatima, Naureen Shams, James T. Tambong, Habtamu Kefale, Sewnet Getahun, Graham Brodie, Mohammed Humayun Kabir, Dorin Gupta, Alexis Pang, Sundip Kumar, Dinesh Kumar Saini, Neeraj Pal, Nilüfer Akci, Ruth Otinga Wanyera, Mercy Wamalwa, Om Prakash Raigar, Mohini Prabha Singh, Mehak Sethi, Kinjal Mondal, Bhallan Singh Sekhon, Archana Mukta, Priyanka, Jasneet Singh, Zahid Manzoor, Muhammad Sheeraz Qadir, Junwei Liu, Muhammad Ahsan Jamil, Zeshan Hassan, Muhammad Shah Jahan, Amir Shakeel, Saba Akram, Maria Ghaffar, Ayesha Wadood, Mian Abdur Rehman Arif, Ashwani Kumar, Hari Kesh, Sunita Devi, Anita Mann, Naresh Kumar, Arvind Kumar, Pooja Dhansu, Parvender Sheoran, Muhammad Amjad Bashir, Muhammad Arshad Hussain, Sadia Afzal, Adeela Haroon, Asad-Ur-Rehman Chaudary, Sagheer Atta, Saqib Bashir, Muhammad Adnan Bodlah, Mahmood-ur-Rahman Ansari, Usman Ali Ashfaq, Abdul Manan

First published in London, United Kingdom, 2022 by IntechOpen
IntechOpen is the global imprint of INTECHOPEN LIMITED, registered in England and Wales, registration number: 11086078, 5 Princes Gate Court, London, SW7 2QJ, United Kingdom

British Library Cataloguing-in-Publication Data
A catalogue record for this book is available from the British Library

Additional hard and PDF copies can be obtained from orders@intechopen.com

Wheat - Recent Advances
Edited by Mahmood-ur-Rahman Ansari
p. cm.
Print ISBN 978-1-80355-522-5
Online ISBN 978-1-80355-523-2
eBook (PDF) ISBN 978-1-80355-524-9

For EU product safety concerns:
IN TECH d.o.o., Prolaz Marije Krucifikse Kozulić 3, 51000 Rijeka, Croatia, info@intechopen.com or visit our website at intechopen.com.

Meet the editor

Mahmood-ur-Rahman Ansari, Ph.D., is an Associate Professor of Molecular Biology at the Department of Bioinformatics and Biotechnology, GC University – Faisalabad, Pakistan. He received his BSc (Hons) in Plant Breeding and Genetics from the University of Agriculture, Faisalabad, Pakistan, in 2003. He obtained an MPhil and Ph.D. in Molecular Biology from the National Centre of Excellence in Molecular Biology, Lahore, Pakistan, in 2006 and 2011, respectively. He has published more than eighty papers in international peer-reviewed journals in the fields of molecular biology, biotechnology, and bioinformatics. He has also published eighteen book chapters and edited five books. His research interest is to understand the molecular mechanisms of stress tolerance in plants. He has been involved in the genetic modification of rice and cotton, their greenhouse and field testing, as well as biosafety studies. He is a member of various national and international professional societies.

Contents

Preface

Wheat is an important food crop grown on 30% of land occupied by cereal crops worldwide. Currently, it feeds about 40% of the global population. The production of wheat is affected by various factors including drought, salinity, water logging, and trends in urbanization. This book discusses strategies to increase wheat yield under adverse conditions. Developing biotic and abiotic stress-tolerant wheat plants may lead to improved production. This book discusses diverse aspects of wheat. It contains five sections.

Section 1 includes an introductory chapter that presents an overview of recent developments for the improvement of wheat. Section 2 describes various diseases of wheat as well as its insect pests. It also describes strategies to develop better plants with biotic stress tolerance. Section 3 presents the latest information on various approaches to developing abiotic stress-tolerant wheat plants. The chapters in this section contain advanced research accomplishments in the field related to drought and salinity. Climate change is the special focus of this section. Section 4 contains chapters on modern and innovative techniques in wheat improvement. The chapters discuss technologies like genome editing and genomics. Section 5 focuses on grain quality and value-added products obtained from the wheat crop.

I would like to thank the staff at IntechOpen, particularly Author Service Manager Ms. Zrinka Tomicic for her valuable help throughout the editing process. I am also grateful to my research student Mr. Abdul Manan for his assistance in preparing the introductory chapter. I am especially thankful to all the authors for their valuable contributions to this book.

Mahmood-ur-Rahman Ansari, Ph.D.
Department of Bioinformatics and Biotechnology,
Government College University,
Faisalabad, Pakistan

Section 1

Introduction

Chapter 1

Introductory Chapter: Advancements in Wheat Research

Abdul Manan, Usman Ali Ashfaq
and Mahmood-ur-Rahman Ansari

1. Introduction

Wheat is the most common food crop we grow. In 2018, it was grown on 214 million hectares, which is approximately 30% of the total land area sown to cereal crops. With an average yield of about 3.4 tonnes per hectare, 734 million tonnes of wheat were produced in 2018 [1]. Wheat's role in human nutrition is another way to show how important it is. Approximately 20% of our protein intake and 20% of our carbohydrate intake come from wheat. Also, wheat represents our most traded cereal [2]. In 2018, over 118 mt of wheat was exported, which was 40% of all cereal exports. Australia, which is a major wheat producer, has average yields of less than 2 tons/ha, whereas yields in the UK are normally somewhere around 7 tons/ha and 8 tons/ha and are well more than 10 tons/ha in several zones [3]. Early in 2020, New Zealand had the maximum wheat yield ever observed: 17.4 t/ha. In many other places, it is hard to get more than 1 t/ha. The wide range of wheat yields shows how different the places where wheat is grown [4]. Due to variable environments, breeding programs usually concentrate on specific target areas. Breeding of wheat has been done mostly by the general public around the world, with help from governmental bodies or farmer groups. However, this has gradually evolved as profit incentive in breeding has grown, particularly in the EU and countries like Australia in which there is a clear way to make money from the benefits of improved varieties [5].

2. Recent developments in wheat for drought stress tolerance

Early on, the breeders realized that making wheat mature at right time for growing season was the most important adaptation trait that had to do with yield. The most important thing that affects yield, or more accurately, yield potential, is how much water is available. To get the most out of the crop, breeders look for patterns that mature and develop throughout the growth period. There is usually, although not always a proper relationship between both grain yield and biomass, growers try to time the development of the plant to coincide with when there is enough water. In cool climates, planting in the fall gives plants time to grow roots before the cold weather of winter, and then they grow quickly in the spring and early summer. If the winters are too cold and there is a chance of freezing damage, growers need to select varieties that can be planted after the danger has receded. They should also endeavor to extend the planting season as late as possible before drought and heat stress slow growth. This is different from hot

IntechOpen

season crops, whose growing season is cut short by killing frosts in temperate zones [6]. If there is enough irrigation water, the season can be extended into the summer. If there is not enough fertigation water, early maturing lines are needed. In Mediterranean-type climates, where it rains in the winter but is hot and dry in the summer, biomass can be built by planting in the fall and letting the plants grow over the winter. However, the plants need to be mature early enough to be harvested before every dry season.

It is well known how important it is to match maturity to the growing season, and so this attribute is usually well organized in existing projects by using genes and loci for earliness, vernalization, and photoperiod response. There might still be ways to change the various growth stages to better match the environment, but generally, the way elite varieties grow and develop makes sure that the crop could indeed take benefit of the times when there is enough water and then flower and have full seed before the end of the season. Even though adjusting development to the atmosphere has been important for increasing the yield of wheat, complications stem during unusual stages when the growth trajectory of elite varieties does not match the patterns of rainfall and temperature [7]. This problem is getting worse because the weather is becoming more unpredictable. Farmers know that some periods will be severe and that they may take a loss. However, good years can make up for this. The fact that bad years are happening more and more often is a big problem, and farmers are looking for varieties that will do well in good years but cause less damage in bad ones. Breeders try to solve this problem by looking for ways to use water more efficiently and reduce the effects of things that might make yield stability less stable.

3. Recent developments in wheat for high temperature tolerance

Wheat production is in danger from many environmental factors. For example, the last 10 years (2010–2019) were the warmest on record, and the steady temperature rise is thought to have caused many changes in the way the climate system works [8]. In its Fifth Assessment Report, the Intergovernmental Panel on Climate Change said that by 2050, the average global temperature could rise by 2–5°C, or even more, and rain trends are likely to become less consistent [9]. "High confidence" says that these changes, like more frequent extreme events, are having an effect on food security. Food insecurity has effects that reach far and wide. Hagel, who used to be the head of the US Department of Defense, said that changes in climate "can add a lot to the problems of global instability, hunger, poverty, and war." In 2018, climate was found to be a cause of "crisis-level acute food insecurity" in 26 of the 33 countries where it happened [10]. Since Russia, India, France, China, and the United States produce 50% of the world's wheat, "any weather shock or external shock to production in these countries will have an immediate effect on global prices and price volatility." Improving how efficiently food is made is seen as a key way to make food production more sustainable in the future. About half of the world's wheat is affected by excessive heat, and 20 million ha or more often have too little water [11].

Models show that there is a chance that crops in global "breadbaskets" could fail at the same time because of heat or drought, and variation in rainfall and temperature (including drought) are indeed blamed for 40% of the variation in the production of wheat from 1 year to the next. By the end of this century, severe water shortages are predicted in approximately 60% of the world's cereal growing regions, and each 1°C temperature rise is expected to decrease yield by a mean of 6% [12]. A few other research and modeling research shows that increasing the level of CO_2 in the atmosphere at least to some extent counteract the harmful effects of drought and heat stress, but the data are not constant [13]. Also, the models do not take into account the harmful effects of rising night-time temperatures, thermal shocks, unsteady rain patterns, and dietary components, which are not helped by higher CO_2 levels.

4. Recent developments in wheat for quality improvement

Wheat can be used for many different things, and each of these needs different qualities. The most important is GPC (grain protein content), which is a key factor how well a grain makes bread and pasta. Farmers usually get more money for grains that are high in protein. Content of protein is the most important quality trait, but the type of grain protein and a number of other qualities, such as grain hardness, also affect how the grain will be used [14]. There are several ways that heat and drought stress can change the quality. Extreme stress can stop starch from forming, which makes the grain smaller. Even mild stress can change the balance between gliadin and glutenin proteins, which changes the quality of the grain. But heat and drought have the most important effect on quality through their relationship to GPC. GPC is a complicated trait that is strongly affected by the environment but also has a clear genetic part. Breeders often choose plants based on how much protein they have since this is such a big part of how much the grain is worth [15]. Not surprisingly, the most important environmental factor that affects GPC is the amount of N fertilizer applied, but there is also a negative relationship between GCP and yield in many environments. GPD stands for "grain protein deviation," which is a way that varieties can be different from the relationship between yield and GPC [16]. It has been suggested that breeders could use GPD as a selection target to get around or lessen the effect of the negative relationship. In a study done in Australia, the negative correlation was found to be the strongest in low-yielding environments that are stressed by heat and drought. Breeders in Australia seem to have chosen high-protein genotypes that do well in low-yield environments based on how well these genotypes limit biomass accumulation to save nitrogen for grain filling. Analysis of a large number of field trials in different environments has shown that breeders need to select under drought stress and limited N supply to maximize both yield and GCP in new varieties, and studies in Europe have shown that selecting based on GPD is also important [17].

5. Future prospects

Way forward in "Climate Change" conditions is to produce wheat cultivars having multiple favorable traits which may better mitigate the crop in stress environment. The wheat plants may be stronger to tolerate various biotic as well as abiotic stresses along with better nutritional value, as food security and safety are both equally important. What we need now is to produce wheat plants having temperature, drought, salinity tolerance as well as more protein contents so that plant may be grown healthy under adverse environments along with more nutrition in terms of protein, carbohydrates, etc. The breeders are highly recommended to include all possible factors contributing toward climate change in their breeding programs while developing new wheat cultivars.

Author details

Abdul Manan, Usman Ali Ashfaq* and Mahmood-ur-Rahman Ansari
Department of Bioinformatics and Biotechnology, G.C. University, Faisalabad, Pakistan

*Address all correspondence to: ashfaqua@gcuf.edu.pk

References

[1] Langridge P, Reynolds M. Breeding for drought and heat tolerance in wheat. Theoretical and Applied Genetics. 2021;**134**(6):1753-1769

[2] Brouns F, van Rooy G, Shewry P, Rustgi S, Jonkers D. Adverse reactions to wheat or wheat components. Comprehensive Reviews in Food Science and Food Safety. 2019;**18**(5):1437-1452

[3] Helman D, Bonfil DJ. Six decades of warming and drought in the world's top wheat-producing countries offset the benefits of rising CO2 to yield. Scientific Reports. 2022;**12**(1):7921. DOI: 10.1038/s41598-022-11423-1

[4] Ladha JK, Jat ML, Stirling CM, Chakraborty D, Pradhan P, Krupnik TJ, et al. Achieving the sustainable development goals in agriculture: The crucial role of nitrogen in cereal-based systems. Advances in Agronomy. 2020;**163**:39-116

[5] Fatima Z, Ahmed M, Hussain M, Abbas G, Ul-Allah S, Ahmad S, et al. The fingerprints of climate warming on cereal crops phenology and adaptation options. Scientific Reports. 2020;**10**(1):1-21

[6] Lippmann R, Babben S, Menger A, Delker C, Quint M. Development of wild and cultivated plants under global warming conditions. Current Biology. 2019;**29**(24):R1326-R1338

[7] Reynolds M, Chapman S, Crespo-Herrera L, Molero G, Mondal S, Pequeno DN, et al. Breeder friendly phenotyping. Plant Science. 2020;**295**:110396

[8] Obembe OS, Hendricks NP, Tack J. Decreased wheat production in the USA from climate change driven by yield losses rather than crop abandonment. PLoS One. 2021;**16**(6):e0252067. DOI: 10.1371/journal.pone.0252067

[9] IPPC. Global Warming of 1.5 °C. 2020. Available from: https://www.ipcc.ch/sr15/

[10] Giraldo P, Benavente E, Manzano-Agugliaro F, Gimenez E. Worldwide research trends on wheat and barley: A bibliometric comparative analysis. Agronomy. 2019;**9**(7):352

[11] Tyagi M, Pandey G. Physiology of heat and drought tolerance in wheat: An overview. Journal of Cereal Research. 2022;**14**(1):13-25. DOI: 10.25174/2582-2675/2022/122868

[12] Gezie M. Farmer's response to climate change and variability in Ethiopia: A review. Cogent Food & Agriculture. 2019;**5**(1):1613770

[13] De Kauwe MG, Medlyn BE, Tissue DT. To what extent can rising [CO2] ameliorate plant drought stress? The New Phytologist. 2021;**231**(6):2118-2124. DOI: 10.1111/nph.17540

[14] De Santis MA, Soccio M, Laus MN, Flagella Z. Influence of drought and salt stress on durum wheat grain quality and composition: A review. Plants. 2021;**10**(12):2599

[15] Rebetzke G, Jimenez-Berni J, Fischer R, Deery D, Smith D. High-throughput phenotyping to enhance the use of crop genetic resources. Plant Science. 2019;**282**:40-48

[16] Fradgley NS, Bentley AR, Swarbreck SM. Defining the physiological determinants of low nitrogen requirement in wheat. Biochemical Society Transactions. 2021;**49**(2):609-616

[17] Hawkesford MJ, Griffiths S. Exploiting genetic variation in nitrogen use efficiency for cereal crop improvement. Current Opinion in Plant Biology. 2019;**49**:35-42

Chapter 2

Breaking Yield Ceiling in Wheat: Progress and Future Prospects

Neeraj Pal, Dinesh Kumar Saini and Sundip Kumar

Abstract

Wheat is one of the most important staple crops that contribute considerably to global food and nutritional security. The future projections of the demand for wheat show significant enhancement owing to the population growth and probable changes in diets. Further, historical yield trends show a reduction in the relative rate of gain for grain yield over time. To maintain future food security, there is a strong need to find ways to further increase the yield potential of wheat. Grain yield is a quantitative trait that is highly influenced by the environment. It is determined by various inter-linked yield component traits. Molecular breeding approaches have already proven useful in improving the grain yield of wheat and recent advances in high-throughput genotyping platforms now have remodelled molecular breeding to genomics-assisted breeding. Hence, here in this chapter, we have discussed various advancements in understanding the genetics of grain yield, its major components, and summarised the various powerful strategies, such as gene cloning, mining superior alleles, transgenic technologies, advanced genome editing techniques, genomic selection, genome-wide association studies-assisted genomic selection, haplotype-based breeding (HBB), which may be/being used for grain yield improvement in wheat and as the new breeding strategies they could also be utilised to break the yield ceiling in wheat.

Keywords: wheat, grain yield, genomics resources, molecular breeding, genomics-assisted breeding

1. Introduction

Wheat (*Triticum aestivum* L.) is the most extensively grown food crop around the world and ranks second after rice [1]. China is the top wheat-growing country which recently in 2020, produced 134,250 thousand tonnes of wheat accounting for approximately 20.66% of the total wheat production around the globe. The top five wheat-growing countries (China, India, Russian Federation, United States of America, and Canada) together account for 63.46% of the world's wheat production (6,499,759 thousand tonnes in 2020) [2]. It accounts for more than 20% of the calorific intake of humans and supplies more protein (approximately 23%) than all animal sources [1]. The progress for the genetic improvement of grain yield in wheat ranged from 0.3% to 1.0% per year during the last century [3]. Nevertheless, it has been decreased in recent years, largely due to the narrow genetic base of the germplasm used for the development of new genotypes and the lack of adoption of novel breeding strategies.

Noticeably, there is a need to increase wheat yield to feed the world population which may be increased from the current 7.5 billion to more than 9 billion by 2050, and this is with the unusual constraints posed by climate change. Under such kind of pressure, wheat breeding programs need to do more to achieve the targeted genetic gain in grain yield ensuring food security in the near future. Many studies have shown that increases in the harvest index (HI), grain weight (GW), grain number per spike (GNPS), and decreases in plant height (PH) are the major traits associated with genetic gain in wheat [4, 5]. Improvement in HI has permitted better partitioning of photosynthetic assimilates to the developing grains, resulting in greater grain yield (GY). The HI of cultivated wheat varieties generally ranges from 0.4 to 0.5 which is already close to the theoretical maximum value of 0.62 [6, 7]. Furthermore, HI values more than 0.5 are very hard to achieve, particularly in unfavourable environmental conditions [8]. This situation again shows that genetic progress in wheat breeding programs may be difficult. Therefore, understanding the changes (either increment or reduction) in yield and related traits is an essential step towards developing new breeding strategies and a further improvement in the grain yield.

Grain yield is the final result of plant growth and development and hence most, if not all, genes are supposed to contribute towards yield either indirectly or directly. Consequently, achieving increased grain yield is a non-trivial task, and the accumulative knowledge from wheat breeding suggests that we would require concurrent improvements of both the 'source' and 'sink' tissues. Traditional breeding largely depends on empirical phenotypic selection, which has already resulted in the development and release of a large number of high-yielding varieties. However, time consumption, labour intensity, environmental dependence, and low efficiency are prime barriers that nowadays hinder conventional/traditional wheat breeding. High-yielding wheat varieties can result from the uncovering of novel genetic variation, better selection techniques, or the identification of superior genotypes with novel or improved characteristics caused by favourable combinations of superior alleles at multiple loci. In recent years, an impressive number of advancements in genetics and genomics have been made in wheat. Owing to the tremendous effort of IWGSC, the 'gold standard' reference genome has become available for wheat cultivar 'Chinese Spring'. The most comprehensive assembly of this reference line has been recently released in 2018 which gave access to a total of 107,891 high-confidence genes [9]. The genome sequences may assist the identification of important genes at an unprecedented level which is a key aspect in wheat breeding. Different types of molecular markers, such as RFLP, AFLP, SCAR, STS, SSR, CAPS, and GBS-SNPs, have been identified and mapped on the different chromosomes of wheat and highly dense genetic maps have also been developed (available at https://wheat.pw.usda.gov/GG3/) which are being utilised in various genetic studies in wheat [10, 11]. To date, more than 15 different high-throughput GBS strategies have been developed and utilised in various crops including wheat [12]. Moreover, several SNP arrays/assays have also been developed which are flexible in terms of data point and sample number customization, which contributes to its high-density scanning and robust call rates compared to PCR and NGS-based markers. Several high-density SNP genotyping arrays have been utilised for genetic dissection of different traits and marker-assisted breeding in wheat namely the Illumina Wheat 9 K iSelect SNP array [13], the Illumina Wheat 90 K iSelect SNP genotyping array [14], the Wheat 15 K SNP array [15], the Wheat 55 K SNP array, the Axiom Wheat 660 K SNP array, the Axiom HD Wheat genotyping (820 K) array [16], the Wheat Breeders' 35 K Axiom array [17], and the Wheat 50 K *Triticum* TraitBreed array [18]. These advancements in genomics have greatly enhanced our understanding of structural and functional aspects of the wheat genome and contributed to wheat improvement in two ways. First, they provided a better understanding of the various biological mechanisms that have led to new or improved screening methods for identifying and selecting superior genotypes more efficiently. Secondly, this new information improved the decision-making process for more efficient breeding strategies. With these advancements, the focus of wheat breeding has gradually switched from phenotype-based to genotype-based selection. Marker-assisted selection (MAS) has improved wheat breeding efficiency to some extent and predominated in breeding programs for decades. Several MAS strategies have been developed—marker-assisted backcrossing (MABC) or introgression of QTL or major genes, selection of complex quantitative traits using molecular markers, and enrichment of favourable alleles in early generations [19]. Availability of high-throughput genotyping platforms and genomics resources now rapidly remodelling marker-assisted breeding to genomics-assisted breeding.

Here in this chapter, we summarise the recent progress in understanding the genetics of grain yield and other related traits together with the new strategies, such as gene cloning and mining of superior alleles, transgenic technologies, genome editing technologies, genomic selection (GS), genome-wide association studies (GWAS)-assisted GS, and haplotype-based breeding (including haplotype-based GWAS and haplotype-based GS), which altogether make it available for genomics-assisted breeding (GAB) in crop improvement and to break the yield ceiling in wheat.

2. Genetics of grain yield and its related traits

Grain yield is a complex polygenic trait, significantly associated with grain number per spike, grain weight, harvest index, number of productive tillers, plant height, days to heading/flowering, etc. The trait is also influenced by the environment and shows a significant level of genotype × environment interaction with low heritability. Previous studies showed that increased yield potential in the major wheat-growing countries was largely associated with increased grains per square meter, harvest index, and biomass, and reduced plant height [4, 5]. Moreover, it has been revealed that the use of dwarfing genes (*Rht1*, *Rht2*, *Rht8,* and *Rht24*), the 1BL.1RS translocation lines [20–23], and positive selection of desirable alleles of major genes including grain size (for instances, *TaGS3-A1*, *TaTGW6*, *TaSus1*, *TaGW2*, *TaGW8*, and so on), vernalization requirements (*Vrn* genes), photoperiod response (*Ppd-1*), etc. resulted into the enormous improvement in wheat grain yield [24, 25]. It is now believed that further improvement in grain yield can be attained only by exploiting untapped genetic variation and depth understanding of its genetic architecture combined with the use of advanced genomics-assisted breeding techniques. QTL mapping has been one of the innovative approaches for understanding the genetic architecture of grain yield and its component traits in wheat. Advancements in molecular marker systems have revolutionised the field of QTL mapping, as hundreds of QTLs for different yield-related traits have been mapped using different bi-parental and multi-parental mapping populations in several countries [26–31]. The QTL regions identified by the standard interval mapping procedure frequently extend to several centimorgans (cM) on linkage map (on the physical map, it may be equivalent to the several Mbp) which may encompass a large number of genes [31]. Therefore, it becomes very hard to pinpoint the causative locus/candidate gene responsible for a specific trait. Furthermore, the introgression of such large QTL regions based on linked or flanking markers might carry several unwanted genes due to linkage drag, thereby negatively affecting the performance of generated cultivars encompassing the introgressed genomic segments. Therefore, the genetic resolution of the mapping procedures must be increased to allow QTL placement within the shortest possible genomic region using advanced strategies. Fine mapping is an important strategy that can be used for refining the QTL region. Three major factors, such as phenotyping, population size, and the number of markers, mainly regulate the success of QTL dissection, fine mapping, and further cloning of desired QTLs. Advances in NGS technologies have dramatically reduced per sample genotyping cost and offered increased throughput. Moreover, with the latest SNP genotyping platforms such as SNP chips or arrays in place, it is now quite possible to genotype tens of thousands of samples in a short period [32]. Moreover, QTL fine mapping occasionally reveals surprises, for instance, the presence of distinct genes whose combined effects contribute to the QTL identified using standard mapping procedures, distinct upstream non-coding enhancer/modifier sequences that contribute to phenotypic effects of a QTL, and substantial genetic differences between the alleles in the QTL region. Identification of the genes or sequence variants that underlie QTL may help in investigating the contribution of specific genes or structural variants to the overall genetic architecture of grain yield and related traits [26, 33].

As discussed above, several studies have reported hundreds of QTLs in different mapping populations evaluated under different environments. An innovative approach i.e., meta-QTL analysis has emerged which helps in refining the QTL positions by combining the QTL results from independent studies and identifying the most stable and consensus QTLs [34]. The power of this approach lies in detecting regions of the genome that are most often involved in trait variation and reducing the QTL confidence intervals, thus facilitating the identification and characterisation of underlying candidate genes. For the first time in 2010, Zhang and his colleagues [35] conducted a meta-QTL analysis of major QTLs for grain yield and yield-related traits and identified 12 significant MQTLs on chromosomes 1A, 1B, 2A, 2D, 3B, 4A, 4B, 4D, and 5A, few of which also included important known genes, such as *Vrn* and *Rht* [35]. Another study reported 16 MQTLs on chromosomes 1B, 2A, 2D, 3B, 4A, 6A, and 6B, related to grain weight [36]. Most recently in 2021, Saini and his colleagues [37] have identified a total of 141 MQTLs responsible for grain yield and related traits, which included 13 breeder's MQTLs and 24 ortho-MQTLs. This study also identified 1202 high-confidence candidate genes within the physical positions of the MQTL flanking markers [37]. Beside these, recently, various other MQTL studies have been also conducted in wheat [38–41]. DNA markers tightly linked to these meta-QTLs (MQTLs) may be used as molecular tools for MAS in wheat breeding. Association mapping or GWAS offers an alternative route for identifying genomic regions that have effects across a wider range of germplasm if false associations that are caused by population structure and relatedness can be minimised. With the advancements in high-throughput genotyping technologies, haplotypes and SNP-sets (instead of single SNPs) are being utilised for GWAS, thereby reducing the detection of false positives via overcoming the limitations of multiple testing and enhancing the identification of underlying candidate genes which in turn facilitate gene-based association mapping. Several GWAS studies have been conducted in wheat for grain yield and related traits, which have also resulted in the identification of hundreds of high-confidence candidate genes governing yield-related traits [42–46]. Combined linkage analysis and joint linkage association mapping (JLAM) have also been used in wheat for genetic dissection of grain yield-related traits. Unlike meta-QTL analysis, meta-GWAS studies have been rare in wheat for yield and related traits. For the first time in 2018, Battenfield and his colleagues [47] described this meta-GWAS approach, which combined GWAS analysis from multi-year unbalanced breeding nurseries and identified the consensus and stable marker-trait associations (MTAs) and underlying candidate genes [47]. The markers, as well as candidate genes identified for grain yield and its component traits, provide important genomic resources for wheat breeding. These genomics resources can be immediately implemented to genomics-assisted breeding in wheat for genetic improvement of grain yield.

3. Gene cloning and allele mining: to be used for MAS

MAS allows a more effective selection of target genotypes which further enable certain traits to be 'fast-tracked', resulting in faster line development and variety release. MAS is a more cost-effective approach that can replace phenotyping and thereby allows selection in off-season nurseries as well. Another advantage of using MAS is that the total number of genotypes that need to be tested can be reduced significantly in early generations which allow more efficient use of field or glasshouse space which is generally limited [48]. MAS remains a valid option for major gene or QTL, whereas QTL cloning or gene cloning may become a more routine activity assisted by increased utilisation of high-throughput phenotyping techniques [49], sequencing [50], and identification of high-confidence candidate genes through 'omics' profiling [51]. Cloned QTL/gene may provide new opportunities for a more targeted search for novel alleles in wild wheat germplasm and mutants (**Table 1**).

At present, tremendous sequence information is available in public databases as a result of the sequencing of diverse wheat crop genomes, including reference lines and wild progenitors. This information can be used for mining the novel and superior alleles of agronomically important genes from gene pools to appropriately deploy for the development of high-yielding cultivars. Allele mining also provides insights into the molecular basis of trait variations and identifies the sequence variants associated with superior alleles. Moreover, it helps in the development of allele-specific molecular markers, assisting the introgression of novel alleles via MAS.

4. Transgenic technologies to boost grain yield

Considerable progress has been made in the past for manipulation of genes from diverse sources, including wild relatives and progenitors, and transferring them into wheat to confer increased grain yield, transgenesis can be employed as a powerful alternative for increasing the grain yield through exploiting the genes/traits which does not occur naturally in the wheat species. Transgenic plants refer to plants that contain a gene(s) that has been artificially inserted from an unrelated plant or a completely different species. The increase in grain yield potential through transgenesis involves an ideotypic detail of potential targets for transformation. In 2017, Nadolska-Orczyk and his colleagues [79] reported potential targets for transgenesis which can result in the increased grain yield in wheat. These include 'transcription factors, regulating spike development, which mainly affect grain number; genes involved in metabolism or signalling of growth regulators—cytokinins, gibberellins, and brassinosteroids—which control plant architecture and consequently stem hardiness and grain yield; genes determining cell division and proliferation mainly impacting grain size; floral

Genes/QTLs	Chromosome	Products/enzymes	Associated yield-related traits	References
TaSus2	2A, 2B, 2D	Sucrose synthase	Endosperm development	[52]
TaCwi-A1	2A	Cell wall invertase	Kernel weight	[53]
TaCWI-5D	5D	Cell wall invertase	Kernel weight	[54]
TaSAP1-A1	7A	Zinc-finger protein	Thousand grain weight, number of grains per spike, spike length, peduncle length and spikelet's per spike	[55]
TaGS1a	6D	Glutamine synthetase	Mineral nutrient and grain size	[56]
TaTGW-7A	7A	Indole-3-glycerol-phosphate synthase	Thousand grain weight	[57]
TaGASR7-A1	7A	Snakin/GASA protein	Grain length	[58]
TaGS-D1	7D	Glutamine synthetase	Thousand grain weight, grain length	[59]
TaCKX6a02	3D	Cytokinin oxidase/ dehydrogenase	Grain size, grain filling rate, grain weight	[60]
Tackx	3A	Cytokinin oxidase	Grain weight and leaf chlorophyll content	[61]
TaTPP-6AL1	6A	Trehalose 6-phosphate phosphatase	Grain weight	[62]
TaFlo2-A1	2A	FLO2 protein	Thousand grain weight, grain size	[63]
TaSnRK2.3	1A, 1B, 1D	Plant-specific protein kinase	Plant height, length of peduncle, penultimate node, thousand grain weight	[64]
TaSnRK2.10	4A, 4B,4D	Sucrose non-fermenting 1-related protein kinases	Thousand grain weight, spike length	[65]
6-SFT-A2	4A	Fructan 6-fructosyltransferase	Thousand grain weight	[66]
TaGW2-6A	6A	E3 ubiquitin ligase	Grain weight, grain size	[25]
TaCKX6-D1	3D	Cytokinin oxidase/ dehydrogenase	Thousand grain weight	[67]
TaGL3-5A	5A	Putative protein phosphatase	Grain length	[68]
TaAPO-A1	7A	F-box protein of 429 amino acids	Total spikelet number per spike	[69]
TaTGW6-A1	3A	Indole-3- acetic acid-glucose hydrolase	Thousand grain weight	[24]
TaGW8-B1a	7B	E3 ubiquitin ligase	Kernel size	[70]

Genes/QTLs	Chromosome	Products/enzymes	Associated yield-related traits	References
TaTAR2.1-3A	3A	Tryptophan amino transferase	Plant height, spike number	[71]
TaNAC2-5A	5A	NAC transcription factor	Spike number, grain number per spike, and thousand grain weight	[72]
TaGS5-3A	3A	Serine carboxypeptidases	Grain size, grain weight	[73]
TaTEF-7A	7A	Transcript elongation factor	Grain number	[74]
TaPPH-A	7A	Pheophytin pheophorbide hydrolase	Thousand grain weight, grain filling	[75]
TaNF-YB4	3B	Histone-like transcription factor	Number of spikes per plant	[76]
TaNFYA-B1	6B	Histone-like transcription factor	Number of spikes per plant	[77]
TaCYP78A3	7A, 7B, 7D	Cytochrome P450 CYP78A3	Seed size	[78]

Table 1.
Cloned genes/QTLs regulating various yield-related traits in wheat.

regulators influencing inflorescence architecture and consequently seed number, and genes involved in carbohydrate metabolism having an impact on plant architecture and grain yield'. Furthermore, modulated expression of flowering genes, which control vernalization and photoperiod-dependent floral induction, may be good for winter or spring wheat varieties [79, 80]. Besides, augmenting photosynthetic rates of laminar and non-laminar organs and the capability to access and utilise a greater amount of resources, such as nutrients or water, may also be potential targets for transgenesis in wheat for grain yield improvement [81, 82]. Besides, information about specific genotypes as well as climatic and agronomic conditions and consideration of the fact that the majority of the genes are members of multigene families is required for successful implementation of selected potential genes in breeding programs [79].

Transgenic wheat has the capacity to transform agriculture, but progress has been very limited as no transgenic wheat cultivar could be commercially approved so far because of consumers' concerns. Few promising reports are available where newly developed transgenic wheat showed a significant grain yield advantage [72, 83]. Over-expression of a nitrate-inducible transcription factor (NAC TF) in wheat enhanced root growth and the ability to uptake nitrogen, therefore, increased nitrogen accumulation and grain yield by 10% (on a single plant basis) [72]. In another study, Gonzalez and his colleagues [83] reported that transgenic wheat lines carrying a mutated version of the sunflower TF (*HaHB4*) can significantly increase grain yield and water use efficiency across a range of environments [83]. Most recently in 2020, Argentina has become the first country to approve a genetically modified wheat variety (HB4). This is a drought-tolerant high-yielding wheat variety jointly developed by Argentine crop inputs manufacturer 'Bioceres' and 'Trigall Genetics' yielding 20% more than other standard wheat varieties in 10 years trials under drought conditions. The commercial approval of this GMO variety solely depends on approval by Brazil, which imports more than

85% of Argentine wheat [84]. Experts have also raised concerns about the growth and marketing of this GMO wheat variety, citing challenges related to food safety, consumer preferences, environmental effects, and socioeconomic issues. More research is required to determine the true safety of this GMO wheat and to decide, whether they are safe for both the consumers and the environment. At least, most would agree that the possible advantage of producing transgenic wheat, which furnishes the human population with cheaper and more food, makes transgenesis a useful invention.

5. Genome-editing technologies

Targeted genome editing has emerged as a powerful tool for studying gene function, correcting defective genes, or introducing novel functionality. Its mechanism involves sequence-specific double-strand breaks (DSBs) in the target DNA, with edits incorporated during the endogenous repair. In the earlier phase of genome editing, to induce the desired double-strand breaks at the target site, the engineering for zinc-finger nucleases (ZFNs) [85] or meganucleases [86] attracted the attention of the researcher community. These genome-editing systems needed specialised competence to produce artificial proteins consisting of customizable DNA-binding domains (sequence-specific), each linked to a non-specific nuclease for target DNA cleavage, and offered researchers with extraordinary tools to perform genetic manipulation. Later, the identification of a novel class of a *Flavobacterium okeanokoites* catalytic domain (FokI) derived from bacterial proteins termed transcription activator-like effectors (TALEs) further offered new possibilities for precisely targeted genome editing [87]. TALE-based programmable nucleases allowed the cleavage of any DNA sequence of interest with comparatively high frequency. Dimerization of FokI nuclease is needed to make an active nuclease, therefore, every time two modules need to be designed to target closely DNA sequences for generating DSBs at target sites. This dimerization requirement limited the use of these two powerful genomes-editing tools, as designing active nucleases was difficult and very expensive [88].

In 2012, an inexpensive, simple, easy to use, and effective genome-editing system that is clustered regularly interspaced short palindromic repeats (CRISPR)/CRISPR-associated protein 9 (CRISPR/Cas9) was introduced which revolutionised the field of genome editing [89]. The use of this powerful tool allows producing genome-edited plants in a very short period. CRISPR technology can be efficiently utilised for both precisely eliminating the negative regulator genes and augmenting the activity of positive-regulator genes that affect the trait of interest. Nevertheless, there are only a couple of reports available for validation of the CRISPR technique in wheat compared to other crops, such as rice [90]. In these reports, different genes were targeted by CRISPR/Cas9 to address the major biotic, and abiotic stresses along with improving a few agronomic traits in wheat [90]. An exciting advantage of using the CRISPR/Cas9 technology is the possibility of simultaneously editing multiple target genes using a single CRISPR construct. For instance, Wang and his colleagues [91] practiced this multiplexed genome editing in hexaploid wheat for targeting three different genes viz. *TaLpx-1*, *TaGW2*, and *TaMLO*. They placed three sgRNAs (each specific to a different gene) in a tRNA polycistronic cassette under the control of a single promoter to produce knockouts. Multiplex genome-editing tools can be efficiently utilised to address more complex traits (such as grain yield) involving multiple genes in a single attempt [91]. Moreover, this CRISPR/Cas9 mediated multiplex genome editing can also be utilised to mimic the domestication process during evolution in a short time frame, with implications for a convenient and rapid generation of high-yielding wheat varieties. Despite the several advantages of using CRISPR/Cas9, one of the prominently associated challenges is off-target effects, that is undesired mutations at unintended sites induced by genome editing. Various methods have been developed to find off-target mutations both *in vitro* and *in vivo*. These include SITE-seq [92], Digenome-seq [93], CIRCLE-seq [94], GUIDE-seq [95], and DISCOVER-seq [96]. In the same way, the engineering of Cas9 proteins has also been performed to enhance the specificity.

5.1 Base editors and prime editing: opening up new avenues for wheat genome engineering

Many crucial agronomic traits are determined by a few base changes or point mutations in a gene [97–99]. CRISPR/Cas9 mediated gene replacements or gene modifications through homology-directed repair (HDR) has been reported as a practicable approach to correct the point mutations in the target DNA/gene and has the capability for accelerating crop improvement [100, 101]. Yet, the low efficiency of template DNA delivery and the rare occurrence of HDR (endogenous) has always been a difficult task in attaining success in plants. Furthermore, the CRISPR/Cas9 system is amenable for gene knock-in or knock-out, but cannot covert base into another. These challenges highlighted the demand for alternative powerful approaches that can result in precise and stable genome editing in crops. In 2016, a novel approach that is 'Base editing' was emerged which allows precise base (nucleotide) substitutions in a programmable manner, without requiring a donor template or disruption of a gene [102]. A base editor is a fusion of catalytically inactive Cas9 domain (Cas9 variants, Cas9 nickase, or dCas9) and an adenosine or cytosine domain that converts one base to another. Nucleotide substitutions or single-base changes may generate elite trait variations in crops which assist in accelerating crop improvement. The base-editing system can revert an SNP or single-base change without gene disruption. In recent years, many adenine and cytosine base editors have emerged as powerful tools for precise genome modifications (A to G or C to T) in eukaryotic genomes [102]. The potential of this approach has been demonstrated in several crops, including wheat [103–106]. As aforementioned, HDR efficiency is comparatively low in plant cells, so knock-ins of DNA fragments to target sites are challenging. Recently in 2019, Anzalone and co-workers developed a more efficient genome-editing technology that is 'Prime editing' which consists of CRISPR-Cas9 nickase–reverse transcriptase fusions programmed with pegRNAs (prime-editing guide RNAs) that enable precise genome editing without inducing DSBs or requiring a donor DNA template (mandatory for genome editing via HDR) in mammalian cells [107]. The prime editors have been adapted for use in wheat via optimization of the codon, promoter, and editing conditions [108]. This optimised suite of prime editors enabled InDels and point mutations in wheat and rice at higher frequencies [108]. Development of new technologies and tools, newly discovered CRISPR/Cas systems, are being continuously reported, inferring that the CRISPR toolbox for wheat genome engineering would expand further in the near future. Researchers have also focused on the development of efficient approaches for eliminating transgenes from genome-edited plants, such as (a) transient expression of DNA and RNA [109], (b) use of CRISPR/Cas9 ribonucleoprotein complexes [110], (c) use of CRISPR-S—an active interference element [111], and (d) programmed self-elimination of the CRISPR/Cas9 constructs [112] to generate transgene-free genome-edited plants. The elimination of transgenes offers the following two advantages—(i) elimination of Cas9 construct from genome-edited plants prevents the induction of genetic changes at undesired loci, (ii) elimination of the transgenes is likely a prerequisite for getting regulatory approval of genome-edited crops for commercial applications. In the future, CRISPR technology may be supposed to accelerate wheat biology research, ultimately facilitating the development of high-yielding wheat varieties.

6. Genomic selection for grain yield improvement

The genetic complexity of grain yield and other yield-related traits limit the power of QTL mapping and association mapping in identifying small effect loci [113]. A powerful breeding strategy that is genomic selection (GS) has been introduced to circumvent this problem which implements whole-genome markers for predictions, and thus can efficiently complement QTL mapping and association analysis in dissecting the complex genetic base of grain yield-related traits in wheat [114, 115]. High-throughput/next-generation genotyping technologies have accelerated the adoption of GS by enabling the development of large sets of DNA marker data at reasonable costs [116]. GS is a potential GAB tool that predicts genomic-estimated breeding values (GEBVs) of individuals (from the breeding population) with genotypic data available via prediction models constructed based on a training population (TP) with available phenotypic and genotypic information [117]. As aforementioned, using the prediction models, the GEBVs of unobserved individuals are predicted, circumventing the omission of the small-effect genomic region (markers) that would fail a threshold (significance) test. Though the effect of each marker is small, a large volume of genotypic information covering the whole genome still has the power to explain all the genetic variance. GS complements conventional breeding approaches and can potentially decrease the requirement of large-scale phenotyping and hasten the rate of genetic gain via shorter breeding cycles [118, 119]. The performance of GS relies mainly on the prediction accuracy, defined as the 'Pearson's correlation between the selection criterion and the true breeding value to select individuals with unknown phenotypes' [120, 121]. Other factors that affect the GS accuracy include gene effects, level of linkage disequilibrium (LD), statistical models, the genetic composition of the TP, relationship between validation population (VP) or selection individuals and TP, and heritability of the target traits [120]. The major objective of GS is to decrease the cost of phenotyping and hasten genetic gains, use of high-throughput phenotyping tools and platforms that enable high-density phenotyping of hundreds to thousands of individuals across time and space using proximal or remote sensing, can increase the intensity and accuracy of selection and, eventually the selection response, as well as reduce phenotyping costs. The main idea of high-throughput phenotyping is to exploit secondary traits, such as canopy temperature, and green normalised difference vegetation index (NDVI) are closely related to grain yield that may be advantageous in early-generation testing of individuals. Data recorded on secondary traits (genetically correlated to grain yield) can be incorporated in multivariate pedigree and GS models, improving indirect selection for GY [122–124]. Moreover, GS can also be applied to gene bank accessions for germplasm enhancement. Accessions stored in germplasm bank represents an under-exploited rich genetic resource for wheat breeders, superior alleles can be extracted from these accessions which may be exploited for grain yield improvement in wheat [125, 126]. In general, lengthy pre-breeding programs are needed to develop lines that possess favourable alleles/genes from the wild accessions with superior agronomic performance and that may be utilised as parents in breeding

Population type and size*	Number of genotyped markers	Traits	Accuracy of GEBV used	References
Advanced breeding lines from CIMMYT (254)	41,371 GBS-SNPs	TGW, DTH, and GY	0.28–0.45	[128]
Two DH populations (165 and 159)	1975 and 1483 SNPs (90K SNP)	GNPS	0.10–0.42	[129]
European winter wheat lines (2325)	12,642 SNPs (9K SNP)	GY	0.5–0.65	[130]
Winter wheat population (273)	40,267 SNPs (90K SNP)	GY, TGW, PH and DTH	0.33–0.67	[131]
Inbred breeding lines (557)	12,083 GBS-SNPs	DTH and GY	0.57	[132]
Advanced elite spring wheat lines (287)	15,000 SNPs (90 K SNP)	GY, TGW and GN	0.38–0.63	[133]
Lines from multiple families (659)	9500 DArT-GBS-SNPs	GY	0.38–0.41	[134]
Winter wheat breeding population from multiple families (861)	6600 DArT-GBS-SNPs	GY	0.39–0.48	[135]
Inbred breeding lines (557)	12,083 GBS-SNPs	GY	0.65–0.76	[136]
Hybrids obtained by crossing 18 males and 667 females (1888)	13,005 SNPs (90 K and 15 K)	GY, DTH and PH	0.5–0.55	[137]
Winter wheat lines (1100)	27,000 GBS-SNPs	GY	0.23–0.55	[138]
European winter and spring cultivars (210)	GBS-SNPs	44 spike morphology traits	0.2–0.5	[139]
Elite wheat lines (4368)	2038 GBS-SNPs	DTH, DTM, PH and GY	0.35–0.44	[140]
Bread wheat lines (10375)	18,101 GBS-SNPs	GY and TGW	0.59–0.98	[141]
Double haploid lines (282)	7426 GBS-SNPs	GY and TGW	0.47–0.54	[142]
Bread wheat lines (3771)	8519 GBS-SNPs	DTH, DTM and GY	0–0.75	[143]
Soft red winter wheat lines (239), Double haploid (100), and Recombinant inbred lines (156)	2721 SNPs (9 and 90K)	GY, DTH, TGW, GNPS, and PH	– 0.14-0.43	[144]
F4:6 generation and double haploid winter wheat breeding lines (1114)	7300 DArT-GBS-SNPs	GY	0.45	[145]
Winter wheat lines (3282)	18,728 GBS-SNPs	GY	0.25	[122]

Population type and size*	Number of genotyped markers	Traits	Accuracy of GEBV used	References
>6400 breeding lines	78,662 GBS-SNPs	GY	0.41	[146]
Advanced breeding lines (456)	11,089 GBS-SNPs	GY	0.33–0.66	[147]
Association mapping panel (456), two F5 populations (61 and 501), two DH populations (447 and 759)	16,233 GBS-SNPs	GY	0.21	[148]
Advanced bread wheat lines (4302)	8443 GBS-SNPs	GY	0.35–0.43	[149]
Winter wheat lines (1325)	11,154 SNPs (15 K)	GY	0.57	[150]

GY, GNPS, DTH, DTM, PH, and TGW refer to grain yield, grain number per spike, days to heading, days to maturity, plant height, and thousand grain weight, respectively. *Figures in parenthesis are the population size.*

Table 2.
Genomic selection studies conducted in wheat for grain yield and related traits.

programs. Using GS, germplasm enhancement breeding programs can be directly started using wild accessions and landraces. In a recent GS-based study, NGS technologies with multi-environment phenotyping were used to study the contribution of exotic genomes to 984 pre-breeding lines. Significant positive contributions of exotic germplasm to pre-breeding lines derived from crosses of CIMMYT's best elite lines with exotics were reported [127]. Genomic selection studies conducted in wheat for grain yield and related traits are presented in **Table 1**. The prediction accuracy of GS for different grain yield-related traits has varied from 0 to 0.98% in wheat (**Table 2**).

6.1 GWAS-assisted GS: making GS more efficient

As discussed above, GWAS estimates marker effects throughout the genome on the target association panel (diverse germplasm) based on prediction models. Based on LD, GWAS may identify new functional variants, including novel MTAs and genes for many agronomically important traits in diverse germplasm. According to a comprehensive simulation study in plants, the use of a few major MTAs/QTLs/genes (each explaining ≥ 10% of the phenotypic variance) as fixed effects in GS models can increase the accuracy of GS for complex quantitative traits [151]. Although, the potential to combine robust and consistent associations identified from GWAS as fixed effects in GS models to increase prediction accuracy for complex traits such as grain yield has not been investigated comprehensively in wheat. The first report of integrating the genetic architecture of GY (revealed through GWAS) into prediction models in wheat has come from the work by Sehgal and co-workers, most recently in 2020 [149]. Firstly, using a haplotype-based genome-wide association study, they identified 58 MTAs for GY. Out of these 58 MTAs, 16 were 'environment-specific' with large effects and eight MTAs were consistent across trials and environments. These consistent MTAs were then used as fixed effects in the prediction models which resulted in a 9–10% increase in prediction accuracy for GY [149]. It is suggested that the utility of GS incorporating GWAS results may be noteworthy for GY when GWAS results detect highly robust and significant genomic regions.

7. Haplotype-based breeding (HBB) for grain yield improvement

Due to low heritability and persistent 'genotype × environment' interactions, improving grain yield (GY) is a difficult task for the global plant breeding community, especially under stressful environmental conditions [152–154]. As discussed earlier, GWAS-assisted GS has proven to be an effective method for deciphering the genetic architecture of complex traits, population improvement, and the development of better varieties with a higher yield. However, the problem of 'missing heritability', which is widespread in single marker-based GWAS, is not addressed by this approach. The alternative approach to boost the power of GWAS is by constructing haplotypes between neighbouring SNPs on a chromosome. As specific sets of alleles are observed on a single chromosome, haplotypes are inherited jointly with the limited probability of contemporaneous recombination. Haplotypes are implemented in crop improvement in two ways—retrospective and prospective [155]. Plant breeders have to choose the advantageous haplotypes that lead to desirable phenotype(s) for the trait(s) of interest during the long-term selection process. As a result, these advantageous haplotypes in elite crop germplasm can be found utilising the genome resequencing technique to sequence an elite gene pool [156]. Later, molecular markers that characterise these beneficial haplotypes can be produced, and all of these haplotype-defining markers can then be utilised to pick the most ideal combination of haplotypes that govern a certain phenotype. Furthermore, by identifying lines with unique recombination in chromosomal blocks of relevance, these haplotype-related markers can be utilised to distinguish between favourable and unfavourable genetic variation. On the other hand, haplotypes can be employed in a prospective approach, in which a vast collection of ancestral and wild germplasm of specific crop species (not just elite breeding pools) is re-sequenced to find haplotypes with a wider range of genetic variation [153, 155]. The genome-wide haplotypes are employed in this strategy to find novel haplotypes in a wide variety of natural germplasm. For the discovery of QTLs/genes, recent GWA studies based on empirical and simulation data (i.e., better p-values) and allelic effect estimation have demonstrated that haplotype blocks have higher mapping accuracy and power than individual SNPs [153, 155–160]. Haplotype superiority can be explained by a number of factors. Stephens and his colleagues [161] showed that haplotype blocks are more informative than SNP markers because of their multi-allelic character in nature. The scientists found that haplotype variants were more common than SNPs, implying that recombination and recurrent mutation events occurred within and among haplotype genes (**Figure 1**). In addition, as compared to individual SNPs, haplotype-based analysis is predicted to reduce the false positives and shows the intricate mechanism of causal haplotypes [162]. Similarly, the haplotype-assisted GS depicts the complex relationships between genotypic information and phenotypes more accurately than individual SNPs. As a result, this method could eventually aid in improving selection gain per unit of time. Because haplotypes can better capture LD and genomic similarities in various lines and may capture local high-order allelic interactions, they may improve the accuracy of genomic prediction [163]. Furthermore, by depicting population structure in the calibration set, prediction accuracy might be enhanced. The superiority of haplotype-based predictions over SNP-based predictions for all studied traits, including yield, test weight, and protein content, was established in a recent GS study that compared the prediction ability computed from haplotypes and SNPs in a set of 383 advanced lines and cultivars of wheat [164]. Based on evidence revealing higher haplotype-assisted genomic prediction efficiency than SNPs, researchers are increasingly embracing haplotype-assisted genomic prediction in crop development programmes.

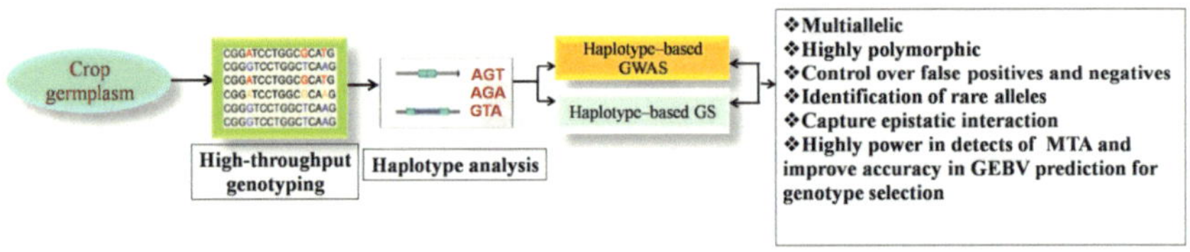

Figure 1.
Flow diagram indicating how haplotype-based GWAS and haplotype-based GS, when combined with high-throughput genotyping, have the potential to improve gene identification precision and accuracy (modified from Bhat et al. [162]).

8. Conclusions

Significant progress has been made in wheat in developing various genomics resources, including high-throughput molecular markers, dense genetic maps, and next-generation genotyping platforms. The availability of high-quality wheat genome information has also enabled many next-generation sequencing-based approaches for genetic mapping, allele mining, and identification of candidate genes which have enhanced the precision, pace, and efficiency of trait mapping. At present, trait-associated markers, high-throughput genotyping platforms, and expertise are available for deploying genomics-assisted breeding in wheat. We believe that in the coming years, extensive deployment of genome editing, transgenic technology, genomic selection, haplotype-based breeding in combination or alone would be undertaken for crop improvement and breaking the yield ceiling. Various steps involved in generating high-yielding wheat genotypes using genomics-assisted breeding technologies are represented in **Figure 2**.

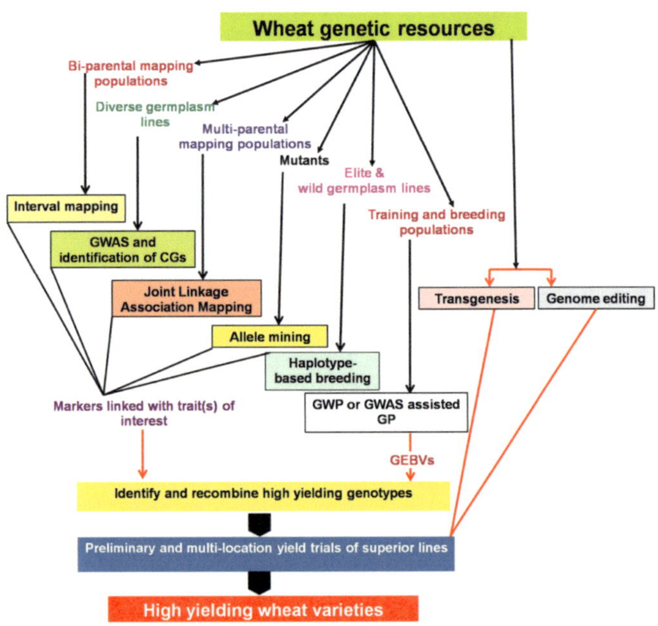

Figure 2.
Flowchart demonstrating the steps involved in generating high-yielding wheat genotypes using different genomics-assisted breeding strategies.

Acknowledgements

Thanks are due to the Head, Department of Molecular Biology and Genetic Engineering, G. B. Pant University of Agriculture and Technology, Pantnagar, (India) for providing necessary facilities.

Funding

This research did not receive any specific grant from funding agencies in the public, commercial, or not-for-profit sectors.

Conflict of interest

The authors declare no conflict of interest.

Author details

Neeraj Pal[1], Dinesh Kumar Saini[2] and Sundip Kumar[1*]

1 Department of Molecular Biology and Genetic Engineering, G.B. Pant University of Agriculture and Technology, Pantnagar, Uttarakhand, India

2 Department of Plant Breeding and Genetics, Punjab Agricultural University, Ludhiana, India

*Address all correspondence to: malik.sundip@gmail.com

References

[1] FAO. Food and Agricultural Organization [Internet]. 2018. Available from: http://www.fao.org/faostat/en/#data/QC [Accessed: January 21, 2022]

[2] Knoema [Internet] 2020. Available from: https://knoema.com/atlas [Accessed: January 21, 2022]

[3] Graybosch RA, Peterson CJ. Genetic improvement in winter wheat yields in the Great Plains of North America, 1959-2008. Crop Science. 2010;**50**:1882-1890

[4] Calderini DF, Reynolds MP, Slafer GA. Genetic Gains in Wheat Yield and Main Physiological Changes Associated with them during the 20th Century. New York: Food Product Press; 1999. pp. 352-377

[5] De Vita P, Nicosia OLD, Nigro F, Platani C, Riefolo C, Di Fonzo N, et al. Breeding progress in morpho-physiological, agronomical and qualitative traits of durum wheat cultivars released in Italy during the 20th century. European Journal of Agronomy. 2007;**26**:39-53

[6] Reynolds MP, Rajaram S, Sayre KD. Physiological and genetic changes of irrigated wheat in the post–green revolution period and approaches for meeting projected global demand. Crop Science. 1999;**39**:1611-1621

[7] Sadras VO, Lawson C. Genetic gain in yield and associated changes in phenotype, trait plasticity and competitive ability of South Australian wheat varieties released between 1958 and 2007. Crop and Pasture Science. 2011;**62**:533-549

[8] Fischer RA, Edmeades GO. Breeding and cereal yield progress. Crop Science. 2010;**50**:S-85

[9] Iwgsc AR, Eversole K, Feuillet C, Keller B, Rogers J, Stein N. Shifting the limits in wheat research and breeding using a fully annotated reference genome. Science. 2018;**361**:7191

[10] Saini DK, Devi P, Kaushik P. Advances in genomic interventions for wheat biofortification: A review. Agronomy. 2020;**10**:1-62

[11] Kaur B, Mavi GS, Gill MS, Saini DK. Utilization of KASP technology for wheat improvement. Cereal Research Communications. 2020;**48**:1-13

[12] Scheben A, Batley J, Edwards D. Genotyping-by-sequencing approaches to characterize crop genomes: Choosing the right tool for the right application. Plant Biotechnology. 2016;**15**:149-161

[13] Cavanagh CR, Chao S, Wang S, Huang BE, Stephen S, Kiani S, et al. Genome-wide comparative diversity uncovers multiple targets of selection for improvement in hexaploid wheat landraces and cultivars. Proceedings of the National Academy of Sciences. 2013;**110**:8057-8062

[14] Wang S, Wong D, Forrest K, Allen A, Chao S, Huang BE, et al. Characterization of polyploid wheat genomic diversity using a high-density 90 000 single nucleotide polymorphism array. Plant Biotechnology Journal. 2014;**12**:787-796

[15] Boeven PH, Longin CFH, Leiser WL, Kollers S, Ebmeyer E, Würschum T. Genetic architecture of male floral traits required for hybrid wheat breeding. Theoretical and Applied Genetics. 2016;**129**:2343-2357

[16] Winfield MO, Allen AM, Burridge AJ, Barker GL, Benbow HR, Wilkinson PA, et al. High-density SNP genotyping array for hexaploid wheat

and its secondary and tertiary gene pool. Plant Biotechnology Journal. 2016;**14**:1195-1206

[17] Allen AM, Winfield MO, Burridge AJ, Downie RC, Benbow HR, Barker GL, et al. Characterization of a wheat breeders' Array suitable for high-throughput SNP genotyping of global accessions of hexaploid bread wheat (*Triticum aestivum*). Plant Biotechnology Journal. 2017;**15**:390-401

[18] Rasheed A, Xia X. From markers to genome-based breeding in wheat. Theoretical and Applied Genetics. 2019;**132**:767-784

[19] Savadi S, Prasad P, Kashyap PL, Bhardwaj SC. Molecular breeding technologies and strategies for rust resistance in wheat (*Triticum aestivum*) for sustained food security. Plant Pathology. 2018;**67**:771-791

[20] He ZH, Liu L, Xia XC, Liu JJ, Pena RJ. Composition of HMW and LMW glutenin subunits and their effects on dough properties, pan bread, and noodle quality of Chinese bread wheat. Cereal Chemistry. 2005;**82**:345-350

[21] Tian XL, Zhu ZW, Xie L, Xu DA, Li JH, Fu C, et al. Preliminary exploration of the source, spread and distribution of *Rht24* reducing height in bread wheat. Crop Science. 2019;**59**:19-24

[22] Zhang XK, Yang SJ, Zhou Y, Xia XC, He ZH. Distribution of *Rht-B1b*, *Rht-D1b* and *Rht8* genes in autumn-sown Chinese wheats detected by molecular markers. Euphytica. 2006;**152**:109-116

[23] Zhou Y, He ZH, Sui XX, Xia XC, Zhang XK, Zhang GS. Genetic improvement of grain yield and associated traits in the northern China winter wheat region from 1960 to 2000. Crop Science. 2007;**47**:245-253

[24] Hanif M, Gao F, Liu J, Wen W, Zhang Y, Rasheed A, et al. *TaTGW6-A1*, an ortholog of rice *TGW6*, is associated with grain weight and yield in bread wheat. Molecular Breeding. 2016;**36**:1

[25] Su Z, Hao C, Wang L, Dong Y, Zhang X. Identification and development of a functional marker of *TaGW2* associated with grain weight in bread wheat (*Triticum aestivum* L.). Theoretical and Applied Genetics. 2011;**122**:211-223

[26] Chen Z, Cheng X, Chai L, Wang Z, Bian R, Li J, et al. Dissection of genetic factors underlying grain size and fine mapping of QTgw.cau-7D in common wheat (*Triticum aestivum* L.). Theoretical and Applied Genetics. 2020a;**133**:149-162

[27] Chen Z, Cheng X, Chai L, Wang Z, Du D, Wang Z, Bian R, Zhao A, Xin M, Guo W, Hu Z. Pleiotropic QTL influencing spikelet number and heading date in common wheat (*Triticum aestivum* L.). Theoretical and Applied Genetics. 2020b;**133**:1-14

[28] Hu J, Wang X, Zhang G, Jiang P, Chen W, Hao Y, et al. QTL mapping for yield-related traits in wheat based on four RIL populations. Theoretical and Applied Genetics. 2020;**133**:917-933

[29] Xin F, Zhu T, Wei S, Han Y, Zhao Y, Zhang D, et al. QTL mapping of kernel traits and validation of a major QTL for kernel length-width ratio using SNP and bulked segregant analysis in wheat. Scientific Reports. 2020;**10**:1-12

[30] Yu M, Liu ZH, Yang B, Chen H, Zhang H, Hou DB. The contribution of photosynthesis traits and plant height components to plant height in wheat at the individual quantitative trait locus level. Scientific Reports. 2020;**10**:1-10

[31] Liu H, Mullan D, Zhang C, Zhao S, Li X, Zhang A, et al. Major genomic

regions responsible for wheat yield and its components as revealed by meta-QTL and genotype–phenotype association analyses. Planta. 2020;**252**:1-22

[32] Jaganathan D, Bohra A, Thudi M, Varshney RK. Fine mapping and gene cloning in the post-NGS era: Advances and prospects. Theoretical and Applied Genetics. 2020;**133**:1-20

[33] Holland JB. Genetic architecture of complex traits in plants. Current Opinion in Plant Biology. 2007;**10**:156-161

[34] Goffinet B, Gerber S. Quantitative trait loci: A meta-analysis. Genetics. 2000;**155**:463-473

[35] Zhang LY, Liu DC, Guo XL, Yang WL, Sun JZ, Wang DW, et al. Genomic distribution of quantitative trait loci for yield and yield-related traits in common wheat. Journal of Integrative Plant Biology. 2010;**52**:996-1007

[36] Tyagi S, Mir RR, Balyan HS, Gupta PK. Interval mapping and meta-QTL analysis of grain traits in common wheat (*Triticum aestivum* L.). Euphytica. 2015;**201**:367-380

[37] Saini DK, Srivastava P, Pal N, Gupta PK. Meta-QTLs, ortho-meta-QTLs and candidate genes for grain yield and associated traits in wheat (Triticum aestivum L.). Theoretical and Applied Genetics. 2022;**5**:1-33

[38] Saini DK, Chopra Y, Pal N, Chahal A, Srivastava P, Gupta PK. Meta-QTLs, ortho-MQTLs and candidate genes for nitrogen use efficiency and root system architecture in bread wheat (*Triticum aestivum* L.). Physiology and Molecular Biology of Plants. 2021a;**27**:2245-2267

[39] Saini DK, Chahal A, Pal N, Srivastava P, Gupta PK. Meta-analysis reveals consensus genomic regions associated with multiple disease resistance in wheat (*Triticum Aestivum* L.). 2021;**42**:11-35. DOI: 10.21203/rs.3.rs-773587/v1

[40] Pal N, Saini DK, Kumar S. Meta-QTLs, ortho-MQTLs and candidate genes for the traits contributing to salinity stress tolerance in common wheat (*Triticum aestivum* L.). Physiology and Molecular Biology of Plants. 2021;**24**:1-20

[41] Kumar S, Singh VP, Saini DK, Sharma H, Saripalli G, Kumar S, et al. Meta-QTLs, ortho-MQTLs, and candidate genes for thermotolerance in wheat (*Triticum aestivum* L.). Molecular Breeding. 2021;**41**:1-22

[42] Akram S, Arif MAR, Hameed A. A GBS-based GWAS analysis of adaptability and yield traits in bread wheat (*Triticum aestivum* L.). Journal of Applied Genetics. 2020;**62**:1-15

[43] Basile SML, Ramírez IA, Crescente JM, Conde MB, Demichelis M, Abbate P, et al. Haplotype block analysis of an Argentinean hexaploid wheat collection and GWAS for yield components and adaptation. BMC Plant Biology. 2019;**19**:553

[44] Gupta PK, Balyan HS, Sharma S, Kumar R. Genetics of yield, abiotic stress tolerance and biofortification in wheat (*Triticum aestivum* L.). Theoretical and Applied Genetics. 2020;**133**:1-34

[45] Muhammad A, Hu W, Li Z, Li J, Xie G, Wang J, et al. Appraising the genetic architecture of kernel traits in hexaploid wheat using GWAS. International Journal of Molecular Sciences. 2020;**21**:5649

[46] Li F, Wen W, Liu J, Zhang Y, Cao S, He Z, et al. Genetic architecture of grain yield in bread wheat based on genome-wide association studies. BMC Plant Biology. 2019;**19**:168

[47] Battenfield SD, Sheridan JL, Silva LD, Miclaus KJ, Dreisigacker S, Wolfinger RD, et al. Breeding-assisted genomics: Applying meta-GWAS for milling and baking quality in CIMMYT wheat breeding program. PLoS One. 2018;**13**:0204757

[48] Collard BC, Mackill DJ. Marker-assisted selection: An approach for precision plant breeding in the twenty-first century. Philosophical Transactions of the Royal Society B: Biological Sciences. 2008;**363**:557-572

[49] Araus JL, Cairns JE. Field high-throughput phenotyping: The new crop breeding frontier. Trends in Plant Science. 2014;**19**:52-61

[50] Berkman PJ, Lai K, Lorenc MT, Edwards D. Next-generation sequencing applications for wheat crop improvement. American Journal of Botany. 2012;**99**:365-371

[51] Shah T, Xu J, Zou X, Cheng Y, Nasir M, Zhang X. Omics approaches for engineering wheat production under abiotic stresses. International Journal of Molecular Sciences. 2018;**19**:2390

[52] Jiang Q, Hou J, Hao C, Wang L, Ge H, Dong Y, et al. The wheat (*T. aestivum*) sucrose synthase 2 gene (*TaSus2*) active in endosperm development is associated with yield traits. Functional & Integrative Genomics. 2011;**11**:49-61

[53] Ma D, Yan J, He Z, Wu L, Xia X. Characterization of a cell wall invertase gene *TaCwi-A1* on common wheat chromosome 2A and development of functional markers. Molecular Breeding. 2012;**2943-52**

[54] Jiang Y, Jiang Q, Hao C, Hou J, Wang L, Zhang H, et al. A yield-associated gene *TaCWI*, in wheat: Its function, selection and evolution in global breeding revealed by haplotype analysis. Theoretical and Applied Genetics. 2015;**128**:131-143

[55] Chang J, Zhang J, Mao X, Li A, Jia J, Jing R. Polymorphism of *TaSAP1-A1* and its association with agronomic traits in wheat. Planta. 2013;**237**:1495-1508

[56] Guo Y, Sun J, Zhang G, Wang Y, Kong F, Zhao Y, et al. Haplotype, molecular marker and phenotype effects associated with mineral nutrient and grain size traits of *TaGS1a* in wheat. Field Crops Research. 2013;**154**:119-125

[57] Hu MJ, Zhang HP, Liu K, Cao JJ, Wang SX, Jiang H, et al. Cloning and characterization of *TaTGW-7A* gene associated with grain weight in wheat via SLAF-seq-BSA. Frontiers in Plant Science. 2016;**7**:1902

[58] Dong L, Wang F, Liu T, Dong Z, Li A, Jing R, et al. Natural variation of *TaGASR7-A1* affects grain length in common wheat under multiple cultivation conditions. Molecular Breeding. 2014;**34**:947

[59] Zhang Y, Liu J, Xia X, He Z. *TaGS-D1*, an ortholog of rice *OsGS3*, is associated with grain weight and grain length in common wheat. Molecular Breeding. 2014;**34**:1097-1107

[60] Lu J, Chang C, Zhang HP, Wang SX, Sun G, Xiao SH, et al. Identification of a novel allele of *TaCKX6a02* associated with grain size, filling rate and weight of common wheat. PLoS One. 2015;**10**:0144765

[61] Chang C, Lu J, Zhang HP, Ma CX, Sun G. Copy number variation of cytokinin oxidase gene *Tackx4* associated with grain weight and chlorophyll content of flag leaf in common wheat. PLoS One. 2015;**10**:0145970

[62] Zhang P, He Z, Tian X, Gao F, Xu D, Liu J, et al. Cloning of *TaTPP-6AL1* associated with grain weight in bread wheat and development of functional marker. Molecular Breeding. 2017a;**37**:1-8

[63] Sajjad M, Ma X, Khan SH, Shoaib M, Song Y, Yang W, et al. *TaFlo2-A1*, an ortholog of rice Flo2, is associated with thousand grain weight in bread wheat (*Triticum aestivum* L.). BMC Plant Biology. 2017;**17**:1-11

[64] Miao L, Mao X, Wang J, Liu Z, Zhang B, Li W, et al. Elite haplotypes of a protein kinase gene *TaSnRK2.3* associated with important agronomic traits in common wheat. Frontiers. Plant Science. 2017;**8**:368

[65] Zhang ZG, Lv GD, Li B, Wang JJ, Zhao Y, Kong FM, et al. Isolation and characterization of the *TaSnRK2.10* gene and its association with agronomic traits in wheat (*Triticum aestivum* L.). PLoS One. 2017b;**12**:0174425

[66] Yue A, Li A, Mao X, Chang X, Li R, Jing R. Identification and development of a functional marker from *6-SFT-A2* associated with grain weight in wheat. Molecular Breeding. 2015;**35**:63

[67] Zhang L, Zhao YL, Gao LF, Zhao GY, Zhou R, Zhang BS, et al. *TaCKX6-D1*, the ortholog of rice *OsCKX2*, is associated with grain weight in hexaploid wheat. New Phytologist. 2012;**195**:574-584

[68] Yang J, Zhou Y, Wu Q, Chen Y, Zhang P, Zhang YE, et al. Molecular characterization of a novel *TaGL3-5A* allele and its association with grain length in wheat (*Triticum aestivum* L.). Theoretical and Applied Genetics. 2019a;**132**:1799-1814

[69] Muqaddasi QH, Brassac J, Koppolu R, Plieske J, Ganal MW, Röder MS. *TaAPO-A1*, an ortholog of rice ABERRANT PANICLE ORGANIZATION 1, is associated with total spikelet number per spike in elite European hexaploid winter wheat (*Triticum aestivum* L.) varieties. Scientific Reports. 2019;**9**:1-12

[70] Yan X, Zhao L, Ren Y, Dong Z, Cui D, Chen F. Genome-wide association study revealed that the *TaGW8* gene was associated with kernel size in Chinese bread wheat. Scientific Reports. 2019;**9**:1-10

[71] Shao A, Ma W, Zhao X, Hu M, He X, Teng W, et al. The auxin biosynthetic TRYPTOPHAN AMINOTRANSFERASE RELATED *TaTAR2.1-3A* increases grain yield of wheat. Plant Physiology. 2017;**174**:2274-2288

[72] He X, Qu B, Li W, Zhao X, Teng W, Ma W, et al. The nitrate-inducible NAC transcription factor *TaNAC2-5A* controls nitrate response and increases wheat yield. Plant Physiology. 2015;**169**:1991-2005

[73] Ma L, Li T, Hao C, Wang Y, Chen X, Zhang X. *TaGS 5-3A*, a grain size gene selected during wheat improvement for larger kernel and yield. Plant Biotechnology Journal. 2016;**14**:1269-1280

[74] Zheng J, Liu H, Wang Y, Wang L, Chang X, Jing R, et al. *TEF-7A*, a transcript elongation factor gene, influences yield-related traits in bread wheat (*Triticum aestivum* L.). Journal of Experimental Botany. 2014;**65**:5351-5365

[75] Wang H, Wang S, Chang X, Hao C, Sun D, Jing R. Identification of *TaPPH-7A* haplotypes and development of a molecular marker associated with important agronomic traits in common wheat. BMC Plant Biology. 2019;**19**:1-2

[76] Yadav D, Shavrukov Y, Bazanova N, Chirkova L, Borisjuk N, Kovalchuk N, et al. Constitutive overexpression of the *TaNF-YB4* gene in transgenic

wheat significantly improves grain yield. Journal of Experimental Botany. 2015;**66**:6635-6650

[77] Qu B, He X, Wang J, Zhao Y, Teng W, Shao A, et al. A wheat CCAAT box-binding transcription factor increases the grain yield of wheat with less fertilizer input. Plant Physiology. 2015;**167**:411-423

[78] Ma M, Wang Q, Li Z, Cheng H, Li Z, Liu X, et al. Expression of *TaCYP78A3*, a gene encoding cytochrome *P450-CYP78A3* protein in wheat (Triticum aestivum L.), affects seed size. The Plant Journal. 2015;**83**:312-325

[79] Nadolska-Orczyk A, Rajchel IK, Orczyk W, Gasparis S. Major genes determining yield-related traits in wheat and barley. Theoretical and Applied Genetics. 2017;**130**:1081-1098

[80] Araus JL, Serret MD, Lopes MS. Transgenic solutions to increase yield and stability in wheat: Shining hope or flash in the pan? Journal of Experimental Botany. 2019;**70**:1419-1424

[81] Sivamani E, Bahieldin A, Wraith JM, Al-Niemi T, Dyer WE, Ho THD, et al. Improved biomass productivity and water use efficiency under water deficit conditions in transgenic wheat constitutively expressing the barley *HVA1* gene. Plant Science. 2000;**155**:1-9

[82] Kulkarni M, Soolanayakanahally R, Ogawa S, Uga Y, Selvaraj MG, Kagale S. Drought response in wheat: Key genes and regulatory mechanisms controlling root system architecture and transpiration efficiency. Frontiers in Chemistry. 2017;**5**:106

[83] González FG, Capella M, Ribichich KF, Curín F, Giacomelli JI, Ayala F, et al. Field-grown transgenic wheat expressing the sunflower gene HaHB4 significantly outyields the wild type. Journal of Experimental Botany. 2019;**70**:1669-1681

[84] Bioceres Crop Solutions [Internet]. 2020. Available from: https://investors.biocerescrops.com/news/news-details/2020/Bioceres-Crop-Solutions-Corp.Announces-Regulatory-Approval-of-Drought-Tolerant-HB4-Wheat-in-Argentina/default.aspx [Accessed: December 19, 2020]

[85] Urnov FD, Rebar EJ, Holmes MC, Zhang HS, Gregory PD. Genome editing with engineered zinc finger nucleases. Nature Reviews Genetics. 2010;**11**:636-646

[86] Silva G, Poirot L, Galetto R, Smith J, Montoya G, Duchateau P, et al. Meganucleases and other tools for targeted genome engineering: Perspectives and challenges for gene therapy. Current Gene Therapy. 2011;**11**:11-27

[87] Joung JK, Sander JD. TALENs: A widely applicable technology for targeted genome editing. Nature Reviews Molecular Cell Biology. 2013;**14**:49-55

[88] Osakabe Y, Osakabe K. Genome editing with engineered nucleases in plants. Plant and Cell Physiology. 2015;**56**:389-400

[89] Jinek M, Chylinski K, Fonfara I, Hauer M, Doudna JA, Charpentier E. A programmable dual-RNA–guided DNA endonuclease in adaptive bacterial immunity. Science. 2012;**337**:816-821

[90] Kumar R, Kaur A, Pandey A, Mamrutha HM, Singh GP. CRISPR-based genome editing in wheat: A comprehensive review and future prospects. Molecular Biology Reports. 2019;**46**:1-13

[91] Wang W, Pan Q, He F, Akhunova A, Chao S, Trick H, et al. Transgenerational

CRISPR-Cas9 activity facilitates multiplex gene editing in allopolyploid wheat. The CRISPR Journal. 2018;**1**:65-74

[92] Cameron P, Fuller CK, Donohoue PD, Jones BN, Thompson MS, Carter MM, et al. Mapping the genomic landscape of CRISPR–Cas9 cleavage. Nature Methods. 2017;**14**:600-606

[93] Kim D, Bae S, Park J, Kim E, Kim S, Yu HR, et al. Digenome-seq: Genome-wide profiling of CRISPR-Cas9 off-target effects in human cells. Nature Methods. 2015;**12**:237-243

[94] Tsai SQ, Nguyen NT, Malagon-Lopez J, Topkar VV, Aryee MJ, Joung JK. CIRCLE-seq: A highly sensitive in vitro screen for genome-wide CRISPR–Cas9 nuclease off-targets. Nature Methods. 2017;**14**:607

[95] Tsai SQ, Zheng Z, Nguyen NT, Liebers M, Topkar VV, Thapar V, et al. GUIDE-seq enables genome-wide profiling of off-target cleavage by CRISPR-Cas nucleases. Nature Biotechnology. 2015;**33**:187-197

[96] Wienert B, Wyman SK, Richardson CD, Yeh CD, Akcakaya P, Porritt MJ, et al. Unbiased detection of CRISPR off- targets in vivo using DISCOVER-Seq. Science. 2019;**364**: 286-289

[97] Doebley JF, Gaut BS, Smith BD. The molecular genetics of crop domestication. Cell. 2006;**127**:1309-1321

[98] Li J, Sun Y, Du J, Zhao Y, Xia L. Generation of targeted point mutations in rice by a modified CRISPR/Cas9 system. Molecular Plant. 2017;**10**:526-529

[99] Sun Y, Zhang X, Wu C, He Y, Ma Y, Hou H, et al. Engineering herbicide-resistant rice plants through CRISPR/Cas9-mediated homologous recombination of acetolactate synthase. Molecular Plant. 2016;**9**:628-631

[100] Ma X, Zhang Q, Zhu Q, Liu W, Chen Y, Qiu R, et al. A robust CRISPR/Cas9 system for convenient, high-efficiency multiplex genome editing in monocot and dicot plants. Molecular Plant. 2015;**8**:1274-1284

[101] Wang M, Lu Y, Botella JR, Mao Y, Hua K, Zhu JK. Gene targeting by homology-directed repair in rice using a geminivirus-based CRISPR/Cas9 system. Molecular Plant. 2017;**10**:1007-1010

[102] Mishra R, Joshi RK, Zhao K. Base editing in crops: Current advances, limitations and future implications. Plant Biotechnology Journal. 2020;**18**:20-31

[103] Li C, Zong Y, Wang Y, Jin S, Zhang D, Song Q, et al. Expanded base editing in rice and wheat using a Cas9-adenosine deaminase fusion. Genome Biology. 2018;**19**:59

[104] Lu Y, Zhu JK. Precise editing of a target base in the rice genome using a modified CRISPR/Cas9 system. Molecular Plant. 2017;**10**:523-525

[105] Tang X, Ren Q, Yang L, Bao Y, Zhong Z, He Y, et al. Single transcript unit CRISPR 2.0 systems for robust Cas9 and Cas12a mediated plant genome editing. Plant Biotechnology Journal. 2019;**17**:1431-1445

[106] Zong Y, Wang Y, Li C, Zhang R, Chen K, Ran Y, et al. Precise base editing in rice, wheat and maize with a Cas9-cytidine deaminase fusion. Nature Biotechnology. 2017;**35**:438

[107] Anzalone AV, Randolph PB, Davis JR, Sousa AA, Koblan LW, Levy JM, et al. Search-and-replace genome editing without double-strand breaks or donor DNA. Nature. 2019;**576**:149-157

[108] Lin Q, Zong Y, Xue C, Wang S, Jin S, Zhu Z, et al. Prime genome editing in

rice and wheat. Nature Biotechnology. 2020;**38**(5):582-585

[109] Zhang Y, Liang Z, Zong Y, Wang Y, Liu J, Chen K, et al. Efficient and transgene-free genome editing in wheat through transient expression of CRISPR/Cas9 DNA or RNA. Nature Communications. 2020;**7**:12617

[110] Liang Z, Chen K, Li T, Zhang Y, Wang Y, Zhao Q, et al. Efficient DNA-free genome editing of bread wheat using CRISPR/Cas9 ribonucleoprotein complexes. Nature Communications. 2017;**8**:14261

[111] Lu HP, Liu SM, Xu SL, Chen WY, Zhou X, Tan YY, et al. CRISPR-S: An active interference element for a rapid and inexpensive selection of genome-edited, transgene-free rice plants. Plant Biotechnology. 2017J;**15**:1371

[112] He Y, Zhu M, Wang L, Wu J, Wang Q, Wang R, et al. Programmed self-elimination of the CRISPR/Cas9 construct greatly accelerates the isolation of edited and transgene-free rice plants. Molecular Plant. 2018;**11**:1210-1213

[113] Korte A, Farlow A. The advantages and limitations of trait analysis with GWAS: A review. Plant Methods. 2013;**9**:29

[114] Mirdita V, He S, Zhao Y, Korzun V, Bothe R, Ebmeyer E, et al. Potential and limits of whole genome prediction of resistance to Fusarium head blight and Septoria tritici blotch in a vast central European elite winter wheat population. Theoretical and Applied Genetics. 2015;**128**:2471-2481

[115] Bentley AR, Scutari M, Gosman N, Faure S, Bedford F, Howell P, et al. Applying association mapping and genomic selection to the dissection of key traits in elite European wheat. Theoretical and Applied Genetics. 2014;**127**:2619-2633

[116] Patel DA, Zander M, Dalton-Morgan J, Batley J. Advances in plant genotyping:Where the future will take us. In: Batley J, editor. Plant Genotyping: Methods and Protocols. New York: Springer New York; 2015. pp. 1-11

[117] Meuwissen THE, Hayes BJ, Goddard ME. Prediction of total genetic value using genome-wide dense marker maps. Genetics. 2001;**157**:1819-1829

[118] Heffner EL, Lorenz AJ, Jannink JL, Sorrells ME. Plant breeding with genomic selection: Gain per unit time and cost. Crop Science. 2010;**50**:1681-1690

[119] Nakaya A, Isobe SN. Will genomic selection be a practical method for plant breeding? Annals of Botany. 2012;**110**:1303-1316

[120] Desta ZA, Ortiz R. Genomic selection: Genome-wide prediction in plant improvement. Trends Plant Science. 2014;**19**:592-601

[121] Sandhu KS, Aoun M, Morris C, Carter AH. Genomic selection for end-use quality and processing traits in soft white winter wheat breeding program with machine and deep learning models. Biology. 2021;**10**:689

[122] Sun J, Poland JA, Mondal S, Crossa J, Juliana P, Singh RP, et al. High-throughput phenotyping platforms enhance genomic selection for wheat grain yield across populations and cycles in early stage. Theoretical and Applied Genetics. 2019;**132**:1705-1720

[123] Crain J, Mondal S, Rutkoski J, Singh RP, Poland J. Combining high-throughput phenotyping and genomic information to increase prediction and selection accuracy in wheat breeding. The Plant Genome. 2018;**11**:1-14

[124] Sandhu KS, Mihalyov PD, Lewien MJ, Pumphrey MO, Carter AH. Combining genomic and phenomic information for predicting grain protein content and grain yield in spring wheat. Frontiers in Plant Science. 2021;**12**:170

[125] McCouch SR, McNally KL, Wang W, Sackville HR. Genomics of gene banks: A case study in rice. American Journal of Botany. 2012;**99**:407-423

[126] Yu X, Li X, Guo T, Zhu C, Wu Y, Mitchell SE, et al. Genomic prediction contributing to a promising global strategy to turbocharge gene banks. Nature Plants. 2016;**2**:1-7

[127] Singh S, Vikram P, Sehgal D, Burgueño J, Sharma A, Singh SK, et al. Harnessing genetic potential of wheat germplasm banks through impact-oriented-prebreeding for future food and nutritional security. Scientific Reports. 2018;**8**:1-11

[128] Poland J, Endelman J, Dawson J, Rutkoski J, Wu S, Manes Y, et al. Genomic selection in wheat breeding using genotyping-by-sequencing. The Plant Genome. 2012;**5**:103-113

[129] Thavamanikumar S, Dolferus R, Thumma BR. Comparison of genomic selection models to predict flowering time and spike grain number in two hexaploid wheat doubled haploid populations. G3: Genes, Genomes, Genetics. 2015;**5**:1991-1998

[130] He S, Schulthess AW, Mirdita V, Zhao Y, Korzun V, Bothe R, et al. Genomic selection in a commercial winter wheat population. Theoretical and Applied Genetics. 2016;**129**:641-651

[131] Huang M, Cabrera A, Hoffstetter A, Griffey C, Van Sanford D, Costa J, et al. Genomic selection for wheat traits and trait stability. Theoretical and Applied Genetics. 2016;**129**:1697-1710

[132] Rutkoski J, Poland J, Mondal S, Autrique E, Pérez LG, Crossa J, et al. Canopy temperature and vegetation indices from high-throughput phenotyping improve accuracy of pedigree and genomic selection for grain yield in wheat. G3: Genes, Genomes, Genetics. 2016;**6**:2799-2808

[133] Sukumaran S, Crossa J, Jarquin D, Lopes M, Reynolds MP. Genomic prediction with pedigree and genotype × environment interaction in spring wheat grown in South and West Asia, North Africa, and Mexico. G3: Genes, Genomes, Genetics. 2017;**7**:481-495

[134] Michel S, Ametz C, Gungor H, Epure D, Grausgruber H, Löschenberger F, et al. Genomic selection across multiple breeding cycles in applied bread wheat breeding. Theoretical and Applied Genetics. 2016;**129**:1179-1189

[135] Michel S, Ametz C, Gungor H, Akgöl B, Epure D, Grausgruber H, et al. Genomic assisted selection for enhancing line breeding: Merging genomic and phenotypic selection in winter wheat breeding programs with preliminary yield trials. Theoretical and Applied Genetics. 2017;**130**:363-376

[136] Sun J, Rutkoski JE, Poland JA, Crossa J, Jannink JL, Sorrells ME. Multitrait, random regression, or simple repeatability model in high-throughput phenotyping data improve genomic prediction for wheat grain yield. The Plant Genome. 2017;**10**:1-15

[137] Basnet BR, Crossa J, Dreisigacker S, Pérez-Rodríguez P, Manes Y, Singh RP, et al. Hybrid wheat prediction using genomic, pedigree, and environmental covariables interaction models. The Plant Genome. 2019;**12**:1-13

[138] Belamkar V, Guttieri MJ, Hussain W, Jarquín D, El-basyoni I, Poland J, et al.

Genomic selection in preliminary yield trials in a winter wheat breeding program. G3: Genes, Genomes, Genetics. 2018;**8**:2735-2747

[139] Guo Z, Zhao Y, Röder MS, Reif JC, Ganal MW, Chen D, et al. Manipulation and prediction of spike morphology traits for the improvement of grain yield in wheat. Scientific Reports. 2018;**8**:1-10

[140] Juliana P, Singh RP, Poland J, Mondal S, Crossa J, Montesinos-López OA, et al. Prospects and challenges of applied genomic selection—A new paradigm in breeding for grain yield in bread wheat. The Plant Genome. 2018;**11**:1-17

[141] Norman A, Taylor J, Edwards J, Kuchel H. Optimising genomic selection in wheat: Effect of marker density, population size and population structure on prediction accuracy. G3: Genes, Genomes, Genetics. 2018;**8**:2889-2899

[142] Hu X, Carver BF, Powers C, Yan L, Zhu L, Chen C. Effectiveness of genomic selection by response to selection for winter wheat variety improvement. The Plant Genome. 2019;**12**:1-15

[143] Krause MR, González-Pérez L, Crossa J, Pérez-Rodríguez P, Montesinos-López O, Singh RP, et al. Hyperspectral reflectance-derived relationship matrices for genomic prediction of grain yield in wheat. G3: Genes, Genomes, Genetics. 2019;**9**:1231-1247

[144] Lozada DN, Mason RE, Sarinelli JM, Brown-Guedira G. Accuracy of genomic selection for grain yield and agronomic traits in soft red winter wheat. BMC Genetics. 2019;**20**:82

[145] Michel S, Löschenberger F, Ametz C, Pachler B, Sparry E, Bürstmayr H. Simultaneous selection for grain yield and protein content in genomics-assisted wheat breeding. Theoretical and Applied Genetics. 2019;**132**:1745-1760

[146] Juliana P, Singh RP, Braun HJ, Huerta-Espino J, Crespo-Herrera L, Govindan V, et al. Genomic selection for grain yield in the CIMMYT wheat breeding program—Status and perspectives. Frontiers in Plant Science. 2020;**11**:1418

[147] Lozada DN, Godoy JV, Ward BP, Carter AH. Genomic prediction and indirect selection for grain yield in US Pacific northwest winter wheat using spectral reflectance indices from high-throughput phenotyping. International Journal of Molecular Sciences. 2020a;**21**:165

[148] Lozada DN, Ward BP, Carter AH. Gains through selection for grain yield in a winter wheat breeding program. PLoS One. 2020b;**15**:0221603

[149] Sehgal D, Rosyara U, Mondal S, Singh R, Poland J, Dreisigacker S. Incorporating genome-wide association mapping results into genomic prediction models for grain yield and yield stability in CIMMYT spring bread wheat. Frontiers in Plant Science. 2020;**11**:197

[150] Tsai HY, Cericola F, Edriss V, Andersen JR, Orabi J, Jensen JD, et al. Use of multiple traits genomic prediction, genotype by environment interactions and spatial effect to improve prediction accuracy in yield data. PLoS One. 2020;**15**:0232665

[151] Bernardo R. Genome wide selection when major genes are known. Crop Science. 2014;**54**:68-75. DOI: 10.2135/cropsci2013.05.0315

[152] Sehgal D, Autrique E, Singh R, Ellis M, Singh S, Dreisigacker S. Identification of genomic regions for

grain yield and yield stability and their epistatic interactions. Scientific Reports. 2017;7:1-2

[153] Sehgal D, Mondal S, Crespo-Herrera L, Velu G, Juliana P, Huerta-Espino J, et al. Haplotype-based, genome-wide association study reveals stable genomic regions for grain yield in CIMMYT spring bread wheat. Frontiers in Genetics. 2020;**11**:1427

[154] Quarrie SA, Pekic Quarrie S, Radosevic R, Rancic D, Kaminska A, Barnes JD, et al. Dissecting a wheat QTL for yield present in a range of environments: From the QTL to candidate genes. Journal of Experimental Botany. 2006;**57**:2627-2637

[155] Bevan MW, Uauy C, Wulff BB, Zhou J, Krasileva K, Clark MD. Genomic innovation for crop improvement. Nature. 2017;**543**:346-354

[156] Qian L, Hickey LT, Stahl A, Werner CR, Hayes B, Snowdon RJ, et al. Exploring and harnessing haplotype diversity to improve yield stability in crops. Frontiers in Plant Science. 2017;**8**:1534

[157] Ledesma-Ramírez L, Solís-Moya E, Iturriaga G, Sehgal D, Reyes-Valdes MH, Montero-Tavera V, et al. GWAS to identify genetic loci for resistance to yellow rust in wheat pre-breeding lines derived from diverse exotic crosses. Frontiers in Plant Science. 2019;**10**:1390

[158] Li F, Wen W, Liu J, Zhang Y, Cao S, He Z, et al. Genetic architecture of grain yield in bread wheat based on genome-wide association studies. BMC Plant Biology. 2019;**19**:1-9

[159] Shokat S, Sehgal D, Liu F, Singh S. GWAS analysis of wheat pre-breeding germplasm for terminal drought stress using next generation sequencing technology. International Journal of Molecular Sciences. 2020;**21**:3156

[160] Brinton J, Ramirez-Gonzalez RH, Simmonds J, Wingen L, Orford S, Griffiths S, et al. A haplotype-led approach to increase the precision of wheat breeding. Communications Biology. 2020;**3**:1-1

[161] Stephens M, Smith NJ, Donnelly P. A new statistical method for haplotype reconstruction from population data. The American Journal of Human Genetics. 2001;**68**:978-989

[162] Bhat JA, Yu D, Bohra A, Ganie SA, Varshney RK. Features and applications of haplotypes in crop breeding. Communications Biology. 2021;**4**:1-2

[163] Hamazaki K, Iwata H. RAINBOW: Haplotype-based genome-wide association study using a novel SNP-set method. PLoS Computational Biology. 2020;**16**:e1007663

[164] Sallam AH, Conley E, Prakapenka D, Da Y, Anderson JA. Improving prediction accuracy using multi-allelic haplotype prediction and training population optimization in wheat. G3: Genes, Genomes, Genetics. 2020;**10**:2265-2273

Section 2

Biotic Stress

Chapter 3

Past, Current and Future of Wheat Diseases in Kenya

Ruth Wanyera and Mercy Wamalwa

Abstract

Wheat (*Triticum aestivum* L.) is an important cereal and is among the crops that contribute significantly to food security in Kenya. However, wheat diseases are among the biotic factors that affect wheat production. Considerable progress has been made to control wheat diseases through host plant resistance breeding and chemical applications. Frequent changes in the pathogens population still present a major challenge to achieving durable resistance. Disease surveillance and monitoring of the pathogens have revealed the changes in virulence across the region, justifying the need to develop and deploy more efficient and sustainable strategies to manage the diseases. Understanding the genetic variability and composition of the diseases is important for variety release with appropriate resistance gene combinations for sustainable disease management. This review highlights the prevalence, distribution of wheat diseases, host plant resistance in the key wheat-growing regions of Kenya, and future prospects in Kenya.

Keywords: wheat, diseases, challenges, control strategies

1. Introduction

Wheat (*Triticum aestivum* L) is the second most important cereal crop in Kenya after maize and is produced mainly under rainfed conditions on 0.4% of the arable land [1, 2]. The crop has greater potential in the country where it is grown in Agro-ecological zones: UH2-UH3; LH2-LH3) [3]. Annual estimated area under production is 150,000 hectares [4] with a production of 320,000MT in 2019 compared to local consumption of 2,450,000MT [5]. The national demand for wheat and wheat consumption is on the increase, partly due to the high population growth, increased urbanization, and changing diet [6, 7]. The local wheat production has not been able to meet this demand leading to the importation of large quantities to fill the gap between supply and demand [8]. However, this is unlikely to be satisfied partly due to pre-harvest sprouting, lodging, losses caused by re-emerging diseases, insect pests, intermittent droughts [6, 9], inadequate seed systems, and poor crop practices under resource-constrained small scale farming conditions. The crop grows in a considerably wide range of altitudes in the country, maturing between 90–145 days depending on the location and cultivars.

There are various wheat diseases such as fungal, which include stem or black rust, caused by *Puccinia graminis* f. sp. *tritici* Erikss and Henning (*Pgt*), yellow/stripe rust, caused by *Puccinia striiformis* f. sp. *tritici* (*Pst*), leaf/brown rust caused by *Puccinia triticina* (*Pt*) and Fusarium caused by *Gibberella zeae* that infect wheat in Kenya. Other diseases include Septoria leaf and glume blotch caused by *S. tritici* and *S. nodorum*, respectively Spot blotch (*Bipolaris sorokiniana*), Loose smut (*Ustilago tritici*), Take All (*Gaeumannomyces graminis* var. *tritici*) and a viral disease, Barley Yellow Dwarf, causal agent Barley Yellow Dwarf Virus (BYDV) [10].

1.1 Wheat rust diseases in Kenya

Among the wheat diseases, rusts have become the most destructive diseases of wheat in Kenya resulting in yield losses of up to 100% in susceptible cultivars [10, 11]. Breeders have been breeding for wheat rust resistance, since 1908, but up to date, there is no permanent solution to the rust diseases as the pathogens keep on evolving rendering the resistant cultivars ineffective [12]. Since the beginning of the wheat breeding program in Kenya in the 1900s, until early 1980s, stem rust was the most serious disease of the three wheat rusts and therefore was given a high research priority by the breeding program. Consequently, many resistant wheat cultivars were developed and the disease seemed to have been controlled. It was until between 1985 and 1988 that trace amounts of the disease were observed in the experimental plots in Njoro; in 1996, it was recorded in some commercial cultivars in Mau-Narok and Molo, and in the year 2000 all the cultivars had become susceptible [10, 12].

Stem or black rust of wheat, caused by *Pgt* is known historically for causing severe losses to wheat production and was the most feared disease in various countries where wheat is grown [13]. The common host is wheat with other small grain cereals, durum wheat (*Triticum durum*), barley (*Hordeum vulgare* L), Rye (*Secale cereael* L.), oats (*Avena fatua* L.), wild barley, goatgrass and forage grasses [14]. Although the disease has been under control through widespread use of resistant cultivars, the re-emergence of a new virulent race, Ug99 [15] first keyed to pathotype TTKS [16] using the North American nomenclature [17] and later as TTKSK after a fifth set of differentials was adapted to further expand the characterization [18]. Prior to the official reporting of the new race, trace amounts of the disease were observed in experimental plots in Njoro between 1985 and 1988, in 1996 the disease was recorded in some commercial cultivars in Mau-Narok and Molo (high altitude areas) in 1996 and in 2000 all the cultivars had become susceptible [10, 12]. This Ug99 race group has evolved and is now composed of 15 races in 14 countries [19–21] with 12 variants (TTKSK, TTKST, TTTSK, PTKSK, PTKST, TTKTT, TTKTK, TTHSK, PTKTK, TTHST, TTKTT+, TTHTT) present in Kenya reversing the gains made by breeders, posing a new and significant threat to wheat production in the Eastern Africa region [16]. In the year 2016, race TKTTF (Digalu race) was genotyped in Kenya for the first time. A new variant TTKTT+ with additional virulence on *Sr*8155B1 was detected in 2019 and another new variant, TTHTT detected in Kenya in 2020 [22]. This is an indication that Ug99 race group is spreading faster specifically, in the areas where close to one billion people reside and the majority of this population consumes wheat and its products [23].

Wheat yellow or stripe rust, caused by *Pst*, is one of the key economical diseases of wheat worldwide [24, 25]. In Kenya, it occurred as early as 1908 and is prevalent in the Rift valley region [12, 26, 27]. Since then it has become a major threat every year as no commercial cultivar is resistant [28, 29]. Serious attacks of the pathogen occur annually and newly introduced resistant cultivars lose their resistance within a short time.

Stripe rust to limits wheat production by affecting the yield and quality of kernels as it develops at an early crop stage when temperatures are favorable for rust development [30]. Stripe rust destroys leaves at jointing to booting growth stages. Consequently, infection of stripe rust on wheat reduces photosynthetic area as early as tillering and jointing stages of development. Stripe rust epidemic has occurred in more than 60 countries in every continent causing yield losses of up to 100% in susceptible cultivars [31]. In East Africa, Kenya, and Ethiopia the epidemics caused yield loss of 67–100% in the year 2010 [25]. In Kenya, wheat is grown throughout the year in different agro-ecological zones, and this increases the concentration of the urediniospores in the air making it difficult to control the disease in susceptible varieties [12, 16]. Yield losses of up to 80% have been estimated but some fields with susceptible cultivars go up to 100% [10, 25].

Stripe rust is a global problem evolving into different races, either from their wild ancestor or their host through introductions [32]. In Kenya and Ethiopia *Yr9* and *Yr27* based cultivars broke down due to evolution of virulent stripe rust races to these genes resulting to yield losses of up to 40% in commercial cultivars like Paa that carried *Yr9* gene [12, 33]. Stripe rust race 134 with virulence for *Yr7*, *6*, *9+* genes were present in Ethiopia, Kenya, Syria, and Yemen [34]. Thirteen races with virulence corresponding to stripe rust resistance genes *Yr1*, *Yr2*, *Yr3*, *Yr6*, *Yr7*, *Yr8*, *Yr9*, *Yr17*, *Yr25*, *Yr27*, and *Avocet S* are present in wheat-growing regions of Kenya [35], these races belong to either strain *Pst1* or *Pst2* which might have been present much earlier than 1982 and 1970.

Wheat leaf rust caused by *Pt* is the most common and widely distributed of the three wheat rusts and occurs in more regions than stem rust and stripe rust [36, 37]. Leaf rust mostly infects wheat in low to medium altitude wheat-growing areas of Kenya [38, 39]. The earliest epidemics of this rust were reported in Kenya as early as 1908 [26]; therefore, it is considered to be among the major three wheat rusts (stem, yellow and leaf) responsible for depressing crop yields drastically depending on the cultivar because of its high frequency and widespread occurrence. Yield losses attributed to leaf rust have been reported to range from 5–16% on average, and up to 40% in epidemic years [40]. Yield losses are usually the result of lower kernel weights and decreased number of kernels as the pathogen may kill wheat seedlings by elevating respiration rate, reducing photosynthetic area on the leaf surfaces, and lessening translocation of carbohydrates [41]. In Kenya, the disease appears sporadically and has not been a problem for the past 20 years, but it has recently emerged in the wheat fields, and experimental plots, including the Kenya Agricultural and Livestock Research Organization (KALRO), Njoro international screening nursery with severity of over 50% [42]. One of the recent studies [43] reported a high reduction in grain yield and kernel weight in some of the Kenyan wheat cultivars.

Highly effective durable resistance to leaf rust has been difficult to achieve due to the high degree of virulence variation in the *Pt* population and the rapid selection of races with virulence to effective *Lr* genes in wheat genotypes [44]. This high degree of specificity has made durable rust resistance in wheat difficult to achieve because the virulence of leaf rust against wheat resistance genes is highly diverse resulting in the existence of many different pathogenic races [37]. For instance, the novel race BBG/BN and its variant BBG/BP overcame the resistance of widely adapted durum cultivars in northwestern Mexico. In Kenya, leaf rust samples collected from wheat-growing areas were found to have virulence for leaf rust resistance genes *Lr*1, *Lr*2b, *Lr*3, *Lr*9, *Lr*11, *Lr*12, *Lr*14a, *Lr*14b, *Lr*18, *Lr*20, Lr22a, *Lr*23, *Lr*24, *Lr*26, and *Lr*27. The race for an isolate collected from Ololulung'a, Narok in the South Rift region was designated as LBBTN [45].

1.2 Other wheat diseases in Kenya

Fusarium diseases, mainly Fusarium head blight of wheat (FHB), also called head scab, are caused mainly by the fungus *Gibberella zeae* (also known as *Fusarium graminearum),* periodically causes significant yield losses and reduced grain quality. *Gibberella zeae* also produces mycotoxins [46]. All Kenya wheat cultivars are susceptible to Fusarium infection [47]. Studies done in Kenya show that the prevalence of FHB and yield loss due to FHB varies from trace to 100% [47, 48].

Septoria diseases are caused by *S. tritici* and *S. nodorum* [49]. Yield losses attributed to heavy incidences of *S. tritici* and *S. nodorum* have been reported to range from 31–53% resulting in shriveled kernels [49]. The occurrence of Septoria diseases in the wheat-growing areas of Kenya is sporadic and severe infection leading to shriveled grain is observed (Wanyera, Personal observation). Some of the foliar fungicides recommended for the control of rust diseases in wheat have been observed to reduce Septoria disease infection when applied at the right time. These foliar fungicides include: (*azoxystrobin* 200 g/L+ *tebuconazole* 300 g/L (Stamina 500 SC); *benzovindiflupy* 30 g/L + *azoxystrobin*114G/L + *propiconazole* 132 g/L (Elatus Arc 265.14 SE); *difenaconazole* 125 g/L + *azoxystrobin*200g/L (Token 325 SC); (trifloxystrobin250g/Kg/L + *tebuconazole* 500 g/Kg (Shadow 750 WG).

Spot blotch caused by *Bipolaris sorokiniana* (Sacc) Shoemaker, perfect stage *Cochliobolus sativus* (S.Ito & Kurib) also causes black point, root rot, and crown rot in wheat. It is known to occur worldwide in warmer environments and is a serious constraint in wheat production in India, Bangladesh, and Nepal [50]. It is also a serious problem on barley. The disease can attack all parts of the wheat plant (seed, roots, shoots and leaves) causing seed-rot, seedling emergence, reduced yield affecting end-use quality of the harvested grain [51, 52]. *Bipolaris* sp. is known to reduce seed viability in wheat and also causes a significant reduction in seed quality and flour [53]. Yield loss estimates of 15–25% and 30% have been reported [54] and on barley in Canada [55].

The sudden upsurge of *Bipolaris* sp. in certain areas of the country has been associated with acid soils. It is estimated that 30% of the soil in the wheat-growing areas is acidic. High infection has been recorded on wheat cultivars Ngamia in Uasin Gishu county (North Rift) and Kenya Nyangumi in Rongai areas of Nakuru county (Central Rift) (Wanyera, Personal observation).

The highland areas of wheat production in Kenya: Molo, Mau Narok (Central Rift), Eldoret and Endebess (North Rift) have a pH of 4.3–5.5 [56]. All wheat cultivars grown in these areas have shown susceptibility to the pathogen but no direct screening has been done. Aluminum toxicity in acid soils has been documented as the primary factor in the reduction of the crop yields [56]. Seed borne nature of the disease has been reported in wheat cultivars in Kenya [57, 58], and studies on disease management revealed that the pathogen can be reduced by the use of seed treatment fungicides [59]. Biological control methods have also been reported [60, 61].

Loose smut caused by *Ustilago tritici* is seed-borne and common in the wheat-growing areas of the world. Infection occurs during flowering through wind-borne spores. In Kenya, the disease occurs rarely and is mostly observed in recycled wheat seeds.

Take All (*Gaeumannomyces graminis* var. *tritici*) is soil and debri-borne, detected mostly in fields that are continuously cultivated with cereals. Both Loose smut and Take All are well managed through use of certified seed, cultural control, and seed treatment fungicides. Some of the recommended seed treatment fungicides are: *prothioconazole* 100 g/L (Redigo FS100), *difenoconazole*92g/L + *metalaxyl-M* 23 g/L

(Dividend Extreme 115 FS), and *azoxystrobin* 141.4 g/L+ *propiconazole*122.4 g/L (Quilt Excel 263.5 SE) [6].

Apart from fungal diseases, another disease that threatens wheat production in Kenya is the Barley Yellow Dwarf Virus (BYDV), which is an important virus disease of cereals globally and has a wide host range that includes wheat, barley, oats, triticale, and over 150 grass species [51]. The disease was first reported in Kenya in 1984 and causes serious damage in barley, wheat, and oats and estimated losses range from 16.5–54.7% [62, 63]. Cereal aphids are vectors of the barley yellow dwarf and five strains have been known to occur in Kenya: RPV (*Rhopalosiphumpadi*), RMV (*R. maidis*), MAV (*Sitobionavenae*), SGV (*Schizaphis graminum*), and PAV (*R. padi*, *S. avenae* [63]. Outbreaks are frequent, and management practices require use of seed dressing insecticides: Gaucho 350FS (*Imidacloprid*), Cruiser 350FS (*Thiamethoxam*), Redigo Deter 350FS (*Clothianidin* + *prothioconazole*), Celest Top 312FS (*Thiamethoxam* + fludioxonil +difenoconozole, and (ii) foliar-applied insecticides (Karate Zeon (*Lambda-cyhalothrin*), Bulldock star 262.5EC (*Betacyfluthrin* + *Chlorpyrifos*), Thunder OD 145 (*Imidacloprid* + *Betacyfluthrin*), Keshet 2.5EC (*Deltamethrin*),Twigathoate 40EC (*Dimethoate*), Nurelle D 50/505 EC (*Cypermethrin* + *Chlorpyrifos*), Alphadime (*Alpha-cypermethrin* + *Dimethoate*), Cyclone 505EC (*Cypermethrin* + *Chlorpyrifos*), and Pirimor 50WG (*Pirimicarb*) [6, 64, 65].

Under favorable environmental conditions, infection of the wheat crop with these diseases can reduce quantity and quality of the grain. Disease surveillance is an epidemiological practice by which the spread is monitored to establish patterns of progression and is key in identifying new diseases and races which can be used in risk assessment and resistance breeding. This review highlights the prevalence, distribution of wheat diseases, host plant resistance in the key wheat-growing regions, and future prospects in Kenya.

1.3 Distribution of diseases in wheat-growing regions of Kenya

Surveys were conducted in the farmer fields in the major wheat-growing regions (Central Rift, South Rift, North Rift, and Mount (Mt) Kenya from 2011 to 2019. The objective was to determine the prevalence and distribution of the wheat diseases and host plant resistance in these regions. Farms were randomly picked along the routes, stopping at every 3 to 5 kilometers. Crops were observed for disease symptoms. An International Standardized survey form was used to keep the records on disease incidence and severity, cultivar grown, production area, and growth stage [66], also any other data that was useful. The Global positioning system (GPS) tool was used to collect precise information on latitude, longitude, and elevation of the sampled farms. Stem, yellow, and leaf rust severities were taken using modified Cobb scale, 0–100% where; 0- immune and 100- susceptible [67]. The host plant response to infection was scored as resistant (R), moderately resistant (MR), moderately susceptible (MS), and susceptible(S) [68]. Incidence and severity of other diseases observed during the surveys were also taken using recommended scales. Septoria diseases were assessed using 0–9 scale [49], where 0 = Free from infection and 9 = Very susceptible/severe infection. Similarly, barley yellow dwarf virus was assessed on a scale of 0–9 [69], where 0 = no symptoms and 9 = full symptom expression, and the Fusarium disease score rating system was 0–5 [70]. **Tables 1** and **2** show the occurrence (percent infection and severity & plant response) of the diseases in all the wheat-growing regions. Rust diseases are common in the wheat fields and stem rust is widespread in all the regions. This explains the importance of stem rust, Ug99 race group, since its detection in

Year	Region	No. of sampled farms	Sr % Infection (%)	Sr % severity and plant response	Yr infection (%)	Yr % severity and plant response	Lr Infection (%)	Lr% severity (%)
2011	Central Rift	62	70.9	0-100S	17.7	TR -60S	17.7	TR-40S
	South Rift	125	68.8	TR-100S	6.4	TR- 20S	17.7	TR-20S
	North Rift	73	48.3	TR-90S	10.9	TR- 20S	10.9	TR-20S
	Mt. Kenya region	67	68.0	TR- 60S	10.4	0 - 40S	13.4	TR-20S
	Total/ Mean	**327**	**63.9**		**11.3**		**14.9**	
2012	Central Rift	67	65.7	TR-80S	4.5	5-50S	11.9	5-30S
	South Rift	71	5.6	TR-20S	—	—	—	—
	North Rift	101	26.7	TR-70S	5.9	5-70S	3.9	10-50S
	Mt. Kenya region	39	58.9	TR-50S	5.1%	5-60S	2.6	30S
	Total/ Mean	**278**	**39.2**		**3.9**		**4.6**	
2013	Central Rift	97	71.0	TR-70S	8.3	TR-50S	6.7	TR-50S
	South Rift	104	68.3	TR-100S	3.8	10S–30S	5.8	TR-50S
	North Rift	78	33.3	TR-70S	10.3	TR-50S	6.5	TR-50S
	Mt. Kenya region	54	25.9	TR-60S	7.4	10S–30S	0	0
	Total/ Mean	**333**	**49.6**		**7.5**		**4.8**	
2014	Central Rift	92	82.5	TR-80S	6.2	TR -50S	6.2	10S–50S
	South Rift	79	72.2	TR-80S	8.9	TR- 60S	—	—
	North Rift	95	55.8	TR-80S	6.3	TR- 40S	5.3	0 - 40S
	Mt. Kenya region	71	57.7	TR- 60S	15.5	5S - 60S	1.4	0-40S
	Total/ Mean	**342**	**67.05**		**4.0**		**4.0**	
2015	Central Rift	66	54.54	TR-80S	5.8	5S–40S	1.5	0-30S
	South Rift	101	35.6	TR-60S	—	—	—	—
	North Rift	106	75.5	TR-50S	8.5	TR-40S	1.9	TR-30S
	Mt. Kenya region	63	71.4	TR-60S	—	—	—	—
	Total/ Mean	**336**	**59.26**		**3.58**		**0.85**	
2016	Central Rift	60	88.3	TR-80S	16.7	TR-60S	3.3	30S–50S
	South Rift	81	76.5	TR-70S	4.9	TR-10S	1.2	0-50S
	North Rift	98	72.4	TR-80S	13.3	TR-40S	10.2	TR-50S
	Mt. Kenya region	61	80.3	TR-90S	1.6	TR	—	—

Year	Region	No. of sampled farms	Sr % Infection (%)	Sr % severity and plant response	Yr infection (%)	Yr % severity and plant response	Lr Infection (%)	Lr% severity (%)
	Total/ Mean	**300**	**79.38**		**9.13**		**3.68**	
2017	Central Rift	54	87.03	TR-70S	8.9	0 -30S	—	—
	South Rift	79	69.2	TR-100S	3.79	TR- 10S	—	—
	North Rift	78	64.1	TR-60S	8.97	TR- 30S	24.4	TR-40S
	Mt. Kenya region	38	44.1	TR- 30S	10.5	TR - 40S	—	—
	Total/ Mean	**249**	**66.11**		**8.04**		**6.10**	
2018	Central Rift	64	74.0	5-50S	10.0	5-60S	10.0	TR-30S
	South Rift	85	42.2	5-70S	3.3	10S–30S	1.1	TR-40S
	North Rift	89	25.84	5-80S	19.1	5S–60S	24.35	5S–70S
	Mt. Kenya region	62	47.9	5-40S	2.81	10S–30S	—	—
	Total/ Mean	**300**	**47.78**		**8.80**		**4.0**	
2019	Central Rift	56	82.2	TR-50S	2.2	0-40S	—	—
	South Rift	87	83.13	TR-80S	1.2	TR	—	—
	North Rift	101	22.77	TR-40S	7.92	TR-40S	4.95	TR-20S
	Mt. Kenya region	46	63.04	TR-50S	10.86	15S–60S	6.5	5S–30S
	Total/ Mean	**290**	**62.79**		**5.62**		**2.86**	

Sr = Stem rust; Yr = Yellow rust; Lr = Leaf rust; TR- trace; S = susceptible; – = no disease observed.

Table 1.
Occurrence of wheat rust diseases in the commercial fields in year 2011–2019.

Uganda and spread to the wheat-growing areas of Kenya, throughout eastern Africa, Yemen, Sudan, Iran, Zimbabwe, Tanzania, South Africa, Mozambique, Zimbabwe, and Iraq [15, 16, 22]. The prediction for the rust diseases to spread towards North Africa, Middle East, Asia and beyond, raises serious concerns of major epidemics that could destroy the world's wheat crop [19].

Yellow rust, which was first described in 1777, and attacked wheat in Kenya as early as 1908 [26], was observed in low incidences but high severities across all the regions (**Table 1**). The disease is also a major threat as no cultivar is resistant [28, 29]. Newly introduced resistant varieties lose their resistance within a short time and farmers are forced to spray to save on yields. Serious attacks of the pathogen occur annually and the disease severity increases with altitude [33]. Serious epidemics also occur in the lower latitudes areas. All the wheat-growing areas are prone to disease in low medium and high altitudes areas.

In Kenya, leaf rust has been sporadic and has not been a problem for the past 20 years, but it has recently emerged in the wheat fields (**Table 1**), and experimental plots, including the international screening nursery with a severity of over 50%. Our

Year	Region	No of sampled farms	Disease incidence (%)		
			Septoria diseases	*Fusarium* sp	BYDV
2011	Central Rift	62	16.1	9.6	0
	South Rift	125	4	1.6	0.8
	North Rift	73	27.4	12.3	0
	Mt. Kenya	67	1.5	0	0
2012	Central Rift	67	42.8	8.9	16.4
	South Rift	71	46.5	2.8	0
	North Rift	101	41.8	0.9	0.9
	Mt. Kenya	39	17.9	2.7	0
2013	Central Rift	97	8.2	6.2	2.1
	South Rift	104	14.4	0.9	0.9
	North Rift	78	28.2	0	0
	Mt. Kenya	54	45.3	1.9	0

Table 2.
Occurrence of Septoria *diseases,* Fusarium *sp. and Barley yellow dwarf virus (BYDV) in commercial wheat field year 2011, 2012, and 2013.*

cultivars are now at risk given the fact that virulences and new races have been identified in Njoro and also South Rift, Ololulung'a areas (data not shown).

The growing of wheat in diverse agro-ecological zones throughout the year [71, 72] in Kenya creates a significant pool of airborne urediniospores, which coupled with favorable climatic conditions and the presence of host plants, favors rapid build up of inoculum and the occurrence of epidemics. This implies that there is a shift in races present each year, which affects different cultivars of wheat. There is continuous attack, due to the presence of wheat crops throughout the year. The breakdown in resistance could also be attributed to mutations [24]. It is, therefore, a problem to reduce the disease infection in susceptible cultivars and also not possible to grow a profitable crop of wheat without the application of fungicides [10, 16]. Septoria diseases, Fusarium spp., Barley yellow dwarf virus are also becoming more prevalent in the commercial fields (**Table 2**), year 2011 to 2013. Disease incidence varied from year to year depending on the chemical/spray applied. Data for the occurrence of these diseases from 2014 to 2019 was not shown because it was similar as shown in **Table 2**.

1.4 Wheat breeding in Kenya

Conventional breeding, which includes testing genotypes in different environments to determine the adaptability of the varieties has been used largely in Kenyan wheat breeding programs to identify resistant varieties [72]. Crop improvement by traditional methods, involves collection, hybridization, and inbreeding that has been practiced since the beginning of 20th Century. However, it has now been realized that these methods are insufficient to make further breakthroughs or cope with the increasing demand for improvement in crop varieties [73]. Some of the limitations of conventional breeding include the exhaustion of the gene pool, low response to biotic and abiotic stress of the introduced materials, and low combining ability, especially with complex characters. In Kenya, diverse agro-ecological zones and favorable environs highly contribute to the emergence of new races. The cultivars grown are at high risk of being infected with diseases, therefore, it is necessary to identify and incorporate genes that confer durable resistance to contain major epidemics [74, 75]. There are various strategies employed to control these diseases in wheat. These include incorporation of genetic resistance into susceptible wheat genotypes, crop management plus use of fungicides. Despite the fact that it takes a long-time, breeding for durable resistance remains to be a cost-effective strategy of minimizing loss due to wheat diseases [76]. Therefore, host resistance is the primary tool to protect wheat crops from wheat fungal rust diseases and other biotic stresses [77]. Breeding for vertical (qualitative) resistance based on major genes and horizontal (quantitative) influenced by several minor genes for wheat disease resistance has been going on in Kenya since wheat introduction in the 19th century. However, due to pathogen evolution, most of the genotypes with qualitative and quantitative resistance become susceptible to the new races, especially wheat rusts pathogens. For instance, wheat cultivars Robin and Eagle 10 released in Kenya as resistant varieties in 2009 and 2010 were overcome by Ug99 variant *Sr*Tmp [78, 79]. Durable resistance by selecting resistant wheat varieties has been going on in Kenya for the past decades, most of the varieties released with resistant genes are now ineffective against the evolving wheat rusts pathogens (**Table 3**).

Kenyan wheat cultivars Robin, NjoroBW2, KS Mwamba, Kwale, Kenya Korongo, Robin, Eagle 10, Kenya Black Hawk 12 (**Tables 3** and **4**), and Kenya Seed Company cultivars were grown by most farmers in the wheat-growing regions of Kenya.

In 2011, KS Mwamba occupied the largest area in Central and South Rift (50.4%), North Rift (45.2%), and in Mount Kenya region 33.8% (**Table 3**). In 2012, the area planted with NjoroBW2 increased: 34.3%, 39.4%, 60.4%, while it decreased for KS Mwamba, 14.9%, 15.5%, and 28.7% in Central, South, and North Rift, respectively (**Table 3**). Cultivar Kwale was highly grown in Central Rift (20.9%), Mt. Kenya (23.1%), and South Rift (20.2%) in 2013. For the cultivars released in 2010 with adult plant resistance (APR) to the wheat stem rust race *Ug99*, Robin occupied 20.6% in Central Rift, 22.2% in Mt. Kenya, 7.7% in North Rift (2013). Cultivar Eagle 10 occupied 1.6% in Mt. Kenya region and 0.9% in South Rift. Mixed and other unknown cultivars were common across the regions and this could be due to the high cost of certified seed.

No	Variety	Region and variety area planted (%)											
		Central Rift			South Rift			North Rift			Mt.Kenya		
		2011	2012	2013	2011	2012	2013	2011	2012	2013	2011	2012	2013
1	NjoroBW2	30.4	34.3	24.7	29.6	39.4	42.3	49.0	60.4	57.7	26.2	—	—
2	KS Mwamba	50.4	14.9	12.3	50.4	15.5	20.2	45.2	28.7	26.9	33.8	25.6	25.9
3	Kwale	14.0	20.9	14.4	14.4	9.9	20.2	4.1	1.9	5.1	6.2	23.1	11.4
4	Robin	—	7.5	20.6	—	1.4	—	—	—	7.7	—	—	22.2
5	Mixed	3.2	8.9	10.3	—	—	1.9	1.3	5.9	1.3	—	17.9	7.4
6	Eagle10	1.6	—	—	—	1.4	0.9		—	—	—	—	1.6
7	Others	0.4	13.5	17.7	5.6	32.4	14.5	0.4	3.1	1.3	33.8	33.1	31.5

-cultivar not planted.

Table 3.
Commonly grown cultivars in the key wheat-growing regions in the year 2011–2013.

Commercial Name	Pedigree	Yield potential tons/Ha	Days to Maturity	Year of release	Resistant status
Robin	BABAX/LR42//BABAX*2/3/TUKURU	8.1	110–120	2009	Overcome by TTKTT race in 2014
Eagle10	EMB16/CBRD//CBRD	6.5	100–110	2010	Good resistance to stem rust (Ug99 strain).
Kenya Wren	THELIN#2/TUKURU	8.5	120–130	2012	APR to both yellow and stem rust diseases.
Kenya Tai	ND643/2*WBLLI	6.5	100–110	2012	Resistant to both stem rust and yellow rust.
Kenya Sunbird	ND643/2*WBLLI	6.5	100–110	2012	Resistance to stem rust,
Kenya Korongo	BABAX/LR42//BABAX*2/4/SNI/TRAP#1/3KAUZ*2/TRAP//KAUZ	8.5	120–130	2012	Overcome by TTKTT race in 2014
Kenya Kingbird	TAM-200/TUI/6/PAVON-76//CAR-422/ANAHUAC-75/5/BOBWHITE/CROW//BUCKBUCK/PAVON-76/3/YECORA-70/4/TRAP-1	6.0	90–110	2012	Developed for Adult plant resistance to both stem rust and yellow rust.
Black Hawk12	URES/JUN//KAUZ/3/BABAX/4/TILHI	8.0	120–130	2012	Overcome by TTKTT race in 2014
Kenya Hornbill	PASTOR//HXL7573/2*BAU/3/SOKOLL/WBLL1	7.5	110–120	2016	High APR to yellow rust and moderate resistance to stem rust.
Kenya Deer	PBW343*2/KUKUNA*2//YANAC	7.8	100–110	2016	High adult plant resistance to stem rust and yellow.
Kenya Weaverbird	PRINIA/3/ALTAR84/AE. SQ //2*OPATA/4/CHEN/AEGILOPS SQUARROSA(TAUS)//BCN/3/BAV92	8.0	110–120	2016	High APR to stem rust.
Kenya Peacock	QUAIU/3/PGO/SERI/BAV92		120–130	2016	High APR to both stem and yellow rusts.
Kenya Falcon	KSW/5/2*ATLAR 84/AE. SQUARROSA (221)//3*BORL95/3/URES/JUN/KAUZ/4/WBLL1	8.0	100–115	2016	Excellent seedling and APR to stem rust. Highly resistant to yellow rust
Kenya Songbird	KSW/5/2*ALTAR 84/AE. SQUARROSA (221)//3*BORL95/3/URES/JUN/KAUZ/4/WBLL1	8.2	110–120	2016	–

Commercial Name	Pedigree	Yield potential tons/Ha	Days to Maturity	Year of release	Resistant status
Kenya Pelican	KSW/5/2*ALTAR84 /AE. AQUARROSA (221)//3*BORL95/3/ URES/JUN/KAUZ/4/WBLL1	8.5	120–130	2016	High APR to stem rust.
Kenya Jacana	KSW/SAUAL//SAUAL/3/REEDLING #1	6.5–8.0	110– 130 Days	2019	Moderately resistant to original Ug99 races. In warmer weather, susceptible to race "TTKTT"
Kenya Kasuko	KSW/SAUAL//SAUAL/3/REEDLING #1	7.0–8.0	110– 120 Days	2019	Moderately resistant to original Ug99 races. In warmer weather, susceptible to race "TTKTT"

APR = adult plant resistant.
Source: http://wheatatlas.org/country/varieties/KEN/0?AspxAutoDetectCookieSupport=1.

Table 4.
Current wheat varieties released in Kenya, their yield potential and resistant attributes.

In 2014, cultivar Robin was highest in Central Rift (43.3%), South Rift (41.8%), and Mt. Kenya region (43.7%) while cultivar NjoroBW2 was highest in North Rift (64.2%), Central Rift (15.5%), Mt. Kenya (12.7%), and South Rift (8.9%). KS Mwamba was highest in North Rift (16.4%), Central Rift (15.5%), South Rift (11.4%), and Mount Kenya (9.9%). The area under cultivar Kwale was highest in Central Rift (10.3%), followed by South Rift and Mt. Kenya region (7.6% and 7.0%), respectively. The area under cultivar Eagle 10 was only noted in South Rift 18.9% and overall occupied only 4.4% across the region. Mixed and other unknown cultivars were common in Mt. Kenya region: Kenya Ibis occupied 1.2%, Duma (0.6%), mixed cultivars (1.8%).

In 2015, the area planted with NjoroBW2 increased in North Rift from 63.2% in 2014 to 70.6% in 2015 cultivar Robin increased in Mt. Kenya region (66.7%) as opposed to 2014 (43.7%), but decreased in Central Rift from to 21.2%. The area under production in North Rift increased from to 16.5% and decreased in South Rift from 35.6%. Cultivar Eagle 10 was only observed in South Rift (20.8%) of the sampled fields. Cultivars Kenya Wren and Kenya Hawk12 were observed only in the South Rift (1.9%).

The area planted on NjoroBW2 decreased in North Rift to 64.3% Mt. Kenya 49.2% in 2016. Cultivar Eagle 10 was only grown in South Rift (9.9%) and in North Rift (2.0%) of the sampled fields. Cultivars Kenya Wren was grown in South Rift (2.5%) and North Rift (1.0%) while Kenya Hawk12 was grown in the South Rift (6.2%). Kenya Korongo was grown in South Rift (8.3%) and North Rift (1.0%).

In 2017, cultivar NjoroBW2 was popular in North Rift (69.2%), Central Rift (40.7%), and South Rift (32.9%). Robin was popular in Mt. Kenya region (32.4%), followed by South Rift (20.3%), Central Rift (10.7%), and North Rift (7.7%). Kenya Korongo was only popular in the Central Rift (27.8%). Kwale was popular in Central Rift (7.4%), South Rift (6.3%), and Mt. Kenya (5.3%) area under production of the sampled fields. Cultivar NjoroBW2 occupied the largest area in North Rift (69.2%) and Central Rift (40.7%). Cultivar Eagle 10 was recorded in South Rift (13.9%), Central Rift (1.9%), and in North Rift (1.3%) of the sampled fields. While variety Duma was popular in Mt. Kenya region (42.1%) area under production of the sampled fields. Kenya Wren was grown in Central Rift 1.9.3%), South Rift (1.3%), and North Rift (1.0%) while Kenya Black Hawk12 was grown in South Rift (6.2%) and North Rift (1.3%). Kingbird was only grown in South Rift (2.5%) and North Rift (1.3%) area under production of the sampled fields.

In 2018, cultivar NjoroBW2 was popular in all the regions: North Rift (70.8%), Central Rift (44.0%), South Rift (34.4%), and Mt. Kenya region (14.08%). Robin was grown in North Rift (16.0%), Mt. Kenya (14.6%), Central Rift (12.0%), and South Rift (11.8). Kenya Korongo was only popular in the Mt. Kenya region (36.6%) while Kwale was grown in Central Rift (8.0%) and South Rift (4.7%). Variety Eagle 10 was only popular in South Rift (14.0%) area under production of the sampled fields. The area under production of variety Eagle 10 remained the same in the South Rift as the previous year. Kenya Wren was only grown in South Rift (3.5%), Kenya Black Hawk12 was grown in North Rift (2.5%). while Kingbird was grown in South Rift (1.2%) and North Rift (1.3%).

In 2019 cultivar NjoroBW2 was popular in Mt. Kenya (43.5%). North Rift (42.5%), Central Rift (39.28%), South Rift (31.0%). Kenya Korongo was grown in Mt. Kenya (23.9%), Central Rift (16.0%), South Rift (11.5%), and North Rift (7.92%). Cultivar Robin was popular in the Mt. Kenya (23.9%), South Rift (13.8%), and Central Rift (5.4%). Kwale was grown in North Rift (9.9.0%), South Rift (8.0%), Mt. Kenya (6.5%), Central Rift (5.4%) area under production of the sampled fields. Cultivar Eagle 10 was only popular in South Rift (14.9%) and Central Rift (7.1%) area under production of the sampled fields. The Kenya Seed Company cultivars were more popular in the North Rift (24.8%) area under production of the sampled fields.

Over fifty percent of the previously released varieties (**Table 4**) are now susceptible to the Ug99 race. Robin, Kenya Black Hawk12, Kenya Korongo, Kenya Jacana, and Kenya Kasuko are susceptible to Ug99 races (TTKTK and TTKTT) that were detected on Robin with virulence to *Sr*Tmp and virulence to *Sr*24, respectively. The resistance in Kwale and other genotypes like Kenya Plume (not included) is due to adult plant resistance (APR) genes and others associated with variable levels of disease symptoms, which show recessive inheritance and is expressed primarily during the APR which has been deployed in a breeding program in Kenya [80]. Stem rust resistance gene *Sr2* is an APR gene present in some of the Kenyan genotypes such as Kwale, Kenya Swara, Kenya Nyangumi, and Kenya Popo together with other APR genes condition resistance to stem rust [11, 81].

There is a long history of wheat breeding in Kenya as early as 1908, however, the use of molecular breeding tools is very limited thereby hampering the rate of genetic gains achieved. As such, the national breeding program has depended on introductions of wheat lines from international wheat breeding programs including CIMMYT and ICARDA. Understanding the composition and diversity of fungal wheat disease resistance in Kenya wheat germplasm is important for defining breeding strategies and prioritizing trait targets for wheat improvement [82].

Biotechnological approaches in wheat breeding such as double haploid (DH) and mutational breeding have been used to speed up breeding by complementing conventional breeding [72]. DH which shortens the breeding period by a single cycle has been used in Kenya to produce varieties such as K. Ibis. Mutation breeding brings about genetic variation and accelerates the outcome of variety release has been applied at KARLO, Njoro to release varieties NjoroBW2 and K. Heroe by irradiation using gamma rays [72, 83]. Conventional method of gene pyramiding is time-consuming, hence, the incorporation of molecular breeding is efficient in breeding for biotic and abiotic stresses in wheat for quick release of resistant varieties. The use of molecular markers enhances phenotypic selection because it makes it more efficient, effective, reliable, and cost-effective compared to conventional plant breeding, hence improving the latter [84]. There has been some concern about the incorporation of DNA marker technology in many plant-breeding institutions and most institutions can now develop their own markers [85, 86]. Molecular markers such as SSR, AFLP, and KASP markers have been developed to evaluate genotypes for biotic stresses such as diseases in Kenyan varieties [7, 82, 87].

1.5 Control of wheat diseases in Kenya

Other than host plant resistance, cultural and chemical methods have been used to control wheat diseases in Kenya. Cultural control techniques such as growing resistant genotypes, late planting, reduced irrigation, avoidance of excessive nitrogen use, and elimination of volunteer and grass plants can reduce stripe rust severities as they limit exposure time to inoculum [25]. Altering planting date and separating the vulnerable crop from the pathogen in either time or space controls certain airborne disseminated pathogens of wheat [88]. Although the cultural techniques are used, they are either not profitable, conflict with conservation farming, or reduce yield potential [89]. Genetic resistance combined with chemical treatments, although expensive to the poor resource farmers may often be very effective in controlling wheat diseases [90]. Some of the fungicides used by farmers in Kenya are listed in **Table 5**. The application of seed treatment chemicals such as *triadimenol* (sterol biosynthesis inhibitors) and *carboxin* (respiratory inhibitors) and the use of moderately resistant cultivars is effective in controlling wheat diseases as it provides the most efficient use of fungicides at the lowest rates [91, 92]. Reduced chemical applications could also minimize the potential development of resistance to the chemicals [90]. Although wheat diseases have been controlled by timely use of effective chemicals, the cost of chemicals and their application creates a huge burden for growers. In Kenya, large-scale farmers are the only ones who can afford to spray chemicals, but it costs about $ 8 million annually [10]. Fungicides can be used to control fungal diseases, but they cause environmental hazards and lead to fungicide tolerant strains [25]. Re-emergence of new virulent races has reversed the gains made by breeders, posing a new and significant threat to wheat breeding in Kenya [16]. Resistance in the commercial wheat cultivars in Kenya, including those released in the last decade, has been overcome by the new races making it impossible to grow a profitable crop of wheat without the use of fungicides [10, 90].

During surveys, we noted that farmers who sprayed following the right recommendations of fungicides in **Table 5** had good yields compared to those who did not spray or sprayed without following the proper recommendations hence losing the crop to the disease. Majority of the farmers sprayed the fungicides to reduce/suppress disease infections, particularly the rusts, but some sprayed farms were noted to have high disease infections. These are farms that either had been sprayed late or the timing/ chemical concentrations were not right.

1.6 Future of breeding for wheat diseases resistance in Kenya

Despite the occurrence of wheat diseases in Kenya, information on the genetic basis of the diseases and wheat cultivars is limited. Molecular genetic markers have been advanced from phenotypic and protein-based markers to DNA sequence polymorphism, this accelerates the process of plant breeding when coupled with conventional breeding [93]. Since many traits valued by plant breeders are complex and polygenic, it is essential to involve the deliberate combination of various genomic regions from many different individuals in the development of an adapted elite variety [94]. Sequencing polymorphism markers are important in identifying genetic diversity in cultivated and wild genotypes, the source of novel genomic regions, alleles, and traits [95].

In crops, marker-assisted selection (MAS) has been made efficient by designation of markers associated with economic importance, for instance, disease resistance (wheat rust), response to abiotic stress and seed quality [96, 97]. The use of molecular markers enhances phenotypic selection because it makes it more efficient, effective, reliable, and cost-effective compared to conventional plant breeding hence improving the latter [84]. There has been some concern about the incorporation of DNA marker technology in many plant-breeding institutions but most institutions can now develop their own markers [85, 86].

In genetic studies of wheat, genetic markers such as amplified fragment length polymorphism (AFLP), restriction fragment length polymorphism (RFLP), random amplified polymorphic DNA (RAPD), simple sequence repeat (SSR), and single nucleotide polymorphism (SNP) have been used but they are limited in their own ways [98]. These limitations are being overcome by improving already available techniques to form next-generation sequencing (NGS) [98]. With next-generation sequencing (NGS) technologies, SNP markers have been discovered in wheat, which is a good choice due to their abundance in the genome as they are distributed across all the wheat chromosomes [99]. These technologies offer easier means to map

No	Chemical name	Common name	Rate L/ha
1	*trifloxystrobin* 100gL+ *tebuconazole* 200 g/Ll	Nativo 300SC	1.0
2	*prothioconazole* 125gL + *tebuconazole* 125gLl	Prosaro 250EC	1.0
3	*epoxiconazole* 250 g/L	Twiga Epox GF	1.0
4	*tebuconazole* 200 g/L	Fezan 250 EW GF	1.0
5	*picoxystrobin* 200 g/L + *cyproconazole* 80 g/L	Acanto Plus	1.0
6	*epoxiconazole* 62.5 g/L + *pyraclostrobin* 62.5 g/L	Abacus SE	1.0
7	*epoxiconazole* 18 g/L + *thiophanate methyl* 310 g/L	Rexduo SE	1.0
8	*metconazole* 27.5 g/L+ *epoxiconazole* 37.5 g/L + 200 g/L *picoxystrobin* + 80 g/L *cyproconazole*	Osiris EC	1.0
9	*propiconazole* 62.5 g/L+ *chlorothalonil* 375 g/L + *cyproconazole* 50 g/L	Cherokee 487.5 SE	1.0
10	*propiconazole* 250 g/L + *cyproconazole* 60 g/L	Menara 410EC	0.5
11	*tebuconazole* 430 g/L	Tebulis 430 SC	0.5
12	*tebuconazole* 200 g/L + *azoxystrobin* 200 g/L	Azimut SC	1.0
13	*bixafen* 75 g/L + *prothioconazole* 100 g/L + *tebuconazole*100g/L	Skyway Xpro 275 EC	1.2
14	*propiconazole* 150 g/L + *difeconazole* 150 g/L	Atlas 300EC	1.0
15	*propiconazole* 172.4 g/L + *azoxystrobin* 141.1 g/L	Quilt Excel 265 SE	1.25
16	*epoxiconazole* 187 g/L + *thiophanate methyl* 310 g/L	Swing Xtra 497 SC	1.0
17	*monopotassium phosphate* 43% + *dipotassiumphosphate* 19%	Fosphite Liquid	4.0
18	*azoxystrobin* 80 g/L+ *chlorothalonil* 400 g/L	Amizoc 480 EC	1.8
19	*bixafen* 50 g/L+ *tebuconazole* 166 g/L	Zantara 216 EC	1.0
20	*fluxapyroxad* 41.6 g/L + *epoxiconazole* 41.6 g/L + *pyraclostrobin* 66.60 g/L	Ceriax 149.8 EC	1.0
21	*azoxystrobin* 200 g/L+ *tebuconazole* 300 g/L	Stamina 500SC	0.9
22	*difenaconazole* 125 g/L + *azoxystrobin*200g/L	Token 325 SC	0.75
23	*benzovindiflupy* 30 g/L + *azoxystrobin*114G/L + *propiconazole* 132 g/L	Elatus Arc 265.14 SE	1.0
24	*tebuconazole/tridimenol*	Silvacur 375 EC	1.0
25	*tebuconazole*	Folicur 250 EC	1.0
26	(*trifloxystrobin* 250g/Kg/L + *tebuconazole* 500 g/Kg))	Shadow 750 WG SC	400 g

** Can control Fusarium Head Blight (FHB) when spayed at flowering** Can control Fusarium Head Blight (FHB) and Septoria diseases GF- Generic fungicide.*

Table 5.
Recommended fungicides for control/reduction of foliar wheat diseases in Kenya.

polymorphic genetic loci and identify genes for important traits [98]. Microsatellite markers have been used to determine the genetic diversity of wheat stem rust races in Kenya ([100]; Wanyera, unpublished data).

1.6.1 Single nucleotide polymorphism (SNP) markers

Single nucleotide variations in genome sequences of individuals of a population are known as SNPs. They result when DNA sequence differs by a single base and are the most abundant molecular markers in the genome [101]. SNPs and flanking sequences are found by library construction and sequencing or through the screening of readily available sequence databases [102]. Genotyping methods, including DNA chips, allele-specific PCR, and primer extension approaches based on SNPs, are particularly attractive for their high data throughput and for suitability for automation [103]. They are used for a wide range of purposes, including rapid identification of crop cultivars and construction of ultra-high-density genetic maps [103, 104]. SNPs markers have been used in wheat in identifying resistance genes for stripe rust *Yr5*, leaf rust *Lr16*, stem rust *Sr6*, the waxy starch gene Wx-D, and Karnal bunt resistance among others [105–107].

1.6.2 Kompetitive allele specific PCR (KASP) markers

Application of modern marker-assisted breeding approaches can help accelerate variety development efforts, single nucleotide polymorphisms (SNPs) markers have emerged as powerful tools for many genetic applications mainly due to their low assay cost, high abundance, co-dominant inheritance, high-throughput, and ease of use [101]. Numerous genotyping platforms have therefore been developed for SNP genotyping [108, 109] including KASP (Kompetitive Allele Specific PCR) which is a gel-free and fluorescent-based genotyping platform. KASP is fast emerging as a global benchmark in SNP genotyping [110, 111] developed and validated 70 KASP assays for functional genes controlling economically important traits such as plant height, disease resistance, yield, and quality in bread wheat. KASP markers have been used to determine alleles for important agronomic traits in wheat in East Africa, Kenya, and Ethiopia [82].

1.6.3 Use of sequence-characterized-amplified region (SCAR)

The application of molecular markers in different epidemiological studies is crucial in developing strain-specific markers such as Sequence-characterized-amplified-region (SCAR) markers [112]. The SCAR markers are codominant, while others are dominant single locus which allows for quick and easy PCR amplification-based detection and hence used in the studies of pathogens [113]. The SCAR markers are efficient in testing large samples and useful in tracing the origin and spread of microbial pathogens with the ability for long-distance disposal and invasion like yellow rust [114]. SCAR markers SCAR1265 and SCAR1400 were developed in wheat to identify powdery mildew (*B. graminis*) gene *Pm21,* which was located on 6AL/6Vs same locus for gene *Yr26* [115]. Species-specific sequence-characterized-amplified-region (SCAR) markers have been used to characterize stripe rust races in Kenya [35].

2. Conclusion

There are high disease incidences and severity of wheat diseases particularly wheat rusts in the farmers' fields, which is attributed to the use of highly susceptible wheat cultivars and also climate change contributing to emerging of new diseases.

For example, the evolution and spread of Ug99 race group and additional races like Digalu race (TKTTF) are spreading very fast causing epidemics subjecting the wheat germplasm to vulnerability.

Other wheat diseases such as *Septoria* and *Fusarium* although sporadic are also a major concern in the wheat-growing regions in the country. The increase in the spread of these diseases is largely due to the widespread of cultivars that are highly susceptible. The favorable climatic conditions and additional costs of fungicides qualify the diseases as damaging with a strong impact on wheat production. Varieties with adequate resistance are now being released and continued monitoring of disease virulences throughout the country is necessary to detect shifts in the pathogen population as early as possible and therefore to effect an appropriate breeding strategy. Effective genetic control of the diseases using the state of the art molecular techniques will require a coordinated effort, including race monitoring, collection, and characterization of sources of resistance and resistance breeding.

In Kenya, different research groups consisting of plant breeders, plant pathologists, agronomists, international partners, and farmers are working towards achieving host plant resistance and ways to combat wheat diseases in order to achieve high yields and contribute to food security.

Acknowledgements

The authors wish to acknowledge the technical staff of the Plant Pathology section KALRO, Njoro, for assistance in collating pertinent information for this article.

Author details

Ruth Wanyera[1*] and Mercy Wamalwa[2]

1 Independent Scientist, Kenya

2 Department of Biological Sciences, Egerton University, Njoro, Kenya

*Address all correspondence to: wanyerar@gmail.com

References

[1] Republic of Kenya. National Food and Nutrition Security Policy. Cathedral Road, Nairobi, Kenya: Agricultural Sector Coordination Unit (ASCU), Kilimo House; 2011

[2] Kenya Agricultural Research Institute. Strategic Plan. Annual Reports, KARI, Nairobi- Kenya; 2009. pp. 2009-2014

[3] Jaetzold R, Schmidt H, Hornetz B, Shisanya C. Farm management Handbook. 2nd ed. Vol. II. Part B Central Kenya. Subpart B1a: Southern Rift Valley. Ministry of Agriculture, Kenya, in Cooperation with the German Agency for Technical Corporation; 2010

[4] Economic Review of Agriculture. Ministry of Agriculture. Nairobi, Kenya; 2013. p. 109

[5] FAOSTAT. 2019. Available at: http://www.fao.org/faostat/en/#home

[6] Kamwaga J, Macharia G, Boyd L, Chiurugwi T, Midgley I, Canales C, et al. Kenya Wheat Production Handbook. Nairobi, Kenya: Kenya Agricultural and Livestock Research Organization; 2016. ISBN 978-9966-30-025-6

[7] Macharia G, Waweru B. Wheat in Kenya: Past and twenty-first century breeding. In: Wanyera R, Owuoche J, editors. Wheat Improvement, Management and Utilization. Rijeka: InTech; 2017. ISBN: 978-953-51-3151-9, Online: ISBN: 978-953-51-3152-6

[8] FAOSTAT. 2020. Statistical Database of the Food and Agriculture of the United Nations. Available from: http://www.fao.org/faostat/en/#home

[9] Kenya Agricultural Research Institute. 2009. Strategic Plan Implementation Frame Work. Annual Reports, KARI, Nairobi, Kenya; 2009-2014

[10] Wanyera R, Macharia JK, Kilonzo SM, Kamundia JW. Foliar fungicides to control wheat stem rust, race TTKS (Ug99) in Kenya. Plant Disease. 2009;**93**:929-932

[11] Wamalwa MN, Owuoche J, Ogendo J, Wanyera R. Multi-pathotype testing of selected Kenyan wheat germplasm and Watkin landraces for resistance to wheat stripe rust (*Puccinia striiformis* f. sp *tritici*) races. Agronomy. 2019;**9**(11):770

[12] Fetch TG, Park RF, Pretorius ZA, Depauw RM. Stem rust: Its history in Kenya and research to combat a global wheat threat. Canadian Journal of Plant Pathology, (just-accepted). 2021;**43**:S275-S297. DOI: 10.1080/07060661.2021.1902860

[13] Singh VK. Pathophenotyping and genome guided characterization of rust fungi infecting wheat and other cereals-A training manual. In: The Cereal Rusts in History. Vol. 10. 2020

[14] Li A, Liu D, Yang W, Kishii M, Mao L. Synthetic hexaploid wheat: Yesterday, today, and tomorrow. Engineering. 2018;**4**(4):552-558

[15] Pretorius ZA, Singh RP, Wagoire WW, Payne TS. Detection of virulence to wheat stem rust resistance gene *Sr31* in *Puccinia graminis* f. sp. *tritici* in Uganda. Plant Disease. 2000;**84**:203

[16] Wanyera R, Kinyua MG, Jin Y, Singh R, Roelfs AP. The spread of stem rust caused by *Puccinia graminis* f. sp. *tritici*, with virulence on *Sr31* in wheat in eastern Africa. Plant Disease. 2006;**90**:113

[17] Roelfs AP, Singh RP, Saari EE. Rust Diseases of Wheat: Concepts and Methods of Disease Management. Mexico, D.F: CIMMYT; 1992

[18] Jin Y, Szabo LJ, Pretorius ZA, Singh RP, Ward R, Fetch T. Detection of virulence to resistance gene Sr24 within race TTKS of *Puccinia graminis* f. sp. *tritici*. Plant Disease. 2008;**92**:923-926

[19] Singh RP, Hodson DP, Jin Y, Lagudah ES, Ayliffe MA, Bhavani S, et al. Emergence and spread of new races of wheat stem rust fungus: Continued threat to food security and prospects of genetic control. Phytopathology. 2015;**105**:872-884

[20] Newcomb M, Olivera PD, Rouse MN, Szabo LJ, Johnson J, Gale S, et al. (2016). Characterization of Kenyan isolates of *Puccinia graminis* f. sp. *tritici* from 2008 to 2014 reveals virulence to *SrTmp* in the Ug99 race group. Phytopathology. 2016;**100**:986-996

[21] Nazari K, Al-Maaroof E, Kurtulus E, Kavaz H, Hodson D, Ozseven I. First report of Ug99 race TTKTT of wheat stem rust (*Puccinia graminis* f. sp. *tritici*) in Iraq. Plant Disease. 2021;**105**(9):2719

[22] RustTracker.org. 2020 Pathotype Tracker – Where Is Ug99? Retrieved from: https://rusttracker.cimmyt.org/?page_id=22

[23] Olson EL. Broadening the wheat gene pool for stem rust resistance through genomic-assisted introgressions from *Aegilops tauschii* [PhD thesis]. Kansas State University: Manhattan, Kansas; 2012

[24] Hovmøller MS. Sources of seedling and adult plant resistance to *Puccinia striiformis* f. sp. *tritici* in European wheat. Plant Breeding. 2007;**126**:225-233

[25] Chen XM. Pathogens which threaten food security: *Puccinia striiformis*, the wheat stripe rust pathogen. Food Security. 2020;**12**(2):239-251

[26] Thorpe HC. Wheat Breeding in Kenya. Canada: First International wheat Genetics Symposium, University of Manitoba; 1959

[27] Wanyera R. The status of rust diseases of wheat in Kenya. In: Breeding for disease resistance with emphasis on durability. In: Daniel DL, editor. Proceedings of a Regional Workshop for Eastern, Central and Southern Africa. Njoro, Kenya. 1994. pp. 222-225

[28] Bonthuis H. Survival of stripe rust (*Puccinia striiformis*) on wheat in the Kenyan highlands and the consequences for virulence. Mededelingen van de Faculteit landbouwwetenschappen. Rijksuniversiteit Gent. 1985;**50**(4): 1109-1117

[29] Kenya Agricultural Research Institute. Annual Reports, KARI, Nairobi, Kenya; 1990-2012

[30] Braun HJ, Singh RP, Huerta-Espino J, Herrera-Foessel SA, Lan CX, Basnet BR. Global status of yellow rust on wheat and strategies for its short- and long-term control. In: Proceedings of the 2nd International Wheat Stripe Rust Symposium, Regional Cereal Rust Research, Center Izmir, Turkey, 28th April – 1st May 2014. 2014. p. 25

[31] Aggarwal R, Kulshreshtha D, Sharma S, Singh VK, Manjunatha C, Bhardwaj SC, et al. Molecular characterization of Indian pathotypes of *Puccinia striiformis* f. sp. *tritici* and multigene phylogenetic analysis to establish inter-and intraspecific relationships. Genetics and Molecular Biology. 2018;**41**:834-842

[32] Hartmann FE, Rodríguez de la Vega RC, Carpentier F, Gladieux P, Cornille A, Hood ME, et al. Understanding adaptation, coevolution, host specialization, and mating system in castrating anther-smut fungi by combining population and comparative genomics. Annual Review of Phytopathology. 2019;**57**:431-457

[33] Danial DL, Stubbs RW, Parlevliet JE. Evolution of virulence patterns in yellow rust races and its implications for breeding for resistance in wheat in Kenya. Euphytica. 1994;**80**:165-170

[34] Ding Y, Cuddy WS, Wellings CR, Zhang P, Thach T, Hovmøller MS, et al. Incursions of divergent genotypes, evolution of virulence and host jumps shape a continental clonal population of the stripe rust pathogen *Puccinia striiformis*. Molecular Ecology. 2021;**00**:1-19

[35] Wamalwa M, Wanyera R, Rodriguez-Algaba J, Boyd L, Owuoche J, Ogendo J, et al. Distribution of *Puccinia striiformis* f. sp. tritici races and virulence in wheat growing regions of Kenya from 1970 TO 2014. Plant Disease. 2021;PDIS-11

[36] Kolmer JA, Long DL, Hughes ME. Physiological specialization of *Puccinia triticina* on wheat in the United States in 2003. Plant Disease. 2005;**89**:1201-1206

[37] Kolmer J. Leaf rust of wheat: Pathogen biology, variation and host resistance. Forests. 2013;**4**(1):70-84

[38] Roelfs AP, Singh RP, Saari EE. Rust Diseases of Wheat: Concepts and Methods of Disease Management. Mexico, D.F: CIMMYT; 1992

[39] Marasas CN, Smale M, Singh RP. The economic impact of productivity maintenance research: Breeding for leaf rust resistance in modern wheat. Agricultural Economics. 2003;**29**:253-263

[40] Bolton MD, Kolmer JA, Garvin DF. Wheat leaf rust caused by Puccinia triticina. Molecular Plant Pathology. 2008;**9**:563-575

[41] Arslan U, Karbulut OA, Yagdi K. Reaction of wheat lines to leaf rust (Puccinia triticina) in Turkey. Bangladesh Journal of Botany. 2007;**36**(2):163-166

[42] Kenya Agricultural Research Institute. Annual Reports. Kenya: Nairobi; 2011

[43] Nyamu GA, Owuoche JO, Ojwang PR, Wanyera R. Evaluation of Kenyan wheat (*Triticum aestivum* L.) genotypes for leaf rust (*Puccinia triticina* Eriks) at adult plant stage. International Journal Agronomic Agriculture Research. 2017;**11**(6):66-78

[44] Jin Y, Kolmer JA, Long DL. Wheat leaf rust and stem rust in the United States. Australian Journal of Agricultural Research. 2007;**58**:631-638

[45] Wamalwa M, Wanyera R. Characterization of wheat (Triticum aestivum L.) leaf rust (Puccinia triticina) races in Kenya. In: Proceedings of EAAPP Mini-Conference and Stakeholders Open Day 12-15 November 2013. Naivasha, Kenya: RDCoE/Morendat Training Centre; 2014

[46] Windels CE. Economic and social impacts of fusarium head blight: Changing farms and rural communities in the northern Great Plains. Phytopathology. 2000;**90**:17-21

[47] Muthomi JW, Oerke EC, Dehne HW, Mutitu EW. Susceptibility of Kenyan wheat varieties to head blight, fungal invasion and Deoxynivalenol

accumulation inoculated with fusarium graminearum. Journal of Phytopathology. 2002;**150**(1):30-36

[48] Wagacha JM, Steiner U, Dehne HW, Zuehlke S, Spiteller M, Muthomi JW, et al. Diversity in mycotoxins and fungal species infecting wheat in Nakuru district, Kenya. Journal of Phytopathology. 2010;**157**:527-535

[49] Eyal Z, Scharen AL, Prescott JM, Van Ginkel M. The Septoria Diseases of Wheat Concepts and Methods of Diseases Management. Mexico, D.F: CIMMYT; 1987. ISBN: 968-6127-06-02

[50] Gupta PK, Chand R, Vasistha NK, Pandey SP, Kumar U, Mishra VK, et al. Spot blotch disease of wheat: The current status of research on genetics and breeding. Plant Pathology. 2018;**67**:508-531

[51] Wiese MV. Compendium of Wheat Diseases. St. Paul, MN: APS Press; 1987

[52] Mathur SB, Cunfer BM. Seed Borne Diseases and Seed Health Testing of Wheat. Copenhagen: Danish Government Institute of Seed Pathology for Developing Countries; 1993. p. 168

[53] Mehta YR. Constraints on the integrated management of spot blotch of wheat. In: Duveiller E, Dubin HJ, Reeves J, McNab A, editors. Helminthosporium Blight of Wheat: Spot Blotch and Tan Spot. Mexico, DF: CIMMYT; 1998. pp. 18-27

[54] Dubin HJ, van Ginkel M. The status of wheat diseases and disease research in the warmer areas. Wheat for the nontraditional warm areas: a proceedings of the International Conference July 29-August 3 1990 Foz do Iguaçu, Brazil. CIMMYT; 1991. pp. 125-45

[55] Ghazvini H, Tekauz A. Virulence diversity in the population of *Bipolaris sorokiniana*. Plant Disease. 2007;**91**(7):814-821

[56] Nyachiro JM. Development of Aluminum-tolerant bread wheat varieties in Kenya. In: Saunders DA, editor. Wheat for the Non-traditional Warm Areas. Mexico, D.F: CIMMYT; 1990. pp. 486-490

[57] Wanyera R, Kinyua MG, Kongsdal O, Hynkemejer A. Detection of seed- bornefungal pathogens infecting some crops in Kenya. In: Korir JK, Gohole LS, Muasya RM, editors. Proceedings of the Stakeholders workshop on Seed Industry in Kenya. Agrotech; 2004. p. 82-88. ISSN: 1727-8155

[58] Wanyera R. Seed-health status in wheat varieties Kenya. Proceedings of the First EgertonUniversity/NPBRC-NJORO Symposium in Agricultural Research. For Development (eds. J.N. Nanua, et al.); 2001. p. 249-256

[59] Wanyera R. Seed-borne pathogens of wheat in Kenya. Regional Wheat Workshop for Eastern, Central and Southern Africa, 10; University of Stellenbosch, South Africa; 14-18 Sep 1998. In: TRegional Wheat Workshop for Eastern, Central and Southern Africa, 10; University of Stellenbosch, South Africa; 14-18 Sep 1998 Addis Ababa (Ethiopia)CIMMYT C1999 (No. Look under series title. CIMMYT.). Centro Internacional de Mejoramiento de Maiz y Trigo (CIMMYT), Addis Ababa (Ethiopia). 1999

[60] Villa-Rodríguez E, Parra-Cota F, Castro-Longoria E, López-Cervantes J, de los Santos-Villalobos S. *Bacillus subtilis* TE3: A promising biological control agent against *Bipolaris sorokiniana*, the causal agent of spot blotch in wheat (*Triticum turgidum* L. subsp. durum). Biological Control. 2019;**132**:135-143

[61] Singh UB, Malviya D, Singh S, Kumar M, Sahu PK, Singh HV, et al. T*richoderma harzianum*- and methyl Jasmonate-induced resistance to *Bipolaris sorokiniana* through enhanced Phenylpropanoid activities in bread wheat (*Triticum aestivum* L.). Frontiers in Microbiology. 2019;**10**:3389

[62] Wangai AW. Effects of barley yellow dwarf virus on cereals in Kenya. In: Burnett PA, editor. World Perspectives on Barley Yellow Dwarf Virus. Mexico, D.F: CIMMYT; 1990. pp. 391-393

[63] Munene M, Nyakwara ZA, Njuguna MN, Maina IN. The occurrence of cereal aphids in Rainfed wheat in Kenya: The problem and possible integrated Pest management strategies. In: Wanyera R, Owuoche J, editors. Wheat Improvement, Management and Utilization. Rijeka: IntechOpen; 2017. DOI: 10.5772/67115. Available from: https://www.intechopen.com/chapters/53906

[64] Wanjama JK. Review of wheat aphid as a pest of cereals in the Rift Valley province Kenya. Entomologist Newsletter. 1979;**9**:2-3

[65] Migui SM, Macharia M, Muthangyia PM, Wanjama JK. Evaluation of insecticides for the control of cereal aphids transmitting barley yellow dwarf virus in barley. Pp-203-210. In: Tanner DG, Mwangi W, editors. Seventh Regional Wheat Workshop for Eastern, Central and Southern Africa. Nakuru, Kenya: CIMMYT; 1992

[66] Zadoks JC, Chang TT, Konzak CF. A decimal code for the growth stages of cereals. Weed Research. 1974;**14**:415-421

[67] Peterson RF, Campell AB, Hannah AE. A diagrammatic scale for estimating rust intensity of leaves and stems of cereals. Canadian Journal of Research. 1984;**26**:496-500

[68] Roelfs AP, Singh RP, Saari EE. Rust Diseases of Wheat: Concepts and Methods of Disease Management. Mexico, D.F: CIMMYT; 1992

[69] Muthangya PM, Macharia M, Wanjama JK. Effects of Imidacloprid seed dressing and foliar aphicide on cereal aphids and BYDV on barley in Kenya. In: Tanner DG, Payne TS, Abdalla OS, editors. The Nineth Regional Wheat Workshop: For Eastern, Central and Southern Africa. 1993

[70] James C. A Manual of Assessment Keys for Plant Diseases. No. BOOK. American Phytopathological Society; 1971

[71] Green GJ, Martens JW, Ribeiro O. Epidemiology and specialization of wheat and oat stem rusts in Kenya in 1968. Phytopathology. 1970;**60**:309-314

[72] Njau PN, Kimurto PK, Kinyua MG, Okwaro HK, Ogolla JB. Wheat productivity improvement in the drought prone areas of Kenya. African Crop Science Journal. 2006;**14**(1):49-57

[73] Kinyua M, Malinga JN, Wanyama J, Karanja L, Njau P, Leo T, Alomba E (2001) Improvement of wheat for resistance to Russian wheat aphid.

[74] Dawit WK, Flath WE, Weber E, Schumann MS, Röderand S, Chen X. Postulation and mapping of seedling stripe rust resistance genes in Ethiopian bread wheat cultivars. Plant Pathology. 2012;**94**:403-409

[75] Kang ZS, Zhao J, Han DJ, Zhang HC, Wang XJ, Wang CF, et al. Status of wheat rust research and control in China. In: BGRI 2010 Technical Workshop, 30-31 May 2010, St Petersburg, Russia. PP; 2010. p. 38

[76] Yuen GY, Schoneweis SD. Strategies for managing fusarium head blight

and deoxynivalenol accumulation in wheat. International Journal of Food Microbiology. 2007;**119**(1-2):126-130

[77] Loughman A, Fitzgerald JR, Brennan MP, Higgins J, Downer R, Cox D, et al. Roles for fibrinogen, immunoglobulin and complement in platelet activation promoted by *Staphylococcus aureus* clumping factor A. Molecular Microbiology. 2005;**57**(3):804-818

[78] Njau PN, Wanyera R, Singh D, Singh R, Gethi M. Release of stem rust resistant wheat varieties for commercial production in Kenya. Journal of Agricultural Science and Technology. 2011;**A1**:587-598

[79] Alemu G. Wheat breeding for disease resistance: Review. Journal of Microbiology and Biotechnology. 2019;**4**(2):000142

[80] Spielmeyer W, Sharp PJ, Lagudah ES. Identification and validation of markers linked to broad-spectrum stem rust resistance gene *Sr2* in wheat (*Triticum aestivum* L.). Crop Science. 2003;**43**:333-336

[81] Njau PN, Jin Y, Huerta-Espino J, Keller B, Singh RP. Identification and evaluation of sources of resistance to stem rust race Ug99 in wheat. Plant Disease. 2010;**94**(4):413-419

[82] Wamalwa M, Tadesse Z, Muthui L, Yao N, Zegeye H, Randhawa M, et al. Allelic diversity study of functional genes in East Africa bread wheat highlights opportunities for genetic improvement. Molecular Breeding. 2020;**40**(11):1-4

[83] M'mboyi F, Njau P, Olembo N, Kinyua M, Malinga MJ. Application of Biotechnology to Wheat Improvement in Sub-Saharan Africa: The Case of Kenya. Nairobi, Kenya: African Biotechnology Stakeholders Forum (ABSF); 2010

[84] Yang Q, Fang T, Li X, Ma C, Yang S, Kang Z, et al. Improving stripe rust resistance and agronomic performance in three elite wheat cultivars using a combination of phenotypic selection and marker detection of *Yr48*. Crop Protection. 2021;**148**:105752

[85] Eagles HA, Barian HS, Ogbonnaya FC, Rebetzke GJ, Hollamby GJ, Henry RJ, et al. Implementation of markers in Australian wheat breeding. Austrialian Journal of Agricultural Research. 2001;**52**:1349-1356

[86] Kaur P, Sachan S, Sharma A. Weed competitive ability in wheat: A peek through in its functional significance, present status and future prospects. Physiology and Molecular Biology of Plants. 2021;**1**:15

[87] Malinga J, Kinyua M, Wanjama J, Kamau A, Karanja L, Awala J, et al. Evidence of a new Russian wheat aphid biotype in Kenya adduced from molecular markers. In: 12th Regional Wheat Workshop for Eastern, Central and Southern Africa, Nakuru, Kenya, 22-26 November, 2004. 2005

[88] Ellis JG, Lagudah ES, Spielmeyer W, Dodds PN. The past, present and future of breeding rust resistant wheat. Frontiers of Plant Science. 2014;**5**:641

[89] Chen XM. Epidemiology and control of stripe rust (*Puccinia striiformis* f sp *tritici*). Canadian Journal of Plant Pathology. 2005;**27**:314-337

[90] Chen XM, Kang ZS, editors. Stripe Rust (719 pages). Dordrecht: Springer; 2017

[91] Boshoff WHP, Pretorius ZA, Van BD. Fungicide efficacy and the impact of stripe rust on spring and winter wheat in South Africa. South African Journal of Plant and Soil. 2003;**20**:11-17

[92] Zobir SAM, Ali A, Adzmi F, Sulaiman MR, Ahmad K. A review on Nano pesticides for plant protection synthesized using the supramolecular chemistry of layered hydroxide hosts. Biology. 2021;**10**(11):1077

[93] Iqbal M, Shahzad R, Bilal K, Qaisar R, Nisar A, Kanwal S, et al. (2021). DNA fingerprinting of crops and its applications in the field of plant breeding. Journal of Agricultural Research. 2021;**59**(1):13-28

[94] Peleman JD, van der Voort JR. Breeding by design. Trends in Plant Science. 2003;**8**:330-334

[95] Li HH, Ribaut JM, Li ZL, Wang JK. Inclusive composite interval mapping (ICIM) for digenic epistasis of quantitative traits in biparental populations. Applied Genetics. 2008;**2008**(116):243-260

[96] Forster BP, Ellis RP, Thomas WTB, Newton AC, Tuberosa R, This D, et al. The development and application of molecular markers for abiotic stress tolerance in barley. Journal of Experimental Botany. 2000;**51**(342):19-27

[97] Rana M, Kaldate R, Nabi SU, Wani SH, Khan H. Marker-assisted breeding for resistance against wheat rusts. In: Physiological, Molecular, and Genetic Perspectives of Wheat Improvement. Cham: Springer; 2021. pp. 229-262

[98] Kumar S, Kumar M, Mir RR, Kumar R, Kumar S. Advances in molecular markers and their use in genetic improvement of wheat. In: Physiological, Molecular, and Genetic Perspectives of Wheat Improvement. Cham: Springer; 2021. pp. 139-174

[99] Hussain B, Alaux M, Sehgal D, Appels R, Batley J, Bentley AR, et al. In: Edwards D, editor. Wheat Genomics and Breeding: Bridging the Gap. AgriRxiv. 2021

[100] Okello OE (2019). Analysis of the population structure 2015 *Pucciniagraminis* f. sp *tritici* (Pgt) in Kenya using simple sequence repeats markers. MSC. Thesis, University of Eldoret, Kenya

[101] Bangarwa SK, Balwan KLS, Choudhary M, Choudhary MK. An introduction to DNA-markers and their role in crop improvement. The Pharma Journal. 2021;**10**(7):638-643

[102] Spooner D, van Treuren R, de Vicente MC. Molecular Markers for Gene Bank Management. IPGRI Technical Bulletin No. 10. Rome, Italy: International Plant Genetic Resources Institute; 2005

[103] Pandurangan S, Workman C, Nilsen K, Kumar S. Introduction to marker-assisted selection in wheat breeding. In: Accelerated Breeding of Cereal Crops. New York, NY: Humana; 2022. pp. 77-117

[104] Mondini L, Noorani A, Pagnotta MA. Assessing plant genetic diversity by molecular tools. Diversity. 2009;**1**:19-35

[105] Kassa MT, You FM, Hiebert CW, Pozniak CJ, Fobert PR, Sharpe AG, et al. Highly predictive SNP markers for efficient selection of the wheat leaf rust resistance gene Lr16. BMC Plant Biology. 2017;**17**(1):1-9

[106] Mourad AM, Sallam A, Belamkar V, Wegulo S, Bowden R, Jin Y, et al. Genome-wide association study for identification and validation of novel SNP markers for Sr6 stem rust resistance gene in bread wheat. Frontiers in Plant Science. 2018;**9**:380

[107] Naruoka Y, Ando K, Bulli P, Muleta KT, Rynearson S, Pumphrey MO. Identification and validation of SNP markers linked to the stripe rust resistance gene *Yr5* in wheat. Crop Science. 2016;**56**(6):3055-3065

[108] Sobrino B, Brio'n M, and Carracedo A. SNPs in forensic genetics: A review on SNP typing methodologies. Forensic Science International. 2005;**154**:181-194

[109] Wang L, Liao B, Gong L, Xiao S, Huang Z. Haploid genome analysis reveals a tandem cluster of four HSP20 genes involved in the high-temperature adaptation of *Coriolopsis trogii*. Microbiology Spectrum. 2021;**9**(1):287-221

[110] Kumpatla SP, Buyyarapu R, Abdurakhmonov IY, Mammadov JA. Genomics-assisted plant breeding in the 21st century: Technological advances and progress. In: Abdurakhmonov I, editor. Plant Breeding. Rijeka: InTech publishers; 2012. Available from: http://www.intechopen.com/books/plantbreeding/genomics-assisted-plant-breeding-in-the-21stcenturytechnological-advances-and-progress

[111] Rasheed A, Wen W, Gao FM, Zhai S, Jin H, Liu JD. Development and validation of KASP assays for functional genes underpinning key economic traits in wheat. Theoretical and Applied Genetics. 2016;**129**:1843-1860

[112] Hodson DP, Hansen JG, Lassen P, Alemayehu Y, Arista J, Sonder K. Tracking the wheat rust pathogen. In: Proceedings from BGRI Technical Meeting, Beijing, China. 2012. p. 20

[113] Naeimi S, Koscsube S. Strain-specific SCAR markers for the detection of *Trichoderma harzianum* AS12- 2, a biological control agent against *Rhizoctonia solani*, the causal agent of rice sheath blight. Acta Biology Hungary. 2011;**62**:73-84

[114] Walter S, Ali S, Kemen E, Nazari K, Bahri BA, Enjalbert J, et al. Molecular markers for tracking the origin and worldwide distribution of invasive strains of *Puccinia striiformis*. Ecology and Evolution. 2016;**6**:2790-2804

[115] Liu Z, Sun Q, Ni Z, Yang T. Development of SCAR markers linked to the *Pm21* gene conferring resistance to powdery mildew in common wheat. Plant Breeding. 1999;**118**:215-219

Chapter 4

Potential of Trichoderma Isolates to Control Plant Pathogen, Leaf Rust on Different Commercial Wheat Varieties/Genotypes

Sadia Afzal, Adeela Haroon, Muhammad Arshad Hussain, Asad-Ur-Rehman Chaudary, Muhammad Amjad Bashir, Sagheer Atta, Saqib Bashir and Muhammad Adnan Bodlah

Abstract

The efficiency in the treatment of leaf rust of wheat was examined for the plant leaf extracts of neem and Moringa at varied concentrations of 50, 100, and 150 ml correspondingly. All treatments decreased fungal growth *in vitro* by greater than 90%. The germination of spores was decreased by 91.99% in the presence of neem leaf extract at 150 ml concentration. The percentage of pustules/leaf was reduced by foliar spray of the same treatments on seedlings of the wheat plant. The wheat plants show the greatest response against the pathogen of leaf rust by plant extract second foliar application on the fourth day of infection. Spray application of 150 ml, 100 mL of neem leaf extracts, and 150 ml of Moringa leaf extracts at wheat seedlings and rust development completely prevented four days after leaf rust inoculation. The application of treatments of all extracts on wheat plants at the mature stage significantly reduced the disease (ACI, average infection coefficient) and increased the efficacy of plant extract application as compared with control but neem 150 ml treatment was most effective in all. There was a higher increase of the chlorophyll and phenol content in wheat plants.

Keywords: *Trichoderma herzianum*, leaf rust, commercial wheat varieties, plant pathogen, biocontrol agent

1. Introduction

Wheat (*Triticum astivum* L.) is an important crop that is grown for years to fulfill the requirements of human hunger. The demand for staple crops increases due to the increase in the world population. In Pakistan, the demand for its product increases therefore growing on a large scale at the government and small agricultural land farmer's level. The tetraploidy and hexaploidy wheat is mostly grown in Asia at 99 million HA (hectares), and overall world production is approximately 215 million HA [1]. In Pakistan, India, and China, the total production is 62 million hectares [2]. In Pakistan, only the wheat sown in an 8.80-million-hectare area produces about 25.09 million tons [3].

The rust fungus gives great loss to the production of wheat all over the world where wheat is grown. Only in Asia does it affect about 43–63% of the growing region if susceptible varieties were grown [4].

2. Background study

In this chapter, a brief review of research work is given in a manner to highlight the contemporary status of findings in leaf rust.

Wheat rust was divided into three types: Of the most common wheat rust is leaf rust caused by *Puccinia triticiana* due to its distribution ability. It has usually fewer losses than the other two types of wheat rust, but the frequent virulence behavior of wheat leaf rust makes it interested to researchers due to its high annual losses. The kernel weight reduction was the major cause of yield loss. The surveillance shows that the rust pathogen has resistance to wheat varieties due to mutation or some migrated genes of rust evolved from other areas. In CIMMYT wheat, the rust pathogen has slow virulence rather than other yielding wheat varieties [5]. It shows that the rust pathogen was resistant against high yielding and low wheat cultivars, disease complexity, and some measures available in practice for control.

2.1 Leaf rust epidemic

In an epidemic situation of leaf rust *Puccinia graminaea* for the wheat susceptible variety, there was a 90% loss in yield. The inoculum that was used against wheat was taken from Research Institute Murree, Pakistan. The fungus was sprayed with five-nozzle sprayer, which was present as a suspension of uredospore [6]. There were about 192 wheat varieties from which the resistibility of wheat genotype was greater in number than the susceptible one. A total of 64 wheat genotypes show resistance, while susceptibility was not shown by any of the genotypes. But some algal species exist in wheat genotypes.

The study purpose of the wheat rust disease damage both qualitatively and quantitatively if the susceptible varieties of wheat line/genotype can be managed by the resistant line development. They evaluate the 30 lines against yellow and leaf rust where they do artificial inoculation and some lines were observed under natural conditions to assess the disease severity [7]. By using cobb's scale method, they observe different rating scales of virulence on the 16 genotypes under a natural condition in comparison with the artificially inoculate rust in wheat lines. The genotypes also show different virulency against leaf rust and yellow rust. The data show that among 30 wheat lines that were inoculated artificially, the resistant and moderately resistant varieties/lines were six in number, while the line/varieties showing MRMS response were 13 and few of the lines showed susceptibility and moderately susceptible. But in natural conditions except from two lines/varieties, others were resistant against leaf and stripe rust. The resistant varieties can be a managed way to manage the leaf and stripe rust so breeders can have a stance on developing resistant varieties.

The surveillance in Pakistan from 2016 to 2018 of leaf rust affects the yield of wheat. A 3-year study design contains 95 districts from which 1202 fields were observed to check the spatial and temporal vigilance of disease severity of leaf rust distributed in the Sindh and Punjab provision of Pakistan. The results of 3-year disease incidence showed the most prevalence of disease in 2017 than in 2016 and 2018. The most affected province is where 60% disease severity occurred in 20% region, while Punjab has only 5% region where south Punjab was most affected and in Khyber-Pakhtunkhwa and Azad Kashmir only 1% disease occurred [8]. Some varieties that show susceptibility were Sehar, Inqlab-91, Shafaqand Morocco.

For the surveillance of rust virulence and disease incidence, their assessment characters are evaluated through the survey of trap plot. The survey is helpful in the seed system, plant breeding, and disease-protecting strategies. Many activities were done at the national level for rust pathogen control but at the global level, the strategies of rust surveillance work very slowly. To make rust surveillance effective at the global level, the Global Cereal Rust Monitoring System GCRMS was recognized to cope with the reinforced problem [9]. The system was a web-based monitoring protocol that will be helpful in testing, disease management, rust virulency, and all the factors that were interrelated for the cause of rust pathogen posing threat to wheat. It also includes surveillance data that will be compared to check the rust pathogen virulency at the global level.

2.2 Susceptibility of commercial wheat varieties

Two-season research was done by Muhammad et al. [10]. During the wheat-growing seasons of 2010–2011 and 2011–2012, 325 genotypes of bread wheat (*Triticum aestivum L.*) were tested for leaf rust resistance against specific pathotypes in the field. In the 2010–2011 growing season, 225 wheat cultivars exhibited no response to leaf rust, 12 genotypes were resistive to leaf rust response, 20 varieties/genotypes showed relatively resistance, 40 wheat lines were moderately susceptible, and 15 were MRMS and 13 genotypes indicated vulnerable response. In total, 233 wheat genotypes did not show any response, 8 genotypes were resistant, 14 genotypes were moderate resistance, 40 wheat lines were moderately susceptible, 8 genotypes were moderately resistant to moderately susceptible, and 22 genotypes were the susceptible response of wheat to leaf rust during the 2011–2012 wheat season. Slow-rusting genotypes had low AUDPC values, whereas high-rusting lines had high AUDPC values. The spread of leaf rust has been strongly impacted by epidemiological variables. Rust responses of different wheat genotypes had shown a strong correlation with environmental factors. Leaf rust reactions were linked to temperature attributes such as maximum and minimum temperature levels, rainfall, and relative humidity. It was also shown that some genotypes responded differently throughout the two crop seasons, which might be due to differences in environmental variables.

How much the severity of the leaf rust disease impacts photosynthetic and grain output in wheat. This was accomplished by calculating the photosynthetic rate, disease severity, chlorophyll content, and wheat reduction in six wheat cultivars grown in uncontrolled and fungicide-treated environments [11]. The mean disease severity level of leaf rust was the greatest on Faisalabad-08 and Galaxy-13 among six wheat varieties/lines such as Faisalabad-08, Galaxy-13, Lasani-08, Millat-11, and two wheat lines NW-3-3341-7 and NW-1-8183-8, although grain yields of wheat were also greater in Millat-11, Galaxy-13, and FSD-08. Fungicide dramatically decreased rust infections and increased chlorophyll concentration and photosynthetic rate, leading to considerably greater production in treated plots. Wheat cultivars FSD-08 and Galaxy-13 were determined to be highly resistant to rust illnesses based on leaf rust severity and yield component assessments. When treated with a rust fungicide, NW-3-3341-7 showed the cheapest and best ratio result. Rust disease drastically reduced grain output, according to the study. When host immunity is combined with little fungicide treatment, the negative effects of leaf rust were reduced and net yields were maximized.

The plants were infected with *Puccinia recondita f.sp.tritici* in natural conditions. Among 197 advanced wheat lines/varieties, based on a measure of disease severity, 89 lines/varieties of leaf rust were clear of any symptoms and 43 lines/varieties showed resistance, 32 MS, 10 mildly susceptible 16, and 7 were susceptible and highly susceptible, respectively, with extremely sensitive in the early planted plots. In late potted plants, 74 were healthy, 28 were immune, 31 were moderately resistant, 8 were moderately resistant, 17 were sensitive, and 39 were moderately sensitive. Either in early sown or late sown nurseries, there were no signs of yellow rust. Most lines/species planted lately exhibited considerably higher rust rates than the comparable genotypes planted earlier. Commercially produced wheat cultivars such as Bahawalpur 97, Inqilab-91, Kohistan-97, MH-97, and Iqbal-99 exhibited no leaf rust signs, suggesting disease resistance. In the early and late sown wheat plants, 89 and 74 lines/varieties were devoid of any disease signs or insect attack, respectively, suggesting their high genetic possibility for improved pest and diseases-resistant breeding [12].

The natural conditions were conducive to the establishment of the wheat leaf rust disease. Out of 150 lines/cultivars tested for brown/leaf rust, 29 lines/cultivars were immune, 57 types exhibited resistance, and the rest were vulnerable [13]. The area under the disease progress curves (AUDPCs) estimates of all types were determined. According to the virulency formula investigated, 57 types of leaf rust were resistant and 49 variations were susceptible to leaf rust fungus. Environmental variables have a significant impact on the progression of wheat leaf rust infections. There was also a relationship between disease severity and environmental factors. Many varieties/lines logically responded to environmental variables. Temperature, humidity levels, wind velocity, and rainfall were found to have a substantial impact on illness severity. Even though pathogenicity incidence did not affect the leaf rust virulence. The occurrence of virulence for them is concerning in situations where the genetic foundation of resistance in currently farmed varieties was stumpy.

The most cultivated varieties were Morocco, Pak-81, Fsd-85, Lylpur-73, Inqlab-91, Fsd-83, and WL-711. From 1991 to 1992 to 2000–2001, these varieties were consistently cultivated in a rust trial. Infection was achieved using both natural inoculum and artificial inoculum of the leaf rust *Puccinia recondita f.sp. tritici*. From the development of the first symptoms until the morphological development of the crop, leaf rust diagnostic evaluations premised on Peterson's scale were collected [14]. A local weather station captured weather parameters such as weekly air temp, humidity levels, and precipitation. Leaf rust prediction models were developed using regression analysis for weather parameters as the independent variable and leaf rust severity as the dependent variable. Leaf rust developed in 10 seasons of wheat at different periods between January and April in different periods. All other environmental factors, except for rainfall, exhibited a significant association with illness severity. Based on 10 years of data, linear regression analysis revealed that the lowest temperature and night-time humidity levels were significant. Except for 1995–1997, the lowest temperature correctly predicted leaf rust in another 8 years.

2.3 Some historic bioagents used for controlling leaf rust

Eight plant extracts were used for the biocontrol of leaf rust and compared the efficacy of these plant extracts, that is, neem, white cedar, clove, garlic, garden quinine, Brazilian pepper, black cumin, and anthi mandhaari. By using foliar spray and *in vitro* evaluation, the spore germination was inhibited in both conditions [15]. In all used plant leaf extract, neem showed a significant result (98.99%) such as fungicide, while other plants extract shows efficacy lower than that of neem. The other method used by them is soaking the seeds in 2 ml/L of plant extract, which inhibits the spore/pustules per leaf by 36.82%. When the soaking method was compared with a foliar spray that reduced the spore's production or pustule per leaf by 100% after 4 days of inoculation, these plants give a maximum result with positive control of Sumi fungicide against the wheat leaf rust pathogen. The most effective extract was neem, clove, and garden quinine. ACI, the average coefficient of infection, shows that at a mature stage of the plant, the foliar spray is more effective than other methods. The foliar spray of the extract also affected test weight, 1000 kernel weight, and grain yield production with one or two sprays after certain days of application. It was a way forward to lessen the use of fungicide and increase the use of plant extract.

Some biological agents against *Bipolaris orzyae* cause rice brown spot disease; its casing agent is a fungus *Cochliobolus miyabeanus*. They used plant extracts, antagonistic organisms, and oil cake. Two plants extracts out of eight plants that extract *Nerium oleander* inhibit 77.4% growth and 8 of 0.3% spores germinate *Pithecolobium dulshows* showing inhibition and spore germination 75.1 and 80%, respectively. These plant extracts were more effective against the *Bipolaris oryzae*, while the oil cake of *Azadirachta indica* inhibition percentage was 80.18% and the percentage of spore germination was 81.13%. The cake extracts of mahua and castors were also effective against the pathogen. The antagonistic organism *Trichoderma viride* showed 62.92% inhibition, and *Trichoderma harzianum* and *Trichoderma reesei* were also effective against the growth of mycelium and spore germination [16]. Under field conditions and glasshouse experiment, *N. oleander*, neem cake extract, and the specie *Trichoderma viride* were effective in controlling the *Bipolars oryzae* when sprayed two times in 15 day of intervals after the appearance of disease in rice plants.

There are three different domains to control leaf rust because the importance of wheat in the world cannot be ignored. The rust species *Puccinia recondita* gives a harsh response to wheat yield. The *in vitro* use of non-systemic fungicides gave such evaluation of pustules/uredospore reduction at different concentrations of 1000 ppm and 500 ppm, which reduce the spore production by 83.33 and 72.31%, respectively, and an average response was 54.85% by mancozeb and 40.89% by chlorothalonil [17]. The response of systemic fungicide to the inhibition percentage of uredospores was 86.03% by propiconazole, of which the highest systemic fungicide was followed by the fungicide hexaconazole with 77.40% and penconazole with 72.29%. While 20 plant extracts with different parts were used for the maximum reduction of uredospores by the bulb of garlic extract at 59.78% and onion bulb extract reduction with 57.70, the rhizome of ginger gives 54.81% reduction of spores.

The extracts of the different plants are affected to control the leaf rust of wheat. The plants used for biocontrol are *Lawsonia Inermis*, *Melia azedarach*, *Acalypha wilkesiana*, *Punica granatum*, and *Lantana camara*. The method of application is before the arrival of disease in wheat plants in two wheat seasons 2016–2018. These plants show the same result as the fungicide used to control wheat leaf rust [18]. It reduces the ACI coefficient of infection while with ACI of non-treated wheat plants. The most effective plant extract was *L. camara* giving 88.88% efficiencies equivalent to the fungicide diniconazole followed by *Lawsonia Inermis*, *M. azedarach*, *A. wilkesiana*, *and P. granatum,* respectively. These plant extracts not only reduce the disease severity but also increase the yield parameters, phenols, chlorophyll content, and the peroxidase and oxidase activity of wheat plants.

In the agricultural field of research, a study was undertaken to assess the allelopathic effects of leaf extracts of daylight neem leaves on wheat production and its constituents. With the treatment of 0, 50, and 100% liquid leaf extract of neem, grain yield, and several yield attributes of wheat, such as the number of viable tillers, grain/spike, 1000-grain weight, and spike length, no meaningful either promotive or negative impact was seen [19]. The leaf extracts of neem, on the other hand, did not affect wheat yield and yield contributing. The use of a natural weedicide (leaf extracts extract of neem) showed no negative impact on wheat quality.

For evaluating the genetic resistance of wheat lines of 45 genotypes/lines against rust, the lines were inoculated at the booting stage with an aqueous inoculum of wheat. It was a field study in Nepal where the results showed different pathogenicity ratios with high variation of wheat varieties [20]. The surveys of 66 production fields where the old varieties are observed have more disease severity rather than the new and resistant varieties. The prevalence of disease on wheat genotype revealed that leaf rust and yellow rust have a high level of prevalence. Leaf rust has more prevalence than yellow rust but low severity, while yellow rust has high a concern with disease severity. The management of disease and farmer literacy about rust makes the varieties more virulent against disease.

Seed priming using leaf extract and compounds has long been utilized to boost agricultural plant development, but the pathways are still unknown. The goal of this study was to figure out how different seed priming methods in greenhouse wheat work. Hydropriming, moringa aqueous extracts priming, and $CaCl_2$ priming were the seed priming methods utilized. The results revealed that all the seed priming treatments were more efficient than the control at enhancing wheat germinating and seed germination. However, *Moringa* was shown to be the most effective technique, followed by $CaCl_2$. The activation of increasing antioxidant and enhanced chlorophyll, soluble phenolics content, and ascorbic acid were all factors in this respect. The findings support the idea that seed priming with *Moringa* is cost effective and may be utilized to promote wheat development in greenhouses. *Moringa oleifera* leaf extract applied on wheat (*T. aestivum* L.) leaves had various results in grain production and yield parameters such as biomass of plant, number of ears, tillers, and 1000 seed weight. A treated field comparing various *M. oleifera* concentrations showed a 19% increase in crop yields at a 10% concentration, but still, no yield potential improvements with higher *M. oleifera* concentrations. Based on the year and data analysis plan, drier field experiments with Moringa doses of 5% and 10% revealed no impact on improvements in crop production. Variations in phytohormone content such as auxin, gibberellins, abscisic acid, jasmonic acid, and salicylic acid are caused by abiotic and biotic stresses before leaf harvest for the disparities in yield component responses. It was discovered that GA4 very probably interaction with auxin is the most important growth promoter. The hormone levels of Moringa are shown to vary significantly on an annual basis, which may have an impact on the biostimulant's prospective application in agriculture.

2.4 Potential of *Trichoderma harzianum*

Wheat (T. aestivum L.) is one of the most important crops for humans worldwide. Stripe rust disease, caused by *Puccinia striiformis f. sp. tritici*, is the most devastating disease, posing a huge danger to wheat production across wide areas, causing significant yield and grain quality losses [21]. This investigation was carried out to evaluate the agents, namely *Trichoderma harzianum*, *T. viridi*, *Chaetomium globosum*, *Saccharomyces cerevisiae*, *Bacillus subtilis*, and *B. chitinosporus*. Under field condition, greenhouse effect, and lab condition, the used biocontrol reduces the coefficient of infection (ACI) of stripe rust *Puccinia striiformis f. sp. tritici*. It is recommended to use a biocontrol agent for plant pathogens.

The biocontrol of wheat stem rust *in vitro* is by using Arbuscular Mycorrhiza fungi and a combination of two species of *Trichoderma harzianum* and *Trichoderma viride* with single use also on the *Puccinia graminis f.sp.tritici* spores. The identification of both species of Trichoderma was done by using a system of GC-MS. The results observed under a scanning electron microscope indicate that when using the Trichoderma species suspension in combination, they were more effective in inhibiting the uredospores than other application methods. The application of bioagents under field condition not only reduces the disease but also increases the phenol content, the peroxidase enzyme activity, yield, and growth parameters [22].

The plant extract of A. indica, bioagent *Trichoderma harzianum* T-2 fungicide Iprodione is against the *Alternaria brassica* causing Alternaria disease radish blight. The inoculation spray of *A. brassica* spore concentration of 5 × 105/ml is on the flowering stage of the Radish field. The bioagent used for treatment purposes was used as a soil amendment in the concentration of 5 × 105 spores ml as a foliar spray after 3 days of spore inoculation, while for seed treatment the concentration was 3 g/Kg. At the flowering stage, the neem leaf extract was applied in the concentration of 10%, and to compare it with fungicide, they applied 200 ppm of Iprodione at 10 days with the four-time application. The results of the three treatments showed that the T8, T7, and T6 were the most effective on disease severity and increase the growth and germination of radish plants [23]. The yield for seed increases due to the nutrient availability by enhancing growth-promoting factors enhanced by the application of treatments.

When compared with the control, all 13 treatments enhanced yield and 1000-grain weight to a larger extent. Fungicide used to has a good effect. Moreover, the leaf extract of neem, *Trichoderma harzianum*, and the Panchgavya were the most efficient treatments for wheat leaf rust in various groups. The yield of fungicides sprayed areas was considerably higher than the control plots, showing that leaf rust had a major impact on yield. In comparison with other procedures, the 1000-kernel weight was beneficial in the abovementioned treated plots [24].

The adoption of resistant varieties is the most cost-effective and efficient means of controlling *Puccinia triticina* Eriks' called leaf rust of wheat. Generally, assessment for rust resistance in wheat cultivars was done in a greenhouse experimental trial and under the natural condition of the field at the seedling stage and mature plants. For this purpose, the varieties were affected by ecological variables that restrict the number of races that may be examined at the same time. In their work, a detached leaf test was used to screen wheat lines for leaf rust resistance. Two senescence compounds that suppress benzimidazole and kinetin were introduced to 5% water-agar as treatments in various doses and combinations [25]. Three leaf rust races were used to verify the chosen medium in 20 wheat genotypes. The media for a treatment having a proportion of 30 mg/L benzamide and 10 mg/L kinetin were injected with the help of a sprayer showing the prominent results in slowing aging and therefore boosting sporulation. There was a positive association (r = 0.9) between the disease types measured by the detached leaf test and the entire seedling test. For detached leaf 0.24 and entire seedling, tests showed 0.3 standard errors. The minimal standard error support uniformity of illness reaction assessment across the tests performed.

The most dangerous disease affecting wheat plants is leaf rust produced by *Puccinia triticina f.sp. tritici*. Bioagents such as *B. subtilis*, *Bacillus pumilus*, *Bacillus chitinospours*, *Trichoderma harzianum*, and *Trichoderma viride* were tested to manage leaf rust. During two wheat-growing seasons, 2016/17 and 2017/18, the bioagents were sprayed pre and post *P. triticina* infection for 24 hours. Our findings revealed that B. subtilis was the most effective treatment, second by *T. viride*, with substantial increases in disease incubation and latent durations, but also an increase in 1000 kernel weight (g) and production (kg). In the contrast, a substantial reduction in the number, length and breadth of spores, the final rust severity percentage, and AUDPC. Furthermore, the treatments increased catalase activity (CAT) and peroxidase (POX), although electrolyte leakage was reduced when compared with the control. The relevance of FRS percentage as a suitable indicator for both the research effects of alternative bioagents in controlling leaf rust was demonstrated by a correlation test [26]. The use of bioagents as a source of disease control is harmless for the environment and the disease produces fungicide-resistant forms.

Under field circumstances, field trials were undertaken on two sowing dates to examine the feasibility of using bioagents to reduce the severity of foliar diseases such as Septoria leaf blotch, powdery mildew, and stem rust. *Trichoderma harzianum*, *B. subtilis*, *Azospirillum brasilense*, *Pseudomonas fluorescens*, and plant shield were tested. When compared with the control group, all therapies lowered the severity of the disease. In general, Giza 171 cv. had the least severe disease, followed by Misr 1 and Gemmiza 12. Furthermore, genotypes seeded sown earlier in November showed less disease severity than many of those sown late in December. Wheat grains sprayed with *T. harzianum* and *B. subtilis* had the greatest impact on the incidence of all diseases, followed by some of the other treatments. Therefore, such sprayed treatments also have the potential to reduce wheat leaf pathogens as a healthy alternative to synthetic treatment with no adverse health effects or environmental pollution [27].

Many environmental factors were the primary hurdles to attaining the maximum productivity possible in the grain yield. Economic losses due to biotic stressors were predicted to be 26–29% in the region. However, physiological processes have a greater negative impact on crop production that accounts for around 70% of the reduction in yield globally. Pesticides and fertilizers are commonly proven as an efficient control mechanism for wheat crop pests and diseases; however, the build-up of synthetic chemicals inside the soil, plant materials, and fungicide-treated kernels harms human and environmental life [27]. Trichoderma is commercially significant as a biocontrol agent, potentially replacing agrochemicals in the fight against biological and chemical stress.

Various methods such as (*B. subtilis*, *Bacillus chitinosporus*, and yeast extract), benzothiadiazole (BTH), salicylic acid, and oxalic acid, as well as the fungicide propiconazole, were developed to optimize the resistance, physiological characters of the wheat variety, and yield production in the vulnerable conditions. Wheat varieties (Gemmiza-7) especially in contrast to the high resistance wheat variety.

In vitro, all treatments reduced fungal growth, disease severity percentage, and the quantity of uredia as compared with the vulnerable cultivar's control infection. In susceptible wheat cultivars treated with bio-agents, salicylic acid, BTH, and oxalic acid treatments, antioxidant enzyme activities catalase, peroxidase, and polyphenol oxidase were significantly increased when compared with the control treatment. In the susceptible-infected cultivar, the concentration of chlorophyll was significantly reduced. The percentage of electrolyte leakage in susceptible treated cultivars was significantly lower than in susceptible infected untreated cultivars [28]. As a result, the treatments were able to boost chlorophyll concentration while also improving yield components such as grain weight 1000 per grain and grain weight in 10 spikes.

The research was carried out during two consecutive fall seasons in 2012/2014 at the El-Kassasein Experimental Farm, Hort. Res. Station, Ismailia Governorate, Egypt. The effect of three different nitrogen fertilizer sources was as follows: namely ammonium sulfate (20.5% N) at 390.2 kg/fed. (fed. = 0.42 ha.), botanical compost at 6.667 tons/fed., and chicken manure at 2.787 tons/fed. (each equaling 80 kg N/fed.), and five biological control (Trichoderma harzianum, Trichoderma viride, the mixture of Trichoderma harzianum + Trichoderma). When compared with Xera cultivar, Paulista cv. provided the maximum dry weight of shoots per plant and final yield. The use of chicken manure enhanced the dry mass of shoots and the overall yield of snap beans. Foliar treatment of a *Trichoderma harzianum* + *Trichoderma viride* combination improved the number of leaves and branches per plant, the dry weight of shoots per plant, pod length, and overall yield. Fertilizer application Paulista cv. with chicken manure and foliar application of *Trichoderma harzianum* + *Trichoderma viride* improved the number of leaves and branches per plant, plant height, dry weight of branches, leaves, and shoots per plant, pod length, and total yield [29]. When compared with Paulista plants, Xera cv. plants had the lowest rating for rust disease severity. Botanical compost application resulted in the lowest rust disease incidence and severity of snap bean plants, followed by chicken manure treatment, and ammonium sulfate at 390.2 kg/fed resulted in the highest results. In comparison with the control, foliar spray of biocontrol agents to snap bean plants reduced the incidence and severity of rust disease on the leaves. The combination of *Trichoderma harzianum* and *Trichoderma viride* reduced rust disease incidence, while *Trichoderma harzianum* reduced rust disease severity.

Plant diseases are one of the most significant restrictions to crop production and productivity, both in terms of quality and quantity. The use of pesticides remains the primary strategy for mitigating agricultural disease threats. However, because of environmental issues, human health problems, and other risks connected with the use of chemicals, the use of bioagents to reduce the disease-causing activities of plant pathogens is gaining popularity. Biocontrol is the intentional use of living organisms, either transferred or indigenous, other than disease-resistant host plants, to decrease the activities or populations of one or more plant diseases. Beneficial organisms, their genes, and/or products, such as metabolites, are used in biological control to lessen the detrimental impacts of plant diseases and stimulate positive plant responses [30]. A variety of commercial products based on diverse fungal and bacterial antagonists have been recognized at both national and international levels in this direction. These commercial products include Biocon, Biogaurd, Ecofit, FStop, Soilgaurd, and others that include *Trichoderma sp*. as an active ingredient, as well as Mycostop, Rhizopus Subilex, and others that contain various *Bacillus* species as active ingredients. Biological control can achieve disease suppression in a variety of methods, including antibiosis.

3. Conclusion

Puccinia triticiana, an obligate parasite, is the cause of wheat leaf rust. Rust is one of the most destructive cereal pathogens and coexisted that developed during grain cultivation. The growth of leaf rust affected and statistically substantial connection with environmental factors were shown in urst responses of various genotypes. Average temperature, maximum temperature, lowest, precipitation, and relative humidity were associated with rust processes. A foliar spray was the most efficient in decreasing leaf rust (ACI) infection and the neem extract at the mature plant stage. This study found several resistant types that may be used for the wheat reproduction programs of various research institutions in Pakistan that contribute to the production of resistance to leaf rust in wheat. The eco-friendly measures used in this study were effective in future research for extracting the secondary substance playing role in controlling leaf rust.

Acknowledgements

We also acknowledge (Lal Hussain Akhtar Director, Regional Agriculture Research Institute, Bahawalpur) for providing their pearls of knowledge with us throughout this research, as well as the reviewers for their contributions.

Conflict of interest

There is no "conflict of interest."

Author details

Sadia Afzal[1], Adeela Haroon[1], Muhammad Arshad Hussain[2], Asad-Ur-Rehman Chaudary[2], Muhammad Amjad Bashir[3]*, Sagheer Atta[3], Saqib Bashir[4] and Muhammad Adnan Bodlah[5]

1 Department of Botany, Women University Multan, Pakistan

2 Regional Agricultural Research Institute Bahawalpur, Punjab, Pakistan

3 Faculty of Agricultural Sciences, Department of Plant Protection, Ghazi University, Dera Ghazi Khan Punjab, Pakistan

4 Faculty of Agricultural Sciences, Department of Soil and Environmental Sciences, Ghazi University, Dera Ghazi Khan Punjab, Pakistan

5 Department of Agricultural Engineering, Khwaja Fareed University of Engineering and Information Technology, Rahim Yar Khan, Pakistan

*Address all correspondence to: abashir@gudgk.edu.pk

References

[1] Singh RP, Hodson DP, Huerta-Espino J, Jin Y, Bhavani S, Njau P, et al. The emergence of Ug99 races of the stem rust fungus is a threat to world wheat production. Annual Review of Phytopathology. 2011;**49**:465-481

[2] Afzal A, Ijaz M, Rafique M. New selection techniques to detect sources of resistance against stripe rust in wheat. Plant Protection. 2021;**5**(3):197-203

[3] Rana IA, Bhatti SS. Lahore, Pakistan–Urbanization challenges and opportunities. Cities. 2018;**72**:348-355

[4] Ekboir J. CIMMYT 2000-2001 World Wheat Overview and Outlook: Developing No-Till Packages for Small-Scale Farmers. 2002

[5] Huerta-Espino J, Singh R, Crespo-Herrera LA, Villaseñor-Mir HE, Rodriguez-Garcia MF, Dreisigacker S, et al. Adult plant slow rusting genes confer high levels of resistance to rusts in bread wheat cultivars from Mexico. Frontiers in Plant Science. 2020;**824**

[6] Yasmeen A, Basra SMA, Wahid A, Nouman W, Rehman HU. Exploring the potential of Moringa oleifera leaf extract (MLE) as a seed priming agent in improving wheat performance. Turkish Journal of Botany. 2013;**37**(3):512-520

[7] Yamin SY, Tariq JA, Raza MM, Bhutto SH, Asif MU. Screening of wheat genotypes against leaf rust under artificial and natural environmental condition. Journal of Applied Research in Plant Sciences (JOARPS). 2021;**2**(1):117-122

[8] Khan MR, Imtiaz M, Munir I, Hussain I, Ali S. Differential distribution of leaf rust across major wheat growing regions of Pakistan revealed through a three years surveillance effort. Pakistan Journal of Botany. 2021;**53**(1):261-266

[9] Gangwar OP, Kumar S, Prasad P, Bhardwaj S, Khan H, Verma H. Virulence pattern and emergence of new pathotypes in Puccinia striiformis f. sp. tritici during 2011-15 in India. Indian Phytopathology. 2016;**69**(4s):178-185

[10] Muhammad S, Khan AI, Aziz-ur-Rehman FSA, Rehman A. Screening for leaf rust resistance and association of leaf rust with epediomological factors in wheat (*Triticum aestivum* L.). Pakistan Journal of Agricultural Sciences. 2015;**52**(3):691-700

[11] Yahya M, Saeed NA, Nadeem S, Hamed M, Saleem K. Effect of leaf rust disease on photosynthetic rate, chlorophyll contents and grain yield of wheat. Archives of Phytopathology and Plant Protection. 2020;**53**(9-10):425-439

[12] Ali Y, Khan MA, Aatif HM, Ijaz M, Atiq M, Bashair M, et al. Research paper (host resistance: Fungi). Plant Protection. 2020;**38**(4):344-353

[13] Mateen A, Khan MA. Identification of yellow rust virulence pattern on wheat germplasm in relation to environmental conditions in Faisalabad. Journal of Biological Agricultural and Health Care. 2014;**4**(13):2224-3208

[14] Khan M, Hussain M, Sajjid M. A two environmental variable model to predict wheat leaf rust based on 10 years data. Pakistan Journal of Phytopathology. 2006;**8**:114-116

[15] Shabana YM, Abdalla ME, Shahin AA, El-Sawy MM, Draz IS, Youssif AW. Efficacy of plant extracts

in controlling wheat leaf rust disease caused by Puccinia triticina. Egyptian Journal of Basic and Applied Sciences. 2017;**4**(1):67-73

[16] Akila R, Mini M. Solvent extraction and antifungal assay of Lawsonia inermis Linn. Against the brown spot fungus Bipolaris oryzae. Journal of Pharmacognosy and Phytochemistry. 2020;**9**(1):05-08

[17] Chaudhary R, Chaudhari M. Effect of fungicides and plant extracts on uredospores germination of Puccinia recondita f. sp. tritici. Bioscan. 2013;**8**(1):59-62

[18] Draz IS, Elkhwaga AA, Elzaawely AA, El-Zahaby HM, Ismail A-WA. Application of plant extracts as inducers to challenge leaf rust of wheat. Egyptian journal of biological. Pest Control. 2019;**29**(1):1-8

[19] Dhanai CS, Nandini D, Panwar G. Allelopathic effect of Azadirachta indica leaf extract on seed germination of different test crops under bioassay. Journal of Pharmacognosy and Phytochemistry. 2019;**8**(6):790-792

[20] Lodhi S, John P, Bux H, Kazi AM, Gul A. Resistance potential of Pakistani wheat landraces (*Triticum aestivum* L.) against stripe rust (*Puccinia striformis*) and Karnal bunt (*Tilletia indica*). Pakistan Journal of Botany. 2018;**50**(2):801-806

[21] ElKazzaz MK, Ghoniem KE, Ashmawy MA, Omar GE, Hafez YM. Suppression of wheat strip rust disease caused by PU a STR Ormis f. sp. Tritici by Ecofriendly Biocontrol Agents Correlated with Yield Improvement. Environmental Bulletin. 2020;**2020**:8385

[22] El-Sharkawy HH, Rashad YM, Ibrahim SA. Biocontrol of stem rust disease of wheat using arbuscular mycorrhizal fungi and Trichoderma spp. Physiological and Molecular Plant Pathology. 2018;**103**:84-91

[23] Arefin MN, Bhuiyan M, Rubayet MT. Integrated use of fungicide, plant extract and bio-agent for management of Alternaria blight disease of radish (*Raphanus sativus* L.) and quality seed production. Research in: Agricultural & Veterinary Sciences. 2019;**3**(1):10-21

[24] Kalappanavar I, Patidar R, Kulkarni S. Management strategies of leaf rust of wheat caused by Puccinia recondita f. sp. tritici Rob. ex. Desm. Karnataka Journal of Agricultural Sciences. 2010;**21**(1)

[25] Peng F, Si M, Zizhu Y, Fu Y, Yang Y, Yu Y, et al. Rapid quantification of fungicide effectiveness on inhibiting wheat stripe rust pathogen (Puccinia striiformis f. sp. tritici). Plant Disease. 2020;**104**(9):2434-2439

[26] Omara RI, El-Kot GA, Fadel FM, Abdelaal KA, Saleh EM. Efficacy of certain bioagents on patho-physiological characters of wheat plants under wheat leaf rust stress. Physiological and Molecular Plant Pathology. 2019;**106**:102-108

[27] El-Mougy NS, Khalil MS, El-Gamal NG, Abdel-Kader MM. Impact of grain treatments with bioagents for suppressing foliar diseases severity of three wheat cultivars under field conditions. Archives of Phytopathology and Plant Protection. 2021;**54**(7-8):431-447

[28] Hafez Y, Emeran A, Esmail S, Mazrou Y, Abdrabbo D, Abdelaal K. Alternative treatments improve physiological characters, yield and tolerance of wheat plants under leaf rust disease stress.

Fresenius Environmental Bulletin. 2020;**29**:4738-4748

[29] Mandour MA, Metwaly HA. Effect of nitrogen fertilizer sources and some biocontrol agents on growth, yield and rust disease incidence of some snap bean cultivars grown in Sandy soil. Egyptian Journal of Horticulture. 2015;**42**(1):591-614

[30] Junaid JM, Dar NA, Bhat TA, Bhat AH, Bhat MA. Commercial biocontrol agents and their mechanism of action in the management of plant pathogens. International Journal of Modern Plant & Animal Sciences. 2013;**1**(2):39-57

Chapter 5

Wheat Stripe, Leaf, and Stem Rust Diseases

Nilüfer Akci

Abstract

Wheat (*Triticum* spp.) is one of the most strategic crops in the world. It provides raw material to the agricultural industry and it is the main source of income for many rural areas. Rust diseases are among the most important biotic factors affecting the yield and quality of wheat plants. Depending on the disease's severity in wheat cultivation fields, the level of yield losses and quality degradation may vary, accordingly, economic losses changes. Wheat rust diseases are categorized into three groups, such as stripe (yellow) (*Puccinia striiformis*), stem (black) (*Puccinia graminis* f.sp. *tritici*), and leaf (brown) (*Puccinia triticina*) rusts. This chapter presents information on the rust symptoms, identification, and management.

Keywords: biotic factors, wheat, wheat rust diseases, stripe rust, leaf rust, stem rust, management

1. Introduction

Wheat is one of the cool climate cereals which is one of the important mineral and energy sources and involved in the nutrition of billions of people due to its suitable nutritional value, easy storage, and processing. Because of its wide adaptation ability, it is in the first place in the world in terms of production amount as well as its cultivation area. In addition, it is also a strategic product in terms of being a raw material for the agricultural industry in the world, contributing to the economy, and being the main source of income for rural areas [1].

There are diseases in wheat that significantly reduce grain yield and grain quality. There are abiotic and biotic factors that affect wheat. Biotic factors causing disease in wheat as in other plants are fungi, bacteria, and viruses. Fungal diseases are wheat leaf diseases, wheat head diseases, and wheat root diseases. The wheat leaf diseases include wheat rust diseases, wheat septoria leaf spot, and wheat powdery mildew.

Rust diseases have been known since ancient times. Especially the ancient Romans considered cereal rust to be very important and accepted it as a punishment given by the God Robigus, and they organized Festivals and sacrificed every year so that this punishment would not be repeated [2].

There are three types of rust diseases in wheat, which are called Stripe (Yellow), Stem (Black), and Leaf (Brown) rust. These rust types got their names from the colors of the pustules they form by tearing the epidermis of the plants. Rust fungi are seen as obligate parasites in nature [2].

2. Stripe rust (Yellow rust) (*Puccinia striiformis*)

It is the earliest and most important rust disease of wheat. Especially in the spring months, there is an increase or decrease in the intensity of the stripe rust disease depending on the climate structure. In the case of abundant and long-term spring rains, it appears suddenly and causes significant yield and quality losses by causing diseases primarily in the leaves. It has been noted that rust diseases are observed even in the early stages with the increase in temperature [3].

Stripe rust disease infects other plants besides wheat. They can infect barley, rye, triticale, and many other related wild wheat crops. Although it is usually seen on the leaves of wheat, it can also occasionally be seen on stems and heads. It can be easily distinguished from other types of rust due to the symptoms it shows on the leaf. It occurs on the upper surface of the leaves, on the leaf sheath, on the head, and even inside the husks. The rust symptoms on the leaves cover the whole leaf and kill the leaf when the disease is severe [3].

Stripe rust disease is named after the color of the disease spores (pustules), which are like a powder of orange-yellow (golden yellow) color. Stripe rust, especially on the upper surface of the leaf, creates yellow pustules like machine stitches. Since the arrangement of these pustules resembles a line, it is also called Line Rust. Summer spores occur inside these pustules which have the form of dots arranged in rows and intra row (**Figure 1**) [4].

High humidity or precipitation in the spring in wheat fields induces the occurrence of the disease. The optimum temperature for the formation and development of the disease is 10–15°C. If the host-pathogen relationship is suitable with the proper environmental conditions for the development of the disease, an epidemic occurs. In the development of the disease, the first infections occurred by urediospores, which can be carried by the wind from long distances. Because rust spores are light they can be spread around even with very little wind and can be drift to the next fields. Millions

Figure 1.
Symptoms of stripe rust in wheat leaves. Photos: Dr. N. Akci.

of summer spores formed from pustules are dispersed by the wind in the spring. The initial infection is initiated by very few spores and is seen in the early stages when plants are just starting to develop. After that, new spores occur every 12 to 15 days, and then every 8 to 10 days, the disease rate increases, and stripe rust disease is suddenly seen everywhere. At the end of the season, winter teliospores are formed from the same pustules. The disease is carried out on wild wheat crops on the edges of the field, which remain alive during the summer, and on the wheat planted in autumn for the winter [4].

With the increase of rust disease, the use of nutrients and water increases, as well as the photosynthesis area of the plant narrows. As a result, the amount of nutrients that will produce grains decreases. It has been determined that rust prevents normal root development and nutrient uptake in the plant to a certain extent. In addition, since rust infection causes plants to reach maturity earlier than usual, it also causes the grain filling period to be shortened thus the damage increases. The severity of the damage caused by the rust disease changes according to the development periods of the plants. Flowering and earlier periods are the most damaging period. The late period is the least damaging period. If the heads are infected with stripe rust, no matter how little rust is on the leaf, the grain yield is greatly reduced. As a result of rust infection, losses occur in the yield and quality of the grain, as well as in the quantity and quality of hay [5].

The losses due to plant deaths can be very large in more severe epidemic conditions. With the decrease in grain size and hectoliter weight, indirect effects such as a decrease in flour yield and quality and even the quality of products obtained from flour occur. Due to the fact that rust diseases slow down plant growth and reduce tillering, they cause large losses in hay yield, as well as large losses in the quality of hay in some toxic substances that occur in plants [6].

3. Leaf rust (brown rust) (*Puccinia triticina*)

It is usually seen on the leaves, so it is also called Leaf Rust. The orange-yellow or burnt brown color pustules are in the form of large and small dots randomly scattered on the leaf surface. Leaf rust can be seen on the upper surface of leaf. The characteristic of this rust is to form smaller pustules in one or two circles around the pustule. This symptom distinguishes brown rust from other types of rust (**Figure 2**) [7].

This rust usually appears on the wheat after the stripe rust before the stem rust. In the spring, summer spores cause infection at 10–18°C and high humidity. The temperature and humidity requirements for the development of leaf rust disease have the ability to spread more easily than stem rust. Thus damage to the product can be very severe. The damage of leaf rust has recently coincided with the maturity period of wheat [8].

Leaf rust disease infects other crops besides wheat. They are barley, triticale, and many other related wild wheat crops. The disease is in the form of uredospores in temperate winter regions. In spring, spores exist on the surface of alternate hosts (*Thalictrum* spp. and *Isopyrum* spp.) leaves. Then they are carried on the leaves of the wheat by the wind and form pustules of rust. It causes significant yield losses by decreasing the number of grains, hectoliter weight, and grain quality in the head. The severity of the damage caused by the rust disease changes according to the development periods of the plants. Flowering and earlier periods are the most damaging period. The late period is the least damaging period [8].

Figure 2.
Symptoms of leaf rust in wheat leaves. Photos: Dr. N. Akci.

4. Stem rust (black rust) (*Puccinia graminis* f.sp. *tritici*)

Stem rust disease is one of the oldest known diseases of wheat and it is also called stem rust because it is usually seen on the stem of the wheat. In the case of an epidemic, it can cause significant yield and quality losses in grain and hay. Stem rust is the last rust disease seen in wheat. Stem rust disease occurs in all parts of the wheat above-ground. It changes in size from 3 mm to 1 cm and is mainly seen on the stem of wheat but can also be seen on the other green parts (**Figure 3**) [9].

The dark red-brown (tile-colored) pustules may occur on the two sides of the leaf, on the stem, and the head. The pustules on the lower surface of the leaf, which are seen on both sides of the leaf, are larger than those on the upper surface. The pustules are sprinkled on the stem and leaves and are large, oval, long, and darker in color than other rust pustules. Their temperature request is higher. Stem rust disease grows well in the temperature between 20 to 25°C with a proportional humidity of over 96%. If

Figure 3.
(a) Symptoms of stem rust in wheat stem (b) Stem rust disease in the wheat field. Photos: Dr. N. Akci.

Figure 4.
(a) Berberis plant and aeciospores on the leaves of it (b) Berberis plant at a flowering period near the wheat field. Photos: Dr. N. Akci.

all environmental conditions are suitable for disease development an epidemic occurs. New urediospores occur every 10–15 days. When plants mature black teliospores occur near the harvest [9].

While the pustules are dispersed in low infections, it can be seen that the pustules may merge in severe infections. The urediospores that form the pustules tear the epidermis and the plant surface takes the form of a whitish collar. These torn pieces of the epidermis are seen very clearly. Stem rust spends the winter on infected plant parts. Spores are found on the underside of the leaves of *Berberis* and *Mahonia* plants, which are alternate hosts in the spring. Then they form rust pustules on the leaves and stems of the wheat by being carried by air and wind. Spores reproduce in suitable conditions and cause great damage to the crop (**Figure 4**) [10].

Stem rust creates new races where *Berberis* plant is present. Thus stem rust can infect wheat varieties that were previously known to be resistant. As a result, new epidemics can occur. Stem rust reduces tillering and decreases grain weight and quality. The whole product can be lost when the appropriate conditions for the disease are formed. Crop loss can change depending on the susceptible varieties, environmental conditions, races of stem rust and it differs from year to year, from region to region [10].

5. Management of wheat rust diseases

The cultural precautions to be taken to prevent rust disease are as follows. Frequent planting should not be done because it prevents ventilation and causes an increase in humidity. Weed control should be done on time and with suitable techniques. Fertilization should be done according to the results of soil analysis. Excessive nitrogen fertilizer should not be given to the field. Activities such as irrigation that will increase the humidity of the air should be avoided [11].

It is important to control weeds and alternate host plants on time. Alternate hosts of rust diseases in the surrounding or on the field edges should be destroyed. The destruction of alternate host plants provides a decrease in the amount of inoculum which causes the first infections in rust disease. Also, the destruction of alternate host plants leads to a decrease in the number of new races that may emerge due to the limited sexual reproduction in rust disease [12].

Resistant cultivars should be used for rust disease. It is of great importance to monitor the races of rust in order to develop resistant varieties for the management of rust disease. In order to determine virulence in rust disease, it is necessary to develop monitoring and prediction warning systems which also include survey studies [13].

Taking into account the climatic conditions, the course of the disease should be monitored, especially for the stripe rust in temperature of 15–20°C and humidity of 90%. In cases where the disease progresses to the upper side of the plant, it is important to prevent the contamination of the disease, especially the flag leaf of the upper leaves. When the first stripe rust symptoms start to appear on the leaves it is recommended to spray the green parts. Spraying should be done so that the surfaces of the leaves and stems are covered with sprayed water. If the climatic conditions are suitable for the development of the disease and an epidemic is possible second spraying should be done, taking into account the effect time of the drug used [14].

With the use of genetic resistance successful progress has been made in the control of wheat stem rust disease in recent years. Resistant varieties can be preferred more reliably because they are economical and environmentally friendly.

6. Conclusion

It is known that climate change causes an increase in disease. It is difficult to predict when and where diseases will spread. With the changing ecosystem the effectiveness of biological factors changes, the distribution of pathogens is affected and the entry of new pathogens is enabled. With climate change, differences can be seen in plant-pathogen systems. Disease development is the result of factors influencing the host and pathogen.

Changes in wind direction and speed affect the spread of rust spores. Therefore, the monitoring of rust diseases is important. Care should be taken against new races, disease surveys and breed analyzes should be done, and identification of the rust races is necessary. Molecular research should be carried out in a variety of breeding studies against the aggressive/virulence of rust diseases that may occur with increasing temperature and other parameters. The pathogen forms new races and these races may overcome the resistance present in wheat cultivars. Besides the increase in the spread of rust diseases is largely due to the widespread of varieties that are susceptible. In addition to this, there may be changes in the number of fungicide applications and doses.

Author details

Nilüfer Akci
Plant Protection Central Research Institute, Yenimahalle, Ankara, Turkey

*Address all correspondence to: niluferakci@hotmail.com

References

[1] Taşçı R, Kabalak S, Bolat M, Acar O, Şanal T, Pehlivan A, et al. Consumer preferences in bread in Ankara province. Journal of Field Crops Central Research Institute. 2017;**26**(1):75-85

[2] Kınacı E. 1983. Wheat Rust Diseases. Central Anatolia Regional Agricultural Research Institute. No:1

[3] CIMMYT. Wheat Diseases and Pests. A Guide for Field Identification1986. p. 135

[4] Kınacı E. 1985. Wheat Diseases. Central Anatolia Regional Agricultural Research Institute. No. 5

[5] Kınacı E, Kınacı G. Winter grain agriculture in Central Anatolia. 2016. ISBN: 978-605-137-564-9

[6] Wheat and Maize Diseases Seminar. Central Anatolia Regional Agricultural Research Institute1985. p. 101

[7] CIMMYT. Common Diseases of Small Grain Cereals. A Guide to Identification. 1985

[8] CIMMYT. Rust Diseaes of Wheat. 1992. p. 81. ISBN: 968-6127-47-X

[9] Compendium of Wheat Diseases and Pests. 2010. ISBN: 978-0-89054-660-4

[10] FAO. Guidelines For Monitoring Diseases, Pests and Weeds in Cereal Crops. 2016. ISBN: 978-92-5-109180-7

[11] Plant Protection Handbook. Ministry of Agriculture and Rural Affairs, Directorate of İzmir Provincial. 2002. p. 2

[12] Agricultural Management Technical Instructions. Cereal Diseases. Ministry of Food, Agriculture and Livestock, Directorate of Agricultural Research and Policies; 2008. pp. 11-15

[13] McIntosh RA, Wellings CR, Park RF. Wheat Rusts, An Atlas of Resistance Genes. Plant Breeding Institute, The University of Sydney; 1995

[14] Cultural Management from Theory to Practice. Republic of Turkiye Ministry of Agriculture and Forestry General Directorate of Food and Control. 2015. p. 170

Chapter 6

Fungal Diseases of Wheat

Mukaddes Kayim, Hira Nawaz and Abdulkreem Alsalmo

Abstract

Wheat is considered the first crop that is grown on earth. It is a staple food in many regions of the world. Due to the increase in the world's population, it is very important to increase wheat production. With an estimate in 2050, almost 50% more production of wheat will be required due to the increase in population. Increased productivity of wheat is the biggest challenge for researchers. It faces several biotic (microbial diseases) and abiotic (water, temperature, and climatic change) limiting factors. But the major threat for wheat is due to a large number of fungal diseased pathogens, which causes massive and destructive loss to the crop. It includes rusts, smuts, Fusarium head blight, Septoria leaf blotch, tan spot, and powdery mildew that cause the most serious losses. It was estimated in 2019 that almost 22% yield loss of wheat was due to diseases. These percentages will increase with time due to mutation and diversity in virulent strains. This chapter includes all major and minor fungal diseases of wheat, symptom, disease cycle, spore identification, disease losses, etiology, and integrated disease management.

Keywords: airborne, management, obligate parasites, root rot, seedborne

1. Introduction

Wheat (*Triticum aestivum*) is a grass commonly grown for its seeds, which are used as a staple food for many countries of the world. Wheat is a good source of carbohydrates and has low protein content. This low protein content help to supply important amino acids to the body. It is also a great source of many other nutrients and dietary fibers [1, 2]. Wheat is cultivated in almost all parts of the world with different ratios. In the world, annually, wheat is grown on an area of 5400 hundred thousand acres. The common forms of wheat that are used for eating are white and red wheat [3]. In the year 2017, 772 million tonnes of wheat were produced around the globe. Global wheat consumption is also increasing, it has gluten protein which helps in producing processed food. Processed food becoming an important part of the modern world [1]. China is the biggest producer of wheat with an annual production of 133.6 million tonnes in the year 2019 [4].

As the world population increases day by day it is estimated that agricultural commodities should be increased by 50% by 2050 to meet the demand and supply chain [5]. But major constrain in this race are abiotic and biotic factors, which affect the production every year. Abiotic factors are generated by the facilities of mankind that are climate change. While biotic factors include major disease pathogens, insects, pests, and weeds. These factors cause a reduction in the yield and quality of grain every year [6]. Serious biotic stress includes major fungal diseases, such as rusts, smuts, bunt, tan spot, fusarium head blight, foot rot, false eyespot, and many more. Three types of rusts and powdery mildews caused major disease epidemics in past and kept on threatening problems to wheat production besides, the development of various fungicidal chemicals and resistant cultivars. Cultivars become vulnerable to pathogens due to variation in pathogen virulence [7].

These effects can be managed by working on resistant cultivars, not against the diseases but also abiotic factors. Cultivars should be best fitted in the environment. Then proper nutrients should be provided to make the crop strong and protect the flag leaf of the plant. Agronomic practices should be done on time and a proper dose of fertilizer should be given to the soil. Explain new and environment-friendly approaches to the farmers to keep the wheat crop healthy and protected from major risks.

This chapter includes major and minor fungal diseases which attack the wheat crop, their mode of action, epidemiology, visual identification, and eradication methods.

2. Major fungal diseases

2.1 Obligate

2.1.1 Loose smut

Loose smut of wheat is a seed-born disease caused by *Ustilago tritici,* an obligate fungal pathogen belonging to division *Basidiomycota*, and the family *Ustilaginaceae*. This disease is reported everywhere, where wheat is cultivated. It was first reported by Romans and given the name *Ustilago* derived from the Latin word which means burn. Correct symptomology was given by Fabricius in 1774 in the book. In this disease, the plant is infected at the flowering stage and produces a sterile kernel-containing seed coat filled with smut spores. So, the disease is rarely spread by man. Its sores can easily spread to long distances with air and rain splashes. Loose smut of wheat was not reported in Australia, America, and South Africa, people from Europe move to these countries to settle down and bring wheat with them which was infected with loose smut. In North America, it was reported in 1832, although resistant genotypes were used against this disease but could not stop it from spreading. It is common in cold and humid regions but in dry areas causes equal yield losses. It does not cause huge economic losses but still, 2% disease in the field can cause yield reduction up to 20% plus make seeds not fit for next sowing. Loose smut is a seedborne pathogen; on maturity, the kernel is filled with dark brown to black color teliospores. The life cycle starts when teliospores enter the ovary through feathery stigma during anthesis [8]. Under favorable conditions when a single spike is infected with two different races, then, it is possible for recombination and genetic diversity. Mycelium survives inside the embryo without showing external symptoms. These spores will germinate when the infected seed is germinated. Pathogen spread systemically from one plant to another as well as inside the plant from cell to cell and reaches the tiller without producing any single external symptom [9]. The external symptom is very clear and easily recognized by black spores on a mature spike. This smut was released from the kernel as soon as the ear emerged out.

2.1.2 Wheat leaf rust or brown rust

Wheat is threatened with several diseases but rust is very dangerous for wheat. It causes huge economic losses all over the world. Wheat leaf rust is caused by *Puccinia triticina.* It was originated in the Middle East [10]. It is widely distributed where wheat is cultivated, therefore adopting a wide range of environments. Rust fungi are biotrophs and obligate parasites, they need living plants to complete their disease and life cycle. Rust fungi are host-specific. It can spread aeciospore, basidiospores, and urediniospores to far-off places by wind [11, 12]. This reason made it more diversified and the biggest reason for wheat economic losses; it adopted a wide range of environmental changes and increased inoculum amounts to cause disease epidemics [13]. It survives in mild temperature and higher moisture conditions. The disease reduces the size and weight of grain in the kernel. Leaf rust shows temporal and geographical variations and causes significant yield losses [14]. In the United States of America, during the years 2000–2004 leaf rust losses reached USD 350 million and in Australia up to AUD 12 million [14, 15].

Puccinia triticina is a macro-cyclic and heteroecious fungus. It requires two different hosts to complete its sexual and asexual life cycle. Its life cycle consists of five different spore stages. When environmental conditions are suitable fungus produce dark brown teliospores, which are diploid and on germination undergo meiosis and produce haploid basidiospores on the surface of a leaf. These basidiospores move by the wind too far-off places and require an alternate host to produce haploid pycnial on germination [10]. Alternate host of *Puccinia triticina* includes Anchusa, Isopyrum, Thalictrum, Barberry, and Clematis. After infection on alternate host rust produce pycniospores disseminate by insects, undergo sexual propagation into two, unlike cells which form plasmogamy. Aeciospores are produced in aecia and liberated by the wind. The life cycle ends when aeciospores germinate into asexual urediniospores and produce visual symptoms on the host plant. External symptoms include pustules circle or slightly oval but smaller than stem rust, these pustules do not merge in each other (**Figure 1A**). These pustules are filled with orange to the brown mass of urediniospores. Infection occurs on the upper side of the leaf surface. All these five types of

Figure 1.
Brown rust of wheat (A), black rust of wheat (B), and stripe rust of wheat (C).

spores are produced either by sexual or asexual reproduction. Asexual reproduction is carried out by urediniospores, which undergo many stages and form haustoria which are used to obtain nutrients from the host plant. It also suppresses the plant defense system and helps in fungal growth and germination [16].

Optimum temperature ranges between 10 and 25°C and free moisture for long period on the leaf surface help in disease infection. If favorable conditions prevail then the uredinial cycle repeats after every 8–14 days. Urediniospores, teliospores, and basidiospores are produced on wheat while pycniospores and aeciospores are produced on the alternate host. Teliospores are produced at the end of the season; so, it helps in overwintering and becomes a source of inoculum for the next growing season. Due to the monocyclic nature of the fungus they produce spores end of season and act as primary inoculum, these spores germinate on wheat individually and cause infection. P. *triticina* produces pustules of 1.5 mm in diameter which contain 20 k spores during advanced growth stages.

2.1.3 Wheat stem rust or black rust

Stem rust of wheat is historically a major disease of wheat caused by *Puccinia graminis* f. sp. *tritici*. It is also known as the black rust of wheat. Due to its historical yield losses of wheat, it comes in the top 10 fungal diseases which can cause devastating losses to crops [17]. It causes severe yield losses up to 50% region wise. On severe attack yield losses reaching up to 90% [18]. It causes yield losses in Canada, South America, Africa, Australia, China, and Indian Subcontinent [19]. Stem rust was a devastating disease for wheat and other cereals for many years. After applying so many management applications, it was eradicated but stem rust re-emerged in Africa and Europe and became a risk for food security. Severe disease attacks occur when a pathogen changes its virulent strain like UG-99 and TKTTF causing 100% yield in Uganda and Ethiopia, respectively [19]. Stem rust destroyed wheat in the United States up to 20% many times during the early 1900s [20]. Airborne pathogens can move from infected fields to healthy in no time, and this connected world increases the chances of spreading too many times. Small size urediniospores can move from one continent to another very smoothly. Like UG-99 moved from Uganda to South Africa, Iran, Russia, and TKTTF moved from Ethiopia to Germany, the UK, Sicily, Sweden, and Denmark [21]. Annually, stem rust causes a loss of 8–54 billion dollars every year globally [22]. Stem rust pathogen is an obligate parasite with a biotrophic mode of nutrition, so, it requires more living hosts to complete its life cycle. Pathogen belongs to phylum *Basidiomycota*, it is different from other fungi because it also produces five different spores to complete its life cycle like leaf rust as discussed earlier. It is also heteroecious, requires two hosts to complete its life cycle. Its primary host is wheat and the alternate host is the barberry plant. However, *Puccinia graminis* can complete its life cycle with or without an alternate host.

After 1–2 weeks of disease, infection urediniospores are produced in uredinia. It is the repeating stage in the whole life cycle. Urediniospores are dikaryotic produced on the separate stalk in the fruiting body. These spores can infect the host plant and produce external symptoms (**Figure 1B**). It produces red rusty spores which become dark, this is the reason it is also known as black rust [23]. When the growing season of crop ends it produces teliospores in telia. Teliospores are also dikaryotic and overwinter in the absence of the host. When teliospores get favorable conditions, they undergo meiosis and produce haploid basidiospores, these are colorless spores capable to infect alternate hosts but not cereal hosts. After germination, it produces pycnidiospore, which produces sticky honeydew which attracts insects and becomes a mode of perpetuation. Pycnidiospores mat and produce dikaryotic aeciospores in aecia. These aeciospores germinate and produce urediniospores but on cereal host not on the alternate host. Urediniospores may pass from winter wheat to spring wheat without infecting the alternate host. The optimum temperature is 30°C and 2 hours of leaf wetness can cause infection [18].

Puccinia graminis goes through meiosis so chances of recombination are many more. The number of strains reported includes UG-99, JRCQC, MCC, QCC, QCCJ, QCCJB, QCCS, QFCS, and TPMK. These all strains were reported at different times from different regions of the world [24, 25].

2.1.4 Stripe rust or yellow rust

All rust species are very destructive but stripe rust is most of all. It is caused by *Puccinia striiformis* f. sp. *tritici*. It belongs to the phylum *Basidiomycota*, and it is also heteroecious. It has a wide range of host plants. It can infect wheat, rye, barley, and other grass species. It can cause 100% yield loss if variety is susceptible and environmental conditions are favorable, but the losses varied between 10 and70% which mainly depends on epidemiology, area, variety under cultivation, and race of pathogen [26]. It was first reported in Sweden in 1794. It is present almost around the globe except in Antarctica. Its losses reached 46% in Asia which makes it a major limiting factor for the wheat crop [27]. The severe outbreak occurred in Turkey, Iran, and Uzbekistan, likewise in 2009-10 diseases occurred in epidemic form in western and central Asia and the North part of Africa [28]. Stripe rust is spreading very fast and changes the strain immediately, so, it is not easy to produce resistant genotype against stripe rust. It causes a wheat loss of 5 million tonnes which is worth equal to 10 million USD annually. It spreads into dry and warmer places as well and due to change in the cropping system and shifting of time of sowing make it more powerful [29].

Disease infection occurs anytime throughout the growing season. It causes symptoms on the green part of the plant which includes leaf, leaf sheath, glumes, and awn (**Figure 1C**). The disease infection cycle is similar to other rust pathogens. Infection caused by urediniospores and optimum temperature for disease infection is 3–20°C and free leaf moisture for 3 hours. The optimum temperature for infection on barberry is 10°C with 32 hours of leaf wetness [30]. When the temperature goes higher urediniospores change into teliospores. If the temperature goes down from −9°C, it completely stops the spores from germination. These two spore phases take place on wheat host remaining carried out on alternate host. *P. striiformis* spores can move by wind and it is assumed that they may travel from Armenia, Georgia, Azerbaijan, Turkey, and then move to Europe [31]. It was reported firstly at higher altitudes with cold weather but now it undergoes environmental adaptations and is reported in Asian countries as well [32]. Weather conditions are very important for disease infection, pathogen viability, growth, and sexual and asexual reproductions, all are linked with weather conditions plus vulnerability of wheat and alternate host [31]. If teliospores germinate at right time on an alternate host, it will complete the infection cycle of *P. striiformis.*

2.1.5 Powdery mildew

After rust and smuts mildew is a more contagious disease for plants. In wheat, powdery mildew is caused by *Blumeria graminis* f. sp. *tritici*. It is an obligate parasite but not heteroecious like a rust pathogen. It is considered as 6th out of 10 fungal yields threatening disease and 8th at the major disease in the world [17]. Powdery mildew is a very important disease in temperate and nautical climates in regions, such as America, Africa, and Europe. It occurred in epidemic form in China in 1980 and caused huge crop yield loss. In the year 2006–2008, it attacked an estimated around 7 million hectares area. It caused yield loss ranging between 5 and 35% [33]. It is an endemic disease and disseminates in different countries of the world. Yield losses up to 22.5% in Egypt on susceptible cultivars [34]. It causes 35% of disease losses in Russia and 62% in Brazil [35]. The pathogen attacks the plant at the lush green vegetative stage, mostly high-yielding genotype and high-nitrogen fertilizer making a plant vulnerable to powdery mildew. Ascospores or conidia germinate and enter the leaf surface to produce hypha with appressoria in 2 hours, this hypha changes into penetration peg, and haustorium allows to enter cell epidermis [36]. Infection reduces photosynthetic rate, leaf assimilation index, and increases the rate of respiration which ultimately affects the quality of grains. It affects plant vigor, tillering, heading, and grain-filling stages. Plant heavily infected gets died completely. When the wheat plant gets infected with powdery mildew at the tillering stage, it affects the booting stage which directly influences yield. It reduces grain size and weight of grain which causes a 40% reduction in crop yield if flag leaf gets infection then losses are much greater [37].

External symptoms include a grayish powdery colony on the leaf and stem of the wheat plant. It is most prevalent on the upper and lower surfaces of the leaf (**Figure 2**). The powder appears in the form of white patches early in the season, but disease duration can be prolonged if favorable conditions prevail. These white cottony pustules produce asexual spores known as conidia which are later on spread by wind. Sexual spores are ascospores that developed in cleistothecia. Both sexual and asexual spores are born in humid weather with a lower temperature range.

On disease progress, it covers the entire plant and turns into yellow color due to chlorosis, black color fruiting bodies appear on a leaf along with the gray powdery mass. Fungus mass can cover the head-on severe attack. The temperature range between 10 and 21°C favors the disease. Infection is reduced at the flowering stage as the temperature rises, so, fungal grow inside the tissues during winter times. When the temperature becomes suitable, it starts germination and spreads from plant to plant and field to field by wind. High humidity and temperature range between 15 and 20°C are most suitable for disease spread, repetition of the life cycle in 7–10 days, and development of new strains [38].

Figure 2.
Visual symptoms of powdery mildew on the wheat leaf (A) and spikes (B).

2.1.6 Karnal bunt

Karnal bunt of wheat is a very common disease everywhere, it is caused by *Tilletia indica* (**Figure 3**). It was first time discovered in India in 1930 [39]. Afterward reported in Iraq, Pakistan, Nepal, Afghanistan, USA, South Africa, and Mexico [40]. Karnal bunt of wheat is an air, seed, and soilborne disease, and spread easily to far-off places. It affects the quality of seeds as well as makes them unfit for eating and sowing. The disease produces a specific smell even 1% contamination occurs. It was assessed that if wheat grains are infected with 3%, it is not suitable for human consumption. Infected seeds also become less fertile [41]. The disease can cause huge yield losses, in India, yield losses due to Karnal bunt up to 40% in severely infected fields; in the year 2014–2015, disease incidence was 15% reported in India [42, 43]. Due to its massive yield losses and easy perpetuation, many countries banned importing wheat from those countries where this disease was common. Almost 30 different countries adopt zero-tolerance quarantine measures to avoid disease and its after-effects. Karnal bunt causes 7 billion US dollars loss in Mexico every year [43]. Disease epidemics mainly depend upon weather conditions, when the temperature and humidity are suitable disease infection occurs on wheat. Teliospores germinate when the temperature is 20–25°C, if 20°C temperature, pH 6–9.5, and moisture for 3 weeks disease outbreak occurs. Teliospores need an 80% moisture rate or free water for germination [40].

Clear symptoms can be seen after threshing as grain in the spike are swollen and fell off with the wind. Spike length and number of spikes per plant also reduce with the disease (**Figure 3B**). It is also found that all ears in the spike are not bunted, however, infected grains are converted into bunt sori (**Figure 3A**). Sori is oval-shaped contains black to brown powdery spores enclosed in the pericarp (**Figure 3C**). In a severely attacked spike, the lining of the seed coat and epidermis is destroyed and spores are enclosed in the lining of the pericarp [40]. This smut fungus has a different pathogenesis method, the fungus infects wheat after diazotization and starts germination and colonization in the epidermis of the plant and infection spreads from cell to cell [44]. Teliospores germinate diploid nuclei which undergo meiosis and several mitoses and produce haploid basidium. Haploid nuclei develop and produce basidiospores. One

Figure 3.
Karnal bunt in wheat spike (A, B) and dark brown teliospores from bunt sori (C).

daughter cell back to basidium and produce 110–185 sporidia on the tip and are sickle-shaped. With rain splashes and wind, teliospores drop on soil and become a source of primary inoculum, it can survive in the soil for many years [45]. Secondary sporidia (allantoid and filiform) are more durable and germinate when getting favorable conditions and play important role in the disease cycle. Allantoid plays role in infection and filiform increases the number of inoculum in the soil. Sporidia are binucleated and on germination produce a germ tube that penetrates newly developing seed in the ovary. Bunt fungus causes infection at the time of anthesis [46]. The disease produces a characteristic fishy smell as teliospores that release trimethylamine [40]. If the conditions are not favorable or the wheat plant does not reach the vulnerable stage when teliospores germinate, it leads to "suicidal germination." [47].

Spores of Karnal bunt can survive in the soil for 3 years, it can spread from one farm to another through farm machinery. It can tolerate very cold temperatures and maintain its viability. Air can spread spores up to 3000 meters [48]. Teliospores are resistant to chloropicrin, hydrogen peroxide, methyl bromide, ozone, and propionic acid [47]. Single pathogen isolates can vary from one another by physical and morphological characteristics, the number of chromosomes, degree of infection

Figure 4.
Flag smut of wheat.

production, and resistance to host barriers. *T. indica* is the most effective and over-breaking smut fungal pathogen from all of the others [40].

2.1.7 Flag smut

Like loose smut, flag smut is also a very important disease; it is caused by *Urocystis agropyri* (**Figure 4**). It is an obligate pathogen like all other rusts and smuts pathogens, it needs a living host to complete its life cycle. Symptoms include twisting and bending of infected seedlings, white patched areas appear with blistered and develop coleoptiles. On the mature plant, symptoms appear on leaves with white areas which turn gray to black afterward. Stunted growth, distorted and twisted tillers may not produce spikes and grains. Poor root development was also observed in the infected plant. If grains produced are also infected and show poor germination if cultivated again. Losses due to flag smut varied between 5 and 20% depending upon the availability of environmental conditions [49].

2.1.8 Common bunt

Common bunt of wheat is caused by *Tilletia caries,* it is the most destructive disease of wheat. It occurs at the grain-filling stage, with spikes filled with bunt spores instead of grains. Spores produce a smell resembled with fish. Spores are airborne and bunt stuck with healthy grains during harvesting and become a source of infection for the next year. These infected seeds spread all around in trading and exportation. Common symptoms appear on spikes that are darker in color as compared to healthy spikes, on maturity these infected spikes turn bluish-gray [50].

3. Non-obligate

3.1 Fusarium head blight

Fusarium head blight [FHB] also known as ear blight and the head scab is caused by different fungal pathogens, including *F. culmorum, F. graminerarum*, and *F. avenaceum.* The most common pathogen is *F. graminearum,* which is found in America, Europe, and Asia [51]. It prevails in crop residues in the form of saproprobes; inoculum consists of conidia and ascospores; ascospores are sexual spores produced in perithecia fruiting body and released by wind [52]. Conidia are asexual spores and produce mycelia which infect plant leaves and spikes. *F. graminearum* produces many mycotoxins, such as nivalenone, zearalenone, deoxynivalenol, and moniliformin, which causes significant food poison for humans and animals [53].

Wheat plant is susceptible to FHB from the anthesis stage till kernel production. Main external symptoms appear on the head, peduncle, spikes, and grains. Yellowing to slight discoloration of infected spikelet starts while healthy spikes are green (**Figure 5**). Infected spikes contain pinkish and orange shade colonies of spores. Spores are produced in cold and humid weather. Infected seed is cultivated in the next season results in red to brown shade colony with poor tiller came out. Reduced size and less vigor and well as affect germination rate of seed. Other morphological characters include the late heading and tiller stage [54]. Its symptoms are mostly confused with root rot and crown rot disease, and incomplete symptoms are confused with glume blotch and black chaff.

Figure 5.
Fusarium graminearum *symptoms on infected wheat nodes and spike.*

Optimum conditions for disease infection are moderate rainfall, and temperature range between 24 and 29°C for 2–3 days is enough for infection. Fungal spores survive in crop residues. High humidity and rainfall not only increase inoculum but also help in dispersal.

3.2 Tan spots

The tan spot of wheat also known as yellow spot or blotch is caused by *Pyrenophora tritici repentis.* Tan spot is very common in the UK, Sweden, Germany, France, and Denmark; this causes serious wheat yield loss.

It has a wild range of host plants, including grass species. Most of them are perennial crops that help in overwintering pathogens and increasing the number of inoculum for disease epidemics. It causes 5–10% yield loss and when environmental conditions are favorable losses reached 50% [49]. Symptoms include small dark brown fleck which turns black spot on basil leaves. Then spots merge and get enlarged into tan and irregular lesions with browning inside and yellow rings surround the lesion. Under humid conditions, lesions produce dark spores, and lesions combine and produce dead tissues. Infected seeds contain pink to red color spores, black points, and low germination.

Fungal survive in the form of dormant mycelium on crop residues. Pseudothecium is the fruiting body and ascospores are produced inside. Ascospores spread with the wind too far-off places, infected seed is also a source of disease spread. Under warm and wet conditions, asexual conidia germinate and spread with rain, it infects the ear, glume, and developing grain. Optimum temperature ranges between 20 and 28°C and disease symptoms appear in 7–14 days [49].

3.3 Septoria disease

Septoria is a disease complex caused by three different pathogens called *P. avenaria triticae, Mycosphaerella graminicola,* and *Phaeosphaeria nodorum.* While diseased caused by *Mycosphaerella graminicola* (anamorph; *Zymoseptoria tritici*) is called Septoria tritici blotch [STB]. This pathogen not only reduces the size but also the quality of grain. When disease occurs in the epidemic form, it causes 30–50% yield losses. It occurs in strong epidemic form in those areas where the temperature is lower and wet humid weather. It is most common in North and South America, North and South Africa, the north part of Europe, and Turkey [55]. External symptoms include chlorotic lesions on leaves appearing in fall and spring, with the disease advancement it becomes darker and produces fruiting bodies on the lesion (**Figure 6**). It produces pseudothecia as a sexual fruiting body and pycnidia as an asexual fruiting body.

The infection starts when airborne ascospores germinate on the plant which is overwintered in plant residues. Infection occurs just after the emergence of seeding. Sexual spores attach with stomatal opening with the help of germ tube and enter into the stomata and produce appressorium. After 7 days, it produces hyphae and mycelium inside the whole plant. The pathogen has both biotrophic and necrotrophic mode or growth when it changes its mode then external lesions appear on leaf and cell collapse. On disease, advancement lesions change from light to dark color. Conidia are produced on the necrotic site which spread with rainwater from infected to healthy plants as well as over winter in residues and become a source of inoculum for the next crop. Pycnidia produce conidia within 15–40 days after infection, it depends upon environmental conditions. Under unfavorable conditions, spores undergo a dormant state and germinate when the temperature and moisture are available [56].

3.4 Common root rot, crown root rot, and black point

The number of diseases caused by *Bipolaris sorokiniana* in different cereal crops causes crown root rot, common root rot, and black point disease in wheat leads to huge yield losses [57]. Root rot occurs everywhere in wheat-cultivated areas. Canada lost 5.7% of wheat due to common root rot which was worth 42 million dollars. Crown root rot is very common in the pacific region. It causes 35% yield losses over there [58]. When seeds are infected with *Bipolaris sorokiniana* causes black points on the wheat plant which on advancement results in rotting and blight disease of seedlings. Black point disease appeared as brown to blackish tips on the embryo of grains, it increased the weight of kernel but grain quality reduced. These infected seeds if cultivated for the next season it will reduce seed germination, increase seedling

Figure 6.
Septoria leaf blotch symptoms on the wheat leaf (A) and sclerotia on the stem and leaf (B).

Figure 7.
Rhizoctonia root rot (A) and fusarium head blight (B).

death, reduce seedling emergence, and reduce photosynthetic rate ultimately affecting normal growth [59]. It is also observed that disease is associated with *Alternaria, Penicillium, and Fusarium spp.* and caused huge losses [60]. The disease causes losses up to 90% if favorable conditions prevail for a longer time. The favorable temperature ranges between 28 and 30°C and relative humidity requires up to 90% [59]. Fungus goes through teleomorph known as *Cochliobolus sativus*, it is the sexual stage and reported first time in Zambia, while sexual reproduction of *C. sativus* has not been reported everywhere. *B. sorokiniana* reproduce asexually through conidia [58]. Common root rot and crown root rot cause significant losses in China, India, Australia, and Europe. Symptoms include necrotic lesions on roots and with the disease advancement lesions become darker [61]. It is very common in dry and warm regions of the world. Disease incidence is affected by soil temperature and moisture, severity increases when the plant is under stress conditions.

Root rot is one of the serious diseases for wheat, especially in Egypt. It can cause significant yield losses because it attacks the seedlings just after germination from the seed. The disease is caused by *Rhizoctonia solani* and *Fusarium graminearum* (**Figure 7**). In combination, these two pathogens are very severe. Fusarium cause wilting and R. solani causes damping-off. Fungal pathogens germinate, colonize, and enter the roots; they block the roots by growing mycelium inside the root. Plant or young seedlings are unable to get nutrition and ultimately die off [62, 63].

4. Minor limiting factors

Wheat is also affected by some minor diseases which are sometimes causing huge crop yields. It includes different types of viruses, bacteria, and nematodes. Wheat is affected by the number of soilborne viruses named as *soilborne wheat mosaic virus* belongs to genus brymovirus. This genus has other members, such as *wheat yellow mosaic virus* and *wheat spindle streak mosaic virus. The soilborne wheat mosaic virus* causes heavy losses in USA and Brazil in the early 90s, 50 and 80% respectively.

Symptoms included were a yellow mosaic appearance on leaves and stunted growth with a poor root system.

Wheat spindle streak mosaic virus causes 30% yield losses to wheat sown in winter. It shows chlorotic as well as necrotic streaks along with leaf veins. It reduces crop height and seed production [64, 65]. Nematodes include cyst nematodes in which 2nd stage juveniles enter the root and move toward the vascular system. Nematode enlarges its size and develops syncytia which withdraw food and nutrition from nearby cells. This serves as an endoparasite. Adult females lay eggs within 3 to 6 weeks of infection and cover these eggs to protect them from unfavorable conditions. Wheat is also infected with *H. avanae*. Another nematode *Anguina tritici* causes seed gall disease in wheat and other important cereals.

5. Management

Wheat is important and staple food all over the world. The main source of carbohydrates and used in a different form. But its production is affected by the number of diseases. Wheat is affected by several fungal diseases. The major threat to wheat is due to rusts, smuts, and mildews. Crop rotation, soil solarization, and zero tillage are important tools for disease management. The use of a resistance cultivar against different pathogens is an effective strategy. Resistant cultivars, certified pure seeds, and seed treatment with strong fungicide are effective to control for these rust and smut diseases [66]. Breeding for resistant varieties to manage loose smut, inheritance of resistance in hexaploid wheat cultivars is examined [67]. Back cross of seven resistant and two susceptible varieties against loose smut disease artificially inoculated in mid of anthesis stage. The segregation ratio showed that resistance against loose smut is controlled by a single dominant gene in wheat [68]. Another study reviled that resistant genes against loose smut are partial and complete resistance which are both dormant and recessive, these resistance genes can stop or hinder the growth of smut inside the plant at different points [68].

One of the powerful and effective tools against disease resistance is host-induced gene silencing in the transgenic plant. It is also helpful in functioning and gene characterization. New and advanced techniques help in contrition of efficient transgenic system and enhance RNAi-driven strategy against the resistant plant. These new and advanced techniques are proven best against biotic and abiotic stresses and another best part of these techniques, they are environmentally friendly and farmer friendly. Genome editing by Cas-9 is very important and helpful in the insertion of resistant genes against pathogens to produce resistant cultivars [69, 70].

The use of biological control agents against pathogens is effective and widely used technique because of the chemical resistance problem. The use of *B. megaterium, T. harzianum, B. amyloliquefaceiens,* and *Epicoccum spp. is* proven effective against different root rots [71]. The use of *T. harzianum* and *T. koningii* are significant against *B. sorokiniana* and *A. alternata* [72]. *Trichoderma spp.* and *Bacillus* give significant inhibition against different root rot pathogens [73].

Chemical fungicides are used for seed treatment and foliar application. Use of Carbendazim, carboxin, triticonazole, thiram, metalaxyl, as a seed treatment for seedborne diseases in wheat [72]. Tubeconazole in combination with imidacloprid and cyproconazole along with difenoconazole are effective chemical fungicides against rust and smut diseases of wheat [74]. Seed treatment with difenoconazole, fludioxonil, homai, and vitavax is proven best against seedborne pathogens.

Author details

Mukaddes Kayim*, Hira Nawaz and Abdulkreem Alsalmo
Faculty of Agriculture, Department Plant Protection, Cukurova University, Adana, Turkey

*Address all correspondence to: kayimukaddes@gmail.com; mkayim@cu.edu.tr

References

[1] Shewry PR, Hey SJ. The contribution of wheat to human diet and health. Food and Energy Security. 2015;**4**(3):178-202

[2] Considine MJ, Siddique KH, Foyer CH. Nature's pulse power: Legumes, food security and climate change. Journal of Experimental Botany. 2017;**68**(8):1815-1818

[3] Liu Q, Qiu Y, Beta T. Comparison of antioxidant activities of different colored wheat grains and analysis of phenolic compounds. Journal of Agricultural and Food Chemistry. 2010;**58**(16):9235-9241

[4] Alam A, Hariyanto B, Ullah H, Salin KR, Datta A. Effects of silicon on growth, yield and fruit quality of cantaloupe under drought stress. Silicon. 2021;**13**(9):3153-3162

[5] Braun H-J, Atlin G, Payne T. Multi-location testing as a tool to identify plant response to global climate change. Climate Change and Crop Production. 2010;**1**:115-138

[6] Bhavani S, Singh PK, Qureshi N, He X, Biswal AK, Juliana P, et al. Globally important wheat diseases: Status, challenges, breeding and genomic tools to enhance resistance durability. Genomic Designing for Biotic Stress Resistant Cereal Crops. 2021:59-128

[7] Beres BL, Rahmani E, Clarke JM, Grassini P, Pozniak CJ, Geddes CM, et al. A systematic review of durum wheat: Enhancing production systems by exploring genotype, environment, and management (G× E× M) synergies. Frontiers in Plant Science. 2020:1665

[8] Kashyap PL, Kumar S, Tripathi R, Kumar RS, Jasrotia P, Singh DP, et al. Phylogeography and Population Structure Analysis Reveal Diversity by Gene Flow and Mutation in Ustilago segetum (Pers.) Roussel tritici Causing Loose Smut of Wheat. Frontiers in Microbiology. 2019;**10**:1072

[9] Menzies J, McLeod R, Tosi L, Cappelli C. Occurrence of a carboxin-resistant strain of Ustilago nuda in Italy. Phytopathologia Mediterranea. 2005;**44**(2):216-219

[10] Bolton MD, Kolmer JA, Garvin DF. Wheat leaf rust caused by Puccinia triticina. Molecular Plant Pathology. 2008;**9**(5):563-575

[11] Mendgen K, Hahn M. Plant infection and the establishment of fungal biotrophy. Trends in Plant Science. 2002;**7**(8):352-356

[12] Kolmer J, Ordonez M, Groth J. The Rust fungi. Encyclopedia of Life Sciences (ELS). Chichester: Wiley; 2009

[13] Abebe W. Wheat leaf rust disease management: A. International Journal of Novel Research in Interdisciplinary Studies. 2021

[14] Huerta-Espino J, Singh R, German S, McCallum B, Park R, Chen WQ, et al. Global status of wheat leaf rust caused by *Puccinia triticina*. Euphytica. 2011;**179**(1):143-160

[15] Murray GM, Brennan JP. Estimating disease losses to the Australian wheat industry. Australasian Plant Pathology. 2009;**38**(6):558-570

[16] Ramachandran SR, Yin C, Kud J, Tanaka K, Mahoney AK, Xiao F, et al. Effectors from wheat rust fungi suppress multiple plant defense responses. Phytopathology. 2017;**107**:75-83

[17] Barton L, Butterbach-Bahl K, Kiese R, Murphy DV. Nitrous oxide

fluxes from a grain–legume crop (narrow-leafed lupin) grown in a semiarid climate. Global Change Biology. 2011;**17**:1153-1166

[18] Huerta-Espino J, Singh RP, Roelfs AP. Rusts fungi of wheat. Fungi from Different Substrates. 2014:217-259

[19] Singh RP, Hodson DP, Jin Y, Lagudah ES, Ayliffe MA, Bhavani S, et al. Emergence and spread of new races of wheat stem rust fungus: Continued threat to food security and prospects of genetic control. Phytopathology. 2015;**105**(7):872-884

[20] Lackermann KV, Conley SP, Gaska JM, Martinka MJ, Esker PD. Effect of location, cultivar, and diseases on grain yield of soft red winter wheat in Wisconsin. Plant Disease. 2011;**95**(11):1401-1406

[21] Lewis CM, Persoons A, Bebber DP, Kigathi RN, Maintz J, Findlay K, et al. Potential for re-emergence of wheat stem rust in the United Kingdom. Communications Biology. 2018;**1**:1-9

[22] Berlin A, Samil SB, Djurle A, Wirsen H, Szabo L, Yuen J. Disease development and genotypic diversity of Puccinia graminis f. sp. avenae in Swedish oat fields. Plant Pathology. 2013;**62**:32-40

[23] Schumann GL, Leonard KJ. Stem rust of wheat. 2011

[24] Chen S, Zhang W, Bolus S, Rouse MN, Dubcovsky J. Identification and characterization of wheat stem rust resistance gene Sr21 effective against the Ug99 race group at high temperature. PLoS Genetics. 2018;**14**(4):e1007287

[25] Olivera P, Newcomb M, Szabo LJ, Rouse M, Johnson J, Gale S, et al. Phenotypicandgenotypiccharacterization of race TKTTF of *Puccinia graminis* f. sp. *tritici* that caused a wheat stem rust epidemic in southern Ethiopia in 2013-14. Phytopathology. 2015;**105**(7):917-928

[26] Begum S, Iqbal M, Ahmed I, Fayyaz M, Shahzad A, Ali GM. Allelic variation at loci controlling stripe rust resistance in spring wheat. Journal of Genetics. 2014;**93**(2):579-586

[27] Curtis BC, Rajaram S, Gómez Macpherson H. Bread Wheat: Improvement and Production. Vialle delle Terme di Caracalla, Rome, Italy: Food and Agriculture Organization of the United Nations (FAO); 2002

[28] Sinha P, Chen X. Potential infection risks of the wheat stripe rust and stem rust pathogens on barberry in Asia and Southeastern Europe. Plants. 2021;**10**(5):957

[29] Beddow JM, Pardey PG, Chai Y, Hurley TM, Kriticos DJ, Braun H-J, et al. Research investment implications of shifts in the global geography of wheat stripe rust. Nature Plants. 2015;**1**(10):1-5

[30] Wang M, Chen X. Barberry does not function as an alternate host for *Puccinia striiformis* f. sp. *tritici* in the US Pacific northwest due to teliospore degradation and barberry phenology. Plant Disease. 2015;**99**(11):1500-1506

[31] Line RF. Stripe rust of wheat and barley in North America: A retrospective historical review. Annual Review of Phytopathology. 2002;**40**(1):75-118

[32] Waqar A, Khattak SH, Begum S, Rehman T, Shehzad A, Ajmal W, et al. Stripe rust: A review of the disease, Yr genes and its molecular markers. Sarhad Journal of Agriculture. 2018;**34**(1):188-202

[33] He Z, Lan C, Chen X, Zou Y, Zhuang Q, Xia X. Progress and

perspective in research of adult-plant resistance to stripe rust and powdery mildew in wheat. Scientia Agricultura Sinica. 2011;**44**(11):2193-2215

[34] Ashmawy M, El-Orabey W, Abu Aly AE-A, Shahin A. Losses in grain yield of some wheat cultivars infected with powdery mildew. Egyptian Journal of Phytopathology. 2014;**42**(1):71-82

[35] Mehta YR. Wheat Diseases and their Management. Switzerland: Springer; 2014

[36] Acevedo-Garcia J, Spencer D, Thieron H, Reinstädler A, Hammond-Kosack K, Phillips AL, et al. Mlo-based powdery mildew resistance in hexaploid bread wheat generated by a non-transgenic TILLING approach. Plant Biotechnology Journal. 2017;**15**(3):367-378

[37] Alam MA, Wang C, Ji W. Chromosomal location and SSR markers of a powdery mildew resistance gene in common wheat line N0308. African Journal of Microbiology Research. 2013;**7**(6):477-482

[38] Piarulli L, Gadaleta A, Mangini G, Signorile MA, Pasquini M, Blanco A, et al. Molecular identification of a new powdery mildew resistance gene on chromosome 2BS from *Triticum turgidum* ssp. *dicoccum*. Plant Science. 2012;**196**:101-106

[39] Mitra M. A new bunt on wheat in India. Annals of Applied Biology. 1931;**18**:178-179

[40] Kumar S, Singroha G, Singh GP, Sharma P. Karnal bunt of wheat: Etiology, breeding and integrated management. Crop Protection. 2021;**139**:105376

[41] Ullah HZ, Haque M, Rauf C, Akhtar L, Munir M. Comparative virulence in isolates of *Tilletia indica* and host resistance against Karnal bunt of wheat. Journal of Animal and Plant Sciences. 2012;**22**(2). Available from: https://link.gale.com/apps/doc/A295848642/AONE?u=anon~de89e5f5&sid=googleScholar&xid=34222a76

[42] Kumari P, Maurya S, Kumar L, Pandia S. Karnal bunt disease a major threatening to wheat crop: A review. IJAR. 2020;**6**(6):157-160

[43] Anonymous. IARI Annual Report (2015-2016). New Delhi, India: Indian Agricultural Research Institute; 2015. p. 83. Available from: www.iari.res.in [Accessed: January 3, 2021]

[44] Aggarwal R, Tripathi A, Yadav A. Pathogenic and genetic variability in Tilletia indica monosporidial culture lines using universal rice primer-PCR. European Journal of Plant Pathology. 2010;**128**(3):333-342

[45] Babadoost M, Mathre D, Johnston R, Bonde M. Survival of teliospores of *Tilletia indica* in soil. Plant Disease. 2004;**88**(1):56-62

[46] Inman A, Magnus H, Riccioni L, Hughes K, Coates M, Barnes A, et al. Survival of Tilletia indica teliospores under European soil conditions. Plant Pathology. 2008;**57**(2):290-300

[47] Rush CM, Stein JM, Bowden RL, Riemenschneider R, Boratynski T, Royer MH. Status of Karnal bunt of wheat in the United States 1996 to 2004. Plant Disease. 2005;**89**(3):212-223

[48] Bonde M, Berner D, Nester S, Peterson G, Olsen M, Cunfer B, et al. Survival of *Tilletia indica* teliospores in different soils. Plant Disease. 2004;**88**(4):316-324

[49] Garima S, Gopal R, Vikas G and Satish K. 2017. Wheat Diseases and their Management. Chapter. Wheat A premier Food Crop_3rd Proof.indb.

[50] Figueroa M, Hammond-Kosack KE, Solomon PS. A review of wheat diseases a field perspective. Molecular Plant Pathology. 2018;**19**(6):1523-1536

[51] Beccari G, Colasante V, Tini F, Senatore M, Prodi A, Sulyok M, et al. Causal agents of fusarium head blight of durum wheat (*Triticumádurum Desf.*) in Central Italy and their inávitro biosynthesis of secondary metabolites. Food Microbiology. 2018;**70**:17-27

[52] Trail F. For blighted waves of grain: *Fusarium graminearum* in the postgenomics era. Plant Physiology. 2009;**149**(1):103-110

[53] Chilaka CA, De Boevre M, Atanda OO, De Saeger S. The status of fusarium mycotoxins in sub-Saharan Africa: A review of emerging trends and post-harvest mitigation strategies towards food control. Toxins. 2017;**9**(1):19

[54] Buerstmayr M, Buerstmayr H. The semidwarfing alleles Rht-D1b and Rht-B1b show marked differences in their associations with anther-retention in wheat heads and with fusarium head blight susceptibility. Phytopathology. 2016;**106**(12):1544-1552

[55] Unal G, Kayim M, Ay T, Yones AM. Evaluation of disease intensity and molecular identification of *Zymoseptoria tritici* causing Septoria leaf blotch on wheat in the eastern Mediterranean region of Turkey. Turkish Agriculture and Forestry. 2017;**41**(6):405-413

[56] Ponomarenko A, Goodwin S, Kema GH. *Septoria tritici* blotch (STB). Plant Health Instr. 2011:10

[57] Singh UB, Malviya D, Singh S, Kumar M, Sahu PK, Singh H, et al. *Trichoderma harzianum*-and methyl jasmonate-induced resistance to *Bipolaris sorokiniana* through enhanced phenylpropanoid activities in bread wheat (*Triticum aestivum* L.). Frontiers in Microbiology. 2019;**10**:1697

[58] Gupta P, Chand R, Vasistha N, Pandey S, Kumar U, Mishra V, et al. Spot blotch disease of wheat: The current status of research on genetics and breeding. Plant Pathology. 2018;**67**(3):508-531

[59] Li Y, Chen M, Hao H, Xu Q. Mechanical stress-initiated signal transductions in vascular smooth muscle cells. Cellular Signalling. 2000;**12**:435-445

[60] Li Q, Niu H, Xu K, Xu Q, Wang S, Liang X, et al. GWAS for resistance against black point caused by *Bipolaris sorokiniana* in wheat. Journal of Cereal Science. 2020;**91**:102859

[61] Touhami O, Hzrbaa D. Study of the fungal complex responsible for root rot of wheat and barley in the north-west of Morocco. Plant Archives. 2019;**19**(2):2143-2157

[62] Moubark MY, Abdel-Monaim MF. Effect of bio-control agents on yield, yield components and root rot control in two wheat cultivars at New Valley region. Notulae Scientia Biologicae. 2011;**3**(4):79-87

[63] Asran M, Amal ME. Aggressiveness of certain *fusarium graminearum* isolates on wheat seedlings and relation with their trichothecene production. Plant Pathology Journal. 2011;**10**(1):590-609

[64] Perovic D, Förster J, Devaux P, Hariri D, Guilleroux M, Kanyuka K, et al. Mapping and diagnostic marker development for Soil-borne cereal mosaic virus resistance in bread wheat. Molecular Breeding. 2009;**23**(4):641-653

[65] Segundo E, Lesemann DE, Martín G, Carmona MP, Ruiz L, Cuadrado IM, et al. Amaranthus leaf mottle

virus: 3′-end RNA sequence proves classification as distinct virus and reveals affinities within the genus Potyvirus. European Journal of Plant Pathology. 2007;**117**(1):81-87

[66] Kumar S, Knox RE, Singh AK, DePauw RM, Campbell HL, Isidro-Sanchez J, et al. High-density genetic mapping of a major QTL for resistance to multiple races of loose smut in a tetraploid wheat cross. PLoS One. 2018;**13**(2):e0192261

[67] Sammour RH, El Shamy MM, Mustafa AE-ZM, Yousef ES. Inheritance of Resistance to Loose Smut *ustilago tritici* of Wheat. 2015;**10**(4):137-146

[68] Muriuki N, Munyua C, Wanga D. Communication channels in adoption of technology with a focus on the use of Purdue improved crop storage (PICS) among small scale maize farmers in Kenya. Journal of Biology, Agriculture and Healthcare. 2016;**6**(18):8-12

[69] Borrelli VM, Brambilla V, Rogowsky P, Marocco A, Lanubile A. The enhancement of plant disease resistance using CRISPR/Cas9 technology. Frontiers in Plant Science. 2018;**9**:1245

[70] Qi T, Guo J, Peng H, Liu P, Kang Z, Guo J. Host-induced gene silencing: A powerful strategy to control diseases of wheat and barley. International Journal of Molecular Sciences. 2019 ;**20**(1):206

[71] El-Gremi SM, Draz IS, Youssef WA-E. Biological control of pathogens associated with kernel black point disease of wheat. Crop Protection. 2017;**91**: 13-19

[72] Mónaco C, Sisterna M, Perelló A, Dal Bello G. Preliminary studies on biological control of the blackpoint complex of wheat in Argentina. World Journal of Microbiology and Biotechnology. 2004;**20**(3):285-290

[73] Akgül DS, Erkilic A. Effect of wheat cultivars, fertilizers, and fungicides on fusarium foot rot disease of wheat. Turkish Journal of Agriculture and Forestry. 2016;**40**(1):101-108

[74] Ishtiaq M, Hussain T, Bhatti KH, Adesemoye T, Maqbool M, Azam S, et al. Management of root rot diseases of eight wheat varieties using resistance and biological control agents techniques. Pakistan Journal of Botany. 2019;**1**(16):1-13

Chapter 7

Bacterial Pathogens of Wheat: Symptoms, Distribution, Identification, and Taxonomy

James T. Tambong

Abstract

Bacterial pathogens are significant biotic factors of wheat, a globally important source of carbohydrates. The diseases caused by these pathogens are reported to reduce annual wheat production by about 10% and up to 40% in severe infections occurring early in the growth period. This chapter presents current information on the symptoms, distribution, identification, and taxonomy of key bacterial pathogens of wheat with a focus on the seed-borne bacterium, *Xanthomonas translucens* pv. *undulosa*, the causative agent of the leaf streak and black chaff disease. Other wheat-pathogenic bacterial pathogens addressed in the chapter are *Pseudomonas syringae* pv. *syringae,* the causal agent of bacterial leaf blight; *P. syringae* pv. *atrofaciens* that cause the basal glume rot; *Pseudomonas fuscovaginae*, the causal agent of the bacterial brown sheath; *Erwinia rhapontici*, the causal agent of the pink seed of wheat; *Pseudomonas cichorii*, the causative agent of wheat stem melanosis; *Clavibacter tessellarius* is responsible for the bacterial mosaic of wheat as well as other minor bacterial pathogens. Finally, the chapter proposed the use of genome-based tools for the accurate identification and classification of bacterial pathogens of wheat.

Keywords: wheat, bacterial pathogens, *Xanthomonas translucens*, *Pseudomonas syringae*, *Clavibacter tessellarius*, bacterial leaf streak and black chaff disease, *Erwinia*, *Pseudomonas fuscovaginae*, *Rathayibacter*, *Pseudomonas cichorii*
Agriculture and Agri-Food Canada

1. Introduction

Wheat is the most important food grain source in the human diet and is considered a global primary commodity ([1]; http://faostat.fao.org/). The "big three" cereal crops, wheat, maize, and rice account for 75% and 50%, respectively, of the carbohydrate and protein intakes by humans [2–4]. Of the three cereal crops, wheat is the most nutritious with a protein content of about 11–14% even though it is low in some essential amino acids, e.g. lysine [1, 5, 6]. Wheat, as an essential staple food, provides about 20% of the calories intake of about 40% of the global population [2].

Notwithstanding the significance of wheat grain and products to humans globally, the worldwide increment in wheat production is third behind maize and rice. Between

1961 and 2013, the total cultivated land area allotted to bread and durum wheat increased from 204 Mha to 218 Mha, an increment of only 6.8%, but recorded a 321% increase in production from 222 MT to 713 MT [4]. In 2017, global wheat production was 757 MT harvested from 220 M ha [4]. An increase of 30% in wheat production to 1 billion tons, as suggested by Bockus et al. [2] is required by 2030 to feed the nine billion estimated population. It is unclear how this can be achieved in the agricultural world threatened by global warming. This is compounded by the fact that annually, about 25–30% of wheat crop yield loss is incurred due to abiotic and biotic stresses.

Biotic stresses are incited by diseases and insect pests of wheat. One of these groups is bacteria. Bacteria are prokaryotic microorganisms that have a nucleus or organelles not bound to a membrane. Plant pathogenic bacteria are unicellular and may be motile due to the presence of one or more flagella; and the mode of reproduction is by binary fission, in which the cell divides into two similar daughter cells with new generations occurring in less than 2 h in some species. Due to their small size, measuring up to about 2 μm, light (at least 400×) and electron microscopes are required for detailed studies and investigations of the cell morphology and structure. Bacteria that are pathogenic to wheat are categorized into two main groups: Gram-positive and Gram-negative based on the reactions to Gram's stain, a reaction during which strains of the former group stains dark purple while the latter group appears red by taking up the counterstain.

Bacterial pathogens are significant biotic factors of wheat, a globally important source of carbohydrates. The diseases caused by these pathogens are reported to reduce annual wheat production by about 10% and up to 40% in severe infections occurring early in the growth period. This chapter presents current information on the typical symptoms, distribution, identification, and taxonomy of key bacterial pathogens of wheat. The pathogens to be profiled include *Xanthomonas translucens*, the causal agent of the bacterial leaf streak and black chaff disease; the causal agent of bacterial leaf blight, *Pseudomonas syringae* pv. *syringae*; *Clavibacter michiganensis* subsp. *tessellarius* (bacterial mosaic of wheat) as well as other minor bacterial pathogens.

2. Wheat bacterial pathogens

2.1 *Xanthomonas translucens* pv. *undulosa* (Xtu), the causal agent of bacterial streak and black chaff disease of wheat

2.1.1 Disease symptoms, importance, and distribution

The leaf streak and black chaff (BLS) is the most important bacterial disease of wheat. Symptoms appear on the leaves and/or spikes of wheat plants. The disease starts with water-soaked necrotic streaks that eventually change to translucent lesions (**Figure 1**). Under favorable temperature and humidity, yellow exudates of the bacteria can be seen oozing out on the surfaces of infected leaves. The black chaff phase of the disease is seed-borne and as such may pose restrictions for germplasm exchange internationally and the trade of wheat grain [7]. In addition, losses in yield resulted mainly from the poor filling of grains and can be as high as 40% [8] but if the heads are heavily diseased no marketable grain is yielded [2]. Black chaff is often confused with the pseudo-chaff or brown melanosis conditions caused by abiotic stress [9, 10] as well as fungal pathogens and genetic factors [7, 11]. In fact, the black chaff was reported to be a composite disease involving three major factors: the bacterial black chaff, alternaria blotch and favorable environmental conditions for melanism [10, 12].

Figure 1.
Symptoms associated with bacterial leaf streak disease of wheat caused by Xanthomonas translucens *pv.* undulosa *(a) typical longitudinal necrotic lesions; (b) completely diseased or dead flag leaves (center of the photo); and (c) typical black chaff symptom of the spikes. Photographs (a) and (b) are courtesy of Dr. M. Harding, Alberta Agriculture and Forestry; and (c) was kindly provided by Dr. S. Wegulo, University of Nebraska-Lincoln.*

BLS disease has been confirmed in 34 countries of wheat-growing regions worldwide. Based on a recent global database of the European Plant Protection Organization (EPPO; https://gd.eppo.int/taxon/XANTTR/distribution, accessed 22 November 2021), the disease is present in Africa (Ethiopia, Kenya, Madagascar, Morocco, South Africa, Tanzania, Tunisia, and Zambia), Australia, South America (Bolivia, Argentina, Brazil, Peru, Paraguay, and Uruguay), North America (Mexico, Canada, and United States,), several European countries (Russia, Romania, Ukraine, Turkey), and Asia (China, India, Iran, Israel, Japan, Kazakhstan, Malaysia, Azerbaijan Pakistan, Syria, Georgia, Yemen). The western part of Europe appears to be shielded from the BLS disease which might be due to unfavorable environmental factors coupled with aggressive quarantine efforts [13, 14]. In North America, the first reports of BLS were in barley and wheat farms in the Midwestern United States [15]. **Figure 2** shows the global distribution of *X. translucens* pv. *translucens* (Xtt). It is worthwhile to note that some plant pathologists

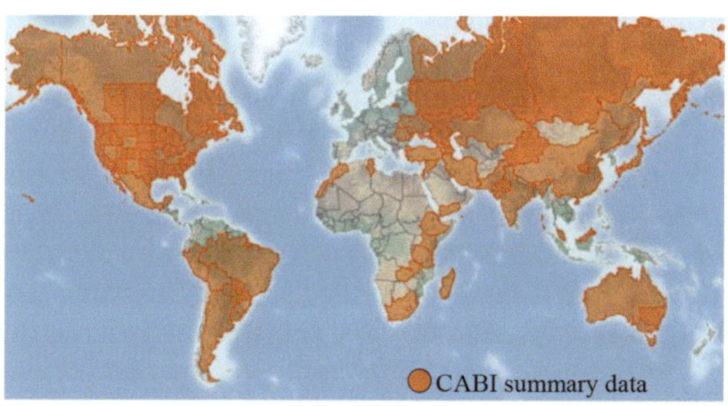

Figure 2.
Global distribution of Xanthomonas translucens *pv.* translucens. *CABI, 2020.* X. translucens pv. translucens *(bacterial leaf streak of barley). In: Crop Protection Compendium. Wallingford, UK: CAB International. https://www.cabi.org/cpc/datasheet/56978#toDistributionMaps [accessed: 17/12/2021]. Note that wheat BLS pathogen has also been referred as* X. t. *pv.* translucens *[2].*

had referred to the BLS disease causal agent of wheat as *X. t.* pv. *translucens* (see the section on taxonomy below).

The disease spread to many other counties with outbreaks and epidemics occurring in southeastern regions of United States [16]. Between 2018 and 2012, the BLS disease incidence increased significantly in the Upper Midwestern states of Minnesota [17], South Dakota [18], and North Dakota [8]. The earliest available record of BLS in Canada was in 1934 [10]. In recent years, no outbreaks have been reported in Canada but the frequency of annual occurrences is becoming high, especially in Alberta and Saskatchewan provinces of Canada. In Mexico, the earliest record of BLS was reported in 1931 in the northeastern part of the country [19]; and currently, high elevation temperate and humid areas of Toluca, 2650m above sea level, seem to have high disease pressure [20].

Although the BLS is one of the most important bacterial diseases of wheat and is considered a potential biothreat, recent data or reports on yield losses do not exist. This is compounded by the fact that the importance of the disease may significantly vary in the different wheat-growing areas which might partly be dependent on the tolerant/resistant levels of the cultivars grown as well as prevalent environmental conditions [14]. Wheat yield grain losses attributed to BLS disease are largely about 10% or lower, but yield losses on highly susceptible cultivars can be as high as 40% [21, 22]. A 20% yield loss as a function of leaf streak severity on bread wheat was reported in Mexico [23]. Duveiller and Maraite [23] and Shane et al. [24] reported that the BLS severity on the flag leaves correlates negatively with the BLS yield loss, and suggested that BLS infection of 50% leaf surface area of flag leaves may lead to a yield reduction of up to 20%. Shane et al. [24] reported that a BLS severity of 100% on flag leaves led to up to 34% yield wheat grain loss. Yield loss due to the severity of BLS disease can be attributed to the reduction in seed weight and the number of seeds per spike [25]. Also, there is a reduction in the quality as BLS infection seems to change the protein content of wheat grains [24].

2.1.2 Taxonomy

The taxonomy of the "translucens" group is not clear due to the pathovar naming system that is based on the first host from which the bacteria was isolated. This section provides some historic and current nomenclature of the group. The bacterial pathogen of the BLS disease on wheat was originally named *Bacterium translucens* pv. *undulosa* [26]. Since then, the nomenclature of the pathogen has evolved through a series of names including *Xanthomonas campestris* pv. *translucens* and *X. campestris* pv. *undulosa* [27]. The pathogen has also been referred in many publications as *X. translucens* pv. *translucens* [2], the valid name for the pathovar that infects barley. The most recent and valid nomenclature for this pathogen is *X. translucens* pv. *undulosa* (Smith et al.) Vauterin, Hoste, Kersters and Swings 1995. Xtu is reported to infect both wheat and barley while Xtt is pathogenic to only barley [8, 28]. The complete current taxonomy of Xtu is Kingdom Bacteria; Phylum *Proteobacteria*; Class *Gammaproteobacteria*; Order *Xanthomonadales*; Family *Xanthomonadaceae*; Genus *Xanthomonas*; and Species *X. translucens*. The translucens group is divided into 10 pathovars: *undulosa*, *translucens*, *cerealis, hordei*, *secalis*, *arrhenatheri*, *graminis*, *poae*, *phlei* and *pistaciae*. Pathovar *cerealis* (Xtc) has also been reported to cause BLS disease symptoms [29, 30], but Xtc has a broad host range than Xtu that includes wheat (*Triticum spp.*), barley (*Hordeum vulgare*), oats (*Avena spp.*), triticale (*Triticosecale*), and rye (*Secale cereale*) [8, 31, 32].

2.1.3 Isolation and identification

The type and quality of the collected BLS diseased leaf sample and culture media are crucial to isolating Xtu strains and could, also, depend on the sample type, leaf, or seeds. Generally, a leaf sample with young BLS lesions is selected, thoroughly washed in running water, surface disinfected, and blotted dry. About 0.25 cm^2 sections of symptomatic leaf tissue are aseptically excised, bisecting, at least, one lesion and immersed in a 100 μl droplet of sterile water, and incubated at room temperature for 5 min [28]. Using a 10-μl loop, the suspensions are streaked onto appropriate non-selective culture media e.g. nutrient agar (NA), Wilbrink's medium, King's B, and Yeast peptone glucose agar plates, and incubated at 28–30°C for up to 72 h. Semi-selective culture media, KM-1 [33], XTS [34] and Wilbrink's boric acid-cephalexin (WBC) [35] have also been developed. KM-1 and XTS have been proposed for isolating Xtt while WBC is a useful medium for isolation of *X. t.* pv. *undulosa* as well as related pathovars. The composition of these culture media is given in **Table 1**.

Xtu can be identified and differentiated from Xtt and *X. translucens* pv. *cerealis* (Xtc) by pathogenicity tests. Application of the bacterial inoculum in the soil and/or seeds is not effective in testing pathogenicity [7, 10]. Artificial inoculation of wheat plants at the 4–5 leaf stage through the injection of bacterial suspension using a hypodermic syringe is the most reliable and effective method [19, 36]. As indicated above, Xtu is reported to infect wheat and barley, Xtt is pathogenic to barley only while Xtc has a broad host range that includes several other cereals. Well-conducted pathogenicity tests could be useful in the identification and differentiation of Xtu.

The pathogen can be distinguished from other pathogenic bacteria of wheat by biochemical and physiological trains. The bacterium causes hypersensitivity on tobacco [37]. Strains of Xtu are non-sporulating, rod-shaped Gram-negative bacteria and use a polar flagellum for motility. This bacterium is oxidative (producing acid from glucose under aerobic conditions) [7]. It does not reduce nitrate to nitrite [38] while esculin hydrolysis is positive but no 2-ketogluconate is produced [39, 40]. Strains of *X. translucens*, unlike *X. campestris*, do not use lactose nor hydrolyze starch [41]. Very few biochemical and physiological tests exist for pathovar differentiation. Metabolic fingerprinting based on BIOLOG MicroPlatesTM on carbohydrates and amino acid utilization may not be effective in the identification of pathovars [7].

Molecular techniques are useful for identification of Xtu but very few records are published. Repetitive element palindromic PCR (rep-PCR) and amplified fragment length polymorphism (AFLP) have been used to partly taxonomically differentiate some *X. translucens* pathovars [32, 42]. Maes et al. [43] reported the development of assay based on rDNA spacer sequences for detection *X. translucens* pathovars causing cereal leaf streak in the seed Iqbal et al. [44] developed a 2-step conventional PCR assay specific for detection and identification of Xtu in wheat in Pakistan. Recently, genome-based and multilocus molecular typing and identification of *X. translucens* pathovars have been reported [17, 28, 45–48]. Langlois et al. [49] characterized the *X. translucens* complex by sequencing and analyzing 16 draft genomes and exploited the genetic variability for the development of diagnostic loop-mediated isothermal amplification (LAMP) assays that can distinguish pathovars (pvs. *undulosa*, *translucens*, *hordei*, and *secalis*) that cause disease on cereals from noncereal pathovars (*arrhenatheri*, *cerealis*, *graminis*, *phlei*, and *poae*). Genome sequencing and comparative genomics have become more reliable and effective

KM-1 [33]		XTS agar [34]		Wilbrink's boric acid-cephalexin (WBC) [35]	
Ingredient	**g/L**	**Ingredient**	**g/L**	**Ingredient**	**g/L**
Lactose	10.0	Glucose	5.0	Bacto peptone	5.0
D (+) trehalose	4.0	Nutrient agar (Difco)	23.0	Sucrose	10.0
Thiobarbituric acid	0.2	Distilled H_2O (ml)	978.0	K_2HPO_4	0.50
K_2HPO_4	0.8			$MgSO_4 .7H_2O$	0.25
$KH_2PO_4 . 7H_2O$	0.8			Na_2SO_3 (anhydrous)	0.05
Yeast extract	1.0			Agar	15.0
NH_4Cl	1.0			Distilled water (ml)	850
Bacto agar (Difco)	15.0				
Distilled water	1.0 L				
Dissolve the ingredients completely and adjust pH to 6.6 using 1 N NaOH before adding agar. Autoclave at 121°C for 20 min., 15 psi; and cool to to 50°C prior to adding: • Cycloheximide (dissolved in 95% ethanol) 100.0 mg • Ampicillin (dissolved in 50% ethanol) 1.0 mg • Tobramycin (dissolved in 50% ethanol) 8.0 mg • Useful for isolation of *X. t.* pv. *Translucens*		Autoclave at 121°C for 20 min, 15 psi and cool to 45°C. Then add: • Cycloheximide (20 ml of a 100 mg/ml 200.0 mg stock solution in 75% ethanol) • Cephalexin (1 ml of a 10 mg/ml 10.0 mg stock solution in 75% ethanol) • Gentamycin (0.8 ml of a 10 mg/ml 8.0 mg stock solution in 75% ethanol) • developed for *X. t.* pv. *translucens*.		Autoclave at 121°C for 20 min. 15 psi and mix with the following solution (autoclaved separately): • Boric acid 0.75 g in 150.00 ml DH2O Allow to cool to 45°C, and then add: • Cycloheximide (in 2 ml of 75% ethanol) 75.00 mg • Cephalexin (1 ml of a 10 mg/ml stock 10.00 mg solution in 75% ethanol) • Developed for isolation of *X. t.* pv. *undulosa* and similar pathovars.	

Table 1.
Semi-selective media for isolation Xanthomonas translucens *pv.* undulosa *and related pathovars.*

tools in the identification of plant pathogenic bacteria including members of the *X. translucens* group (see Section 3).

2.1.4 Management strategies

Since the BLS pathogen is seed-borne, one of the best control strategies is to avoid planting infected seeds. It is, thus, important to do a simple seed wash test [7]. The test is conducted by adding 120 g of seeds in 120 ml aqueous saline solution containing 0.02% v/v Tween 20, shake vigorously for 1 minute, and perform serial tenfold dilutions up to 10^{-3} before plating on semi-selective media e.g. KM-1 and WBC, described above. Levels of 1000 colony forming units per gram or less in seed washes have been reported to result in little or no disease [34]. Avoid using infested seeds for germplasm exchange [7] to minimize the spread of the pathogen.

Crop rotations have been reported as another management strategy but little data exist on how this measure works in reducing the BLS and black chaff epidemics. Given that the main source of spread of the pathogen is through infested seeds, crop rotation may not be a key management strategy. However, the bacterium can survive in wheat straw and can induce initial field infections. Boosalis [50] noted that the viable number of the pathogen is low in overwintered straw and greatly reduced when the straw are buried in the soil. Also, pathogen survival seems to be almost impossible under non-host rotations as well as the high susceptibility of the bacteria to antagonistic bacteria [34].

2.2 *Pseudomonas syringae* pathovars

2.2.1 Pseudomonas syringae *pv.* syringae, *the causal agent pathogen of the leaf blight*

2.2.1.1 Disease symptoms, distribution, and importance

P. syringae pv. *syringae* is the pathogen that causes the bacterial leaf blight [51] or leaf necrosis [52] disease of wheat. During the initial infection, the symptoms are small, water-soaked spots that expand into characteristic large lesions, blotches, or streaks often observed under rainy and high humidity conditions [2]. Under low humidity, the water-soaked lesions quickly turn into characteristic gray-green. Within 3 or 4 days, the lesions coalesce to form irregularly shaped blotches and become tanned or white. Whole leaves might be necrotic while the heads and lower leaves are without any symptoms [2].

The disease was first reported in the USA in the following states, South Dakota, North Dakota, and Nebraska [53], Minnesota [51] and Montana [54]; and 2 Canadian provinces of Saskatchewan and Alberta [52]; Italy [55], South Africa [56]; Pakistan [57] and Argentina [58]. Besides wheat, *Pseudomonas syrinage* pv. *syringae* is a widespread plant epiphytic bacterium with a very wide host range consisting of many herbaceous and woody plants including corn (Holcus spot), sorghum (bacterial leaf blight), and lilac (bacterial blight). Otta [52] found that strains of *P. syringae* pv. *syringae* isolated from corn, foxtail, sorghum, and peach are pathogenic on wheat seedlings. This opportunistic bacterium has been reported to cause disease in various host plants of agricultural importance as well as many weeds [59, 60] in Europe, Asia, Oceania, South America, Caribbean, and North and Central America. Its worldwide distribution is well documented as the causal agent of the bacterial canker or blast disease of stone and pome fruits (**Figure 3**).

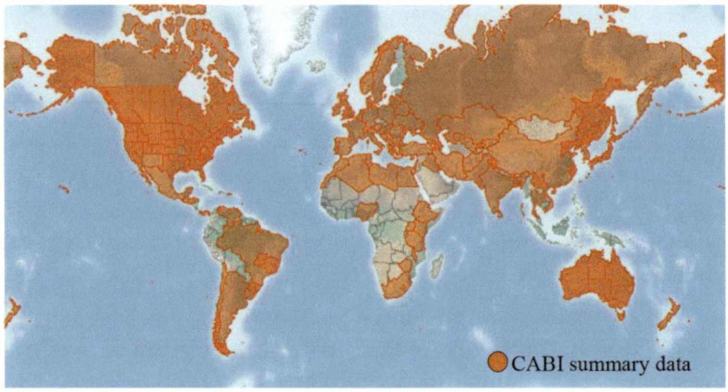

Figure 3.
Distribution of Pseudomonas syringae pv. syringae Van Hall worldwide. CABI, 2020. P. syringae pv. syringae (bacterial canker or blast (stone and pome fruits)). In: Crop Protection Compendium. Wallingford, UK: CAB International. https://www.cabi.org/cpc/datasheet/45014#toDistributionMaps [accessed: 17/12/2021] [61]).

The importance of the disease on wheat is dependent on specific weather conditions. The disease is more often observed during several days of high humidity, cool temperatures (15–25°C), and heavy rainfall [62]; and plants are more susceptible at the boot stage. The disease is generally considered to be of minor importance with minimal impact on yield losses. However, Otta [52] reported high yield losses in South Dakota where an epidemic outbreak resulted in very high infection severity in fields with 75% or more necrotic leaves. Also, in North America, the cultivation of susceptible cultivars, e.g. Chris, Era, Scout 66, and Winoka, led to significant yield losses. These susceptible cultivars have been replaced with resistant wheat cultivars and as such the disease is only sporadic in North America [2].

The current and complete taxonomic standing of the bacterium is Kingdom Bacteria; Phylum *Proteobacteria*; Class *Gammaproteobacteria*; Order *Pseudomonadales*; Family *Pseudomonadaceae*; Genus, *Pseudomonas*; and Species, *P. syringae*; pathovar syringae.

There are no traditional management strategies for the bacterial leaf blight disease caused by *P. syringae* pv. *syringae*. It is, however, recommended to avoid planting very susceptible cultivars.

2.2.2 Pseudomonas syringae *pv.* aatrofaciens, *the incitant of the basal glume rot disease*

2.2.2.1 Disease symptoms, distribution, and importance

P. syringae pv. *atrofaciens* (McCulloch) Young, Dye, & Wilkie is the incitant of basal glume rot of wheat. Typical and characteristic symptoms of the disease are dull brownish to blackish discoloration, usually, on the lower part of the wheat glume (**Figure 4a**). It is more visible in the inner than the outer side of the glume. At times, water-soaked sections can be seen around the lesions [63]. Seeds in florets showing typical symptoms of the disease are often brownish to blackish in color (**Figure 4b**). The grains may be shriveled when wet darkish green lesions appear on the peduncle and surrounding it fully [2]. Also, small (2–10 mm) wet, darkish green lesions may appear on the leaves which then could quickly become necrotic [62]. Symptoms on the glume and peduncles can be confused with black-chaff caused by *X. translucens*

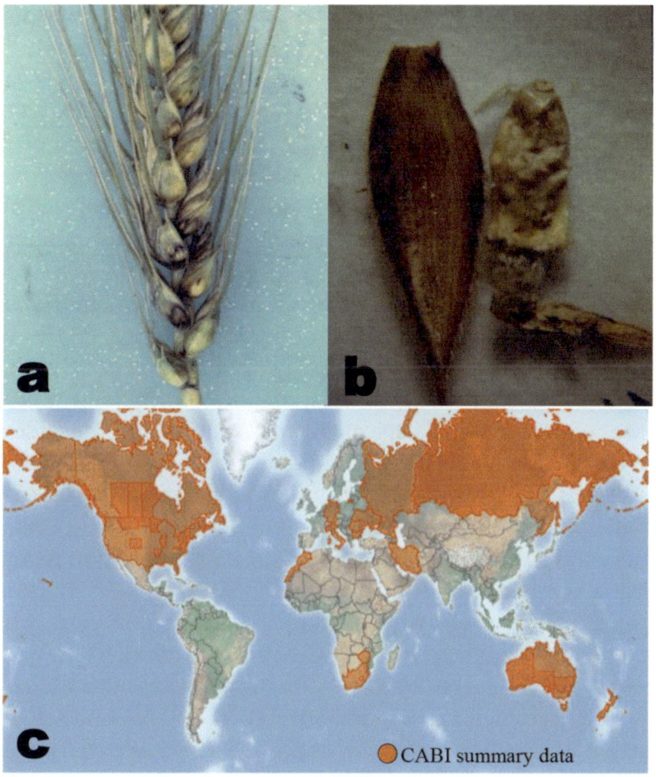

Figure 4.
(a) Typical dark brown to black discoloration and (b) diseased seed symptoms caused by Pseudomonas syringae *pv.* atrofaciens*; and (c) global distribution of the causal agent of the wheat basal glume rot disease. Photos: Centro Internacional de Mejoramiento de Maíz y Trigo (CIMMYT) (http://wheatdoctor.org/basal-glume-rot). Distribution map from CABI, 2020. In: Crop Protection Compendium. Wallingford, UK: CAB International. https://www.cabi.org/cpc/datasheet/44934#toDistributionMaps [accessed: 17/12/2021].*

or pseudo-chaff, a disorder associated with abiotic stress factors. Seed symptoms are often confused with the black point disease. Wet climatic conditions after heading favor the development of the disease.

The basal glume rot disease caused by *P. s.* pv. *atrofaciens* was first described in America [64]. It has since been reported in almost all wheat-growing regions of the world [65]. The basal glume rot disease of wheat has been reported in the USA and Canada [64]; Mexico [66]; Ukraine [67]; Bulgaria [68]; Australia [69]; New Zealand [70]; South Africa [56, 71]; Germany and Denmark [72]; and Belgium [62]. **Figure 4c** shows the global distribution of the disease.

The basal glume rot occurs only sporadically. It is generally thought to be a minor disease and as such, the impact on wheat grain yield has not been well studied. However, *P. syringae* pv. *atrofaciens* is reported to cause yield losses exceeding 50% in marshy soils of Germany [73]. Also, 7-year outbreaks of leaf necrosis led to leaf infections of over 75% [52] and can cause a severe reduction in grain quality [74, 75].

Since the bacterium is seed-borne, contaminated seeds remain the most important source of infections to new wheat plantings. No effective seed treatment exists for the management of *P. s.* pv. *atrofaciens*. Also, wheat genotypes may vary in resistance but little is known of the level of resistance of present-day cultivars [2]. Wheat growers are recommended to avoid planting seeds harvested from infected fields.

2.2.3 Isolation and differentiation of P. syringae *pv.* syringae *and* P. syringae *pv.* *atrofaciens*

Epiphytic populations of *P. syringae* pv. *atrofaciens* and *P. syringae* pv. *syringae* are extensive and easily isolated from wheat plants. As such, isolating these bacteria is not definite prove that the observed symptoms are induced by *P. syringae* pv. *atrofaciens* or *P. syringae* pv. *syringae* [62]. However, if the concentrations of the bacteria in diseased leaf tissues are high, e.g. 10^8 colony forming units per gram fresh weight, then they might play somewhat a likely causal role [62].

These two pathovars cannot be differentiated based on colony morphology or physiological traits [53, 62, 76], nor by using serological methods [52, 77] while Iacobellis et al. [76] reported that *P. syringae* pv. *syringae* and *P. syringae* pv. *atrofaciens* could not be differentiated by genetic features. However, advances in whole-genome sequencing and comparative genomics analysis have made it possible to accurately identify and differentiate these two pathovars (see Section 3). Genome-based DNA-DNA relatedness calculations between *P. syringae* pv. *syringae* and *P. syringae* pv. *atrofaciens* revealed 82.2% hybridization values, suggesting a genetic difference of about 18% (Tambong, unpublished). Accurate and reliable pathovar identification/differentiation is only possible based on the disease symptoms induced on cereals [78, 79].

The isolation of *P. syringae* pv. *syringae* and *P. syringae* pv. *atrofaciens* has been routinely achieved using King's medium B. Colonies are whitish-gray in color and are circular and convex in shape after 24 h of incubation; and exhibit blue fluorescence under ultra-violet light, which turns green after 48 h [62]. No semi-selective media have been reported for use in the isolation of these *P. syringae* pathovars. However, KBC medium, King's B medium amended with boric acid, cephalexin, and cycloheximide [80] showed high selectivity for *P. syringae* pv. *atrofaciens* even though the bacterium seemed to grow slightly slower compared to King's medium B.

2.3 Pathogens of the genus "*Clavibacter*"

2.3.1 Clavibacter tessellarius, *the causative agent of bacterial mosaic of wheat*

The wheat bacterial mosaic disease was first reported in Nebraska (USA) in 1976 and taxonomically described in 1982 [81]. The disease is caused by *Clavibacter michiganensis* subsp. tessellarius (Carlson & Vidaver) Davis et al.. The taxonomy has evolved from *Corynebacterium michiganese* subsp. *tessellarius* through *C. michiganensis* subsp. *tessellarius* to *Clavibacter tessallarius* based on whole-genome sequence analysis [82]. The geographic distribution is restricted to North America: USA [81] and Canada [13]. The pathogen is specific to wheat but seems to be related to *C. michiganensis* [81]. It can, however, be differentiated from other plants pathogenic corynebacterium by bacteriocin typing. The disease is sporadic, annually, and also occurs in triticales. The economic significance of the disease is yet to be documented.

The bacterial mosaic disease caused by *C. tessellarius* is a foliar disease characterized by a mosaic of small yellow lesions that resemble infections by viral pathogens. The small lesions on the leaf may coalesce to form streaks [7]. Under greenhouse temperature conditions of 20–22°C, artificially inoculated seedlings develop typical mosaic-like symptoms in three to 5 days [81]. The pathogen is seed-borne [83] and as such control strategies should include removing contaminated seeds and the development of tolerant/resistant genotypes. The available wheat genotypes seem to respond differentially to the pathogen. This could be an indication that genetic improvement is possible [7].

2.3.2 Clavibacter iranicus, *the pathogen of gumming disease of wheat spikes*

C. iranicus is the causative agent of the gumming disease of wheat spikes and has only been reported in Iran [84]. The taxonomic name of the pathogen was initially *Corynebacterium iranicum* [84] followed by *C. iranicus* (ex Scharif) Davis, Gillaspie, Vidaver & Harris 1984 while Zgurskaya et al. [85] proposed the renaming of this bacterial pathogen as *Rathayibacter iranicus*. The economic significance of the pathogen to wheat is unknown.

There are no control or management strategies proposed in the literature for the gumming disease of wheat spikes.

C. iranicus (Syn. *R. iranicus*) exhibits a close relatedness with *Rathayibacter tritici* but they are different [2].

2.3.3 R. tritici, *causative agent of spike blight of wheat*

R. tritici is the causal agent of spike blight of wheat and was first reported in India in 1917 as the causal agent of the tundu disease [2]. It is also known as the yellow ear rot or yellow slime rot disease; and also pathogenic to several grasses, e.g. barley. The spike blight is considered a disease complex involving the bacterium *R. tritici* and a nematode, *Aguina tritici*, the causal agent of the seed galls, also known as ear cockle, in some wheat cultivars [86].

The current taxonomic nomenclature of the pathogen is *R. tritici* (Carlson & Vidaver) Zgurskaya, Evtushenko, Akimov & Kalakoutskii. Previous scientific names, from latest, included *Clavibacter tritici* (Carlson & Vidaver) Davis, Gillaspie, Vidaver & Harris, *Corynebacterium michiganense* pv. *tritici* (Hutchinson) Dye & Kemp *Corynebacterium tritici* (Hutchinson) Burkholder, *Phytomonas tritici* (Hutchinson) Bergey and *Pseudomonas tritici* Hutchinson.

The geographic distribution includes 14 countries: Egypt, Ethiopia, Morocco, Zambia, Afghanistan, China, India, Iran, Iraq, Pakistan, Cyprus, Australia [60, 86–89].

Initial field symptoms of the spike blight disease include parallel white or yellow streaks generally along the veins of leaves. Later, this is transformed into a sticky mass, yellow gummosis on wheat spikes. The spikes and peduncles (necks) are often distorted when they emerge from the whorl. Also, early leaves may also be twisted or wrinkled. When the sticky mass is dry, the gummois becomes pale yellow-colored flecks on the spikes and the adaxial leaf surface [2]. Since the bacterial sticky mass is watery under wet weather conditions and dry when the RH is low, the hardened gummy substance mechanically causes the leaves, spikes, and peduncles to be distorted. The symptoms that are caused by *A. tritici* (nematode) are part of the spike blight disease complex.

R. tritici is reported to persist in crop residues in moist soils. To facilitate the colonization of wheat, the pathogen has to be carried by the nematode *A. tritici* into the whorl enclosures. Generally, the juveniles of *A. tritici* are contaminated by the cells of *R. tritici* in the soil. This helps with the dissemination of the bacterial cells of *R. tritici* on seeds, in seed galls, and in soil. The *A. tritici* and *R. tritici* can survive in seed galls for over 5 years [90].

The management of spike blight disease is not well studied. However, growing wheat on well-drained soils significantly reduces spike blight disease. Also, management strategies used to control *A. tritici* would be helpful in reducing the spike blight incidence. A 2- or 3-year crop rotation of wheat with non-grass crops is another control method.

2.4 Other bacterial pathogens

This section focuses on other bacterial causal agents of wheat diseases that are rarely observed and are pathogenic to a variety of host plants. These are considered to be less specialized pathogens some of which are epiphytes or opportunistic organisms that cause disease to the wheat only under favorable unusual conditions [91]. Since the diseases caused by these bacteria have not been extensively studied very limited information exists. The pathogens described here include *Pseudomonas fuscovaginae*, *Pseudomonas cichorii*, and *Erwinia rhapontici*.

2.4.1 Pseudomonas fuscovaginae*, causal agent of the bacterial brown sheath*

P. fuscovaginae (ex Tanii et al.; Miyajima et al.) is the causal agent of the bacterial sheath rot, also referred to as brown sheath rot, of wheat. Tanii et al. [92] were the first to report this pathogen on rice but have since been isolated from several other cereal crops including wheat [93]. A recent genome-based taxonomic study re-classified *P. fuscovaginae* as a later heterotypic synonym of *Pseudomonas asplenii* [94].

The bacterial brown sheath rot disease caused by *P. fuscovaginae* (heterotypic synonym of *P. asplenii*) after the first report in Japan in 1976 on rice, it has been reported in 33 other countries, e.g. Burundi, Mexico, (**Figure 5**) on a variety of crops and grasses including wheat, barley, maize, oats, bentgrasses, bromegrass, perennial ryegrass, smooth meadow-grass, rye, and sorghum (CABI, accessed November 17, 2021). Little information exists on the effects of the pathogen on the yield of wheat. But severity seems to vary with the genetics of the host plants. Two wheat cultivars, Anahuac and Seri 82 suffered severe damage in Mexico in 1990 with 18–20% infections of the tillers [2, 93]. In Nepal, the disease was not observed on genotype RR21 while four other cultivars (WK685, Annapurna-1, Annapurna-2, and Annapurna-3) were heavily infested [95].

The symptoms are characterized by black brown lesions of angular to irregular shapes on the leaves. The lesions have blackish-purple water-soaked discolored borders [96]. The adaxial surface of the leaf sheath is often where the initial infection starts [97]. Plants with severe incidences may show poor spike emergence and even sterility. *P. fuscovaginae* is disseminated by seed and plant susceptibility is dependent on the developmental stage. If the pathogen is present, infections are highly favorable

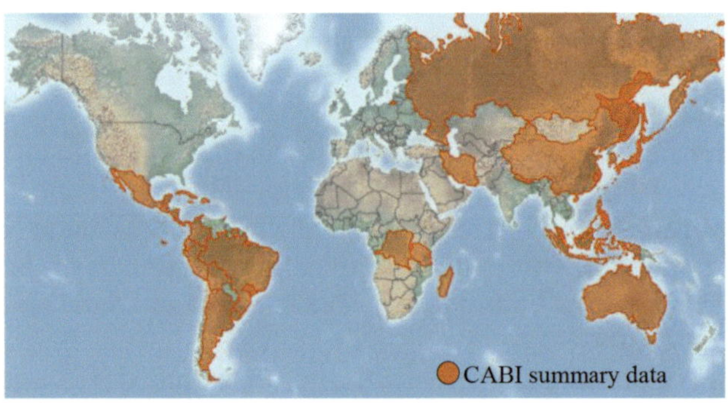

Figure 5.
Global distribution of Pseudomonas fuscovaginae. *CABI, 2021.* P. fuscovaginae *(sheath brown rot). In: Crop Protection Compendium. Wallingford, UK: CAB International. https://www.cabi.org/cpc/datasheet/44957#toDistributionMaps [accessed: 17/12/2021].*

during the flowering stage at a temperature range of 17–25°C with 100% relative humidity (RH) [2].

Management of the disease is very difficult since some of the factors involved cannot be controlled easily. Given that the pathogen is seed-borne [93], preventive measures are key to reducing the incidence of the disease. Avoid sowing contaminated seeds or susceptible cultivars, especially in growing areas under low temperatures and high RH microclimates.

2.4.2 Pseudomonas cichorii, *the causative agent of wheat stem melanosis*

The Gram-negative bacterium *P. cichorii* is a ubiquitous organism that is pathogenic to several host plants worldwide [60]. The only report of *P. cichorii* on wheat is that of Piening et al. [98] isolated from the Park spring wheat cultivar grown in copper-deficient soil in Canada (Alberta). No recent report of its occurrence on wheat exists and data on the economic importance of the disease is not available.

The symptoms were first observed in 1965 in central Alberta as irregularly shaped, sharply defined, dark patches in fields of cv. Park spring wheat at the milky stage of growth [98]. The initial symptoms start at the milky stage with the development of small lesions of light brown coloration under the lower two nodes which later darken and coalesce on the stem, rachis, and peduncle [91]. The spikes (also called the ear or head) are bleached and the grains are shriveled [2]. Even though the epidemiology is not understood, high temperatures (29°C) and relative humidity are conducive for the spread of the pathogen; and the disease has, also, been associated with soils that are cooper-deficient [99, 100]. There is no known management strategy for stem melanosis. However, the application of Cu chelate as amendments at the rate of 2.4 kg/ha to copper-deficient soils has been reported to reduce the severity of stem melanosis and improve wheat grain yield [99].

The current taxonomy of the pathogen is Kingdom, Bacteria; Phylum, *Proteobacteria*; Class, *Gammaproteobacteria*; Order, *Pseudomonadales*; Family, *Pseudomonadaceae*; Genus, *Pseudomonas*; and Species, *P. cichorii*.

2.4.3 Erwinia rhapontici, *the causal agent of the pink seed of wheat*

E. rhapontici (Millard) Burkholder emend. Hauben et al., a homotypic synonym to *Pectobacterium rhapontici* (Millard) Patel & Kulkarni, *Erwinia carotovora* var. *rhapontici* (Millard) or *Pseudobacterium rhapontici* (Millard) Krasil'nikov, is the causal agent of the pink seed of wheat. Heterotypic synonymic names include *Bacterium rhaponticum*, *Phytomonas rhapontica*, *Aplanobacter rhaponticum*, and *Xanthomonas rhapontica*.

It has been reported in Canada [101], France [102], England [103], USA [104], Belgium [105] and Russia and Ukraine [106]. It is also reported to be pathogenic to pea [107], onion [108], garlic [109], common bean [110], lentil [111], rhubarb [7] and several other plant species.

This pathogenic bacterium is opportunistic as it affects mainly injured kernels caused by the gall midge (family *Cecidomyiidae*). Seeds/kernels infected by *E. rhapontici* exhibit pinkish discoloration and slightly shriveled compared normal kernels. The germination vigor of infected seeds is poor. The market value of pink seeds is significantly low and often not used in pasta production [104].

The infrequent and somewhat random occurrence of pink seed on wheat makes it difficult to understand the disease and devise effective prevention strategies.

3. Genome-based identification of key bacterial pathogens of wheat

With advances in whole-genome sequencing (wgs) and bioinformatics tool developments, genome-based methods of classification and identification of prokaryotes including wheat pathogens are fast replacing traditional methods such as colony morphology, conventional polymerase chain reaction-based assays, multilocus sequence analysis, and wet-lab DNA-DNA hybridization similarity values. Some of the traditional approaches, e.g., wet-lab DDH have inherent drawbacks such as irreproducibility between laboratories [112]. Whole-genome sequencing provides complete and draft chromosome data that can be used to better understand the evolutionary and taxonomic relationships in bacteria, in general [113, 114]. Genome sequencing and comparative genomics are powerful technologies in the accurate identification and classification of bacterial pathogens of wheat.

Table 2 shows the number of publicly available whole-genome sequences (wgs) of each of the bacterial pathogens of wheat profiled in this chapter. Application of taxogenomics approach to identifying bacterial pathogens of wheat can be achieved by computing genome-to-genome distance (GGDC; [115]); average nucleotide identity (ANI; [116]), MuMmer-based average nucleotide identity (ANIm; [117]; tetranucleotide usage patterns (TETRA; [118]; and codon usage [119] as well as supertree analysis and other genomic signatures [116]. The three most commonly used genome-based parameters in bacterial identification are GGDC, ANI, and TETRA. GGDC outputs a pairwise genome-based digital DNA-DNA hybridization (dDDH) value between two bacterial strains with a species-level cut-off value of 70%; the ANI similarity value based on the BLAST or MuMmer approach has a species delineation threshold of 95–96%; and TETRA cut-off value of 0.998 for same species assignment of two bacterial strains. It is key that the whole genome sequence of the unknown strain is compared to the type strain or pathotype of the target bacterial species.

Pathogen	Wheat disease	Number of sequenced genomes	Genome size (Mb)	G + C content (%)	Number of Protein-coding sequences (CDS)
X. t. pv. *undulosa*	Leaf streak and spike black chaff	12 (8)	4.6	67.93	3406
P. s. pv. *syringae*	Leaf blight	76 (8)	5.99	59.08	4652
P. s. pv. *atrofaciens*	Basal glume rot	7 (1)	5.91	59.1	5109
P. fuscovaginae	Brown sheath	9 (1)	6.54	61.88	5693
P. cichorii	Stem melanosis	51 (3)	5.89	58.28	4990
Erwinia rhapontici	Pink seed	8 (5)	5.31	54.04	4648
C. tessellarius	Bacterial mosaic	2 (0)	3.31	73.45	3028
R. iranicus	Gumming	6 (1)	3.33	67.18	3066
R. tritici	Spike blight	6 (1)	3.23	69.78	2982

Table 2.
Number of publicly available whole-genome sequences (wgs) of each of the bacterial pathogens of wheat. X. t., Xanthomonas translucens*; P. s.,* Pseudomonas syringae*; C.* Clavibacter; *and R.,* Rathayibacter. *wgs presented are from wheat as well as other hosts, and the number in brackets indicate genomes having complete chromosomes instead of draft genomes.*

The procedure can be summarized into 5 crucial steps: (1) bacteria isolation from plant tissues. Single colony purification, by repetitive streaking, is important to ensure that only one bacterial species is present. The presence of two or more species will result in poor genome assembly; (2) genomic DNA extraction and quantification. High purity genomic DNA is required as well as the quantity which is dependent on the genome sequencing method to be used; (3) library preparation. During this step, DNA is fragmented into short-reads, end-repaired, and adapter-ligated. Adapters allow the attachment of sequences to the flow cell, to identify samples, and to permit multiplexing; (4) Next-generation genome sequencing (NGS). Also referred to as high-throughput sequencing, enables sequence profiling of genomes. In a relatively short time, NGS generates large amounts of short or long sequence data; (5) Genome assembly and analysis. This step involves the use of a specialized Bioinformatics tool to assemble the raw reads into contigs, scaffolds, or chromosomes. These tools can be user-friendly or command-line accessed. The assembled whole-genome sequences (contigs, scaffolds, or complete chromosomes) can then be utilized as input in other bioinformatics tools to compute dDDH, ANI, or ANI similarity values between the type strain of the target bacterial species and the newly isolated or unknown bacterial strains. There are several user-friendly bioinformatics pipelines that can be routinely used by biologists, plant pathologists, or other scientists with basic or little bioinformatics knowledge. Examples of these pipelines are PATRIC ([120]; https://patricbrc.org/) and galaxy@pasteur [121]; https://galaxy.pasteur.fr/).

4. Conclusion

The purpose of this chapter was to summarize the disease symptoms, distribution, taxonomy, and identification methods of key bacterial pathogens of wheat. Occurrences of wheat diseases caused by bacteria are sporadic or confined to limited ecological niches due to favorable weather conditions. However, with the advancement of climatic change some of these diseases might become more prevalent and as such the information provided here should be helpful to agriculturists, biologists, plant pathologists as well as wheat breeders with limited experience to recognize and identify these pathogens. This is particularly important since some of these pathogens, e. g., *X. translucens* pv. *undulosa*, are seed-borne and efforts are being made to minimize spread through germplasm exchange. It is highly recommended to confirm the pathogenicity of bacterial isolates prior to using appropriate determinative tests to validate species-level identification. Accurate identification of the causal pathogen(s) of field observed symptoms is key to the development or selection of appropriate management strategies. With significant reductions in bacterial genome sequencing costs, as low as $220 per strain, and the availability of user-friendly bioinformatics tools, biologists and plant pathologists are encouraged to adopt the genome-based approach in the identification of pathogens of wheat. Whole-genome sequences parameters provide reproducible data leading to reliable and accurate identification of bacteria, in general, and wheat pathogens, in particular.

Acknowledgements

Funding for this work was provided by Agriculture and Agri-Food Canada through projects # J-002272 and # J-002749. The author is grateful to Dr. M. Harding and Dr. S. Wegulo for providing some of the photographs. Also, the author is thankful to CAB International (England) for authorizing the use of distribution maps generated in Plantwise.

Author details

James T. Tambong
Agriculture and Agri-Food Canada, Ottawa, Ontario, Canada

*Address all correspondence to: james.tambong@agr.gc.ca

References

[1] Shewry PR. Wheat. Journal of Experimental Botany. 2009;**60**:1537-1553. DOI: 10.1093/jxb/erp058

[2] Bockus WW, Bowden RL, Hunger M, Morrill WL, Murray TD, Smiley RW. Compendium of Wheat Disease. 3rd ed. St Paul, Minnesota, U.S.A.: The American Phytopathological Society; 2010. p. 117

[3] Shewry PR, Hey SJ. The contribution of wheat to human diet and health. Food and Energy Security. 2015;**4**:178-202

[4] Igrejas G, Branlard G. The importance of wheat. In: Igrejas G, Ikeda TM, Guzmán C, editors. Wheat Quality for Improving Processing and Human Health. Switzerland: Springer Nature; 2020. pp. 1-7. DOI: 10.1007/978-3-030-34163-3_1

[5] Nigro D, Gadaleta A, Mangini G, Colasuonno P, Marcotuli I, Giancaspro A, et al. Candidate genes and genome-wide association study of grain protein content and protein deviation in durum wheat. Planta. 2019;**249**(4):1157-1175. DOI: 10.1007/s00425-018-03075-1

[6] FAO/UN. FAO Food and Nutrition Paper 92. Dietary protein quality evaluation in human nutrition (PDF). Rome, Italy: Food and Agriculture Organization of the United Nations; 2013 [Accessed: November 22, 2021]

[7] Duvieller E, Bragard C, Mariate H. Bacterial leaf streak and black chaff caused by *Xanthomonas translucens*. In: Duvieller E, Fucikovsky L, Rudolph K, editors. The Bacterial Diseases of Wheat: Concepts and Methods of Disease Management. Mexico: CIMMYT; 1997. pp. 25-47. Available from: https://repository.cimmyt.org/handle/10883/1227

[8] Adhikari TB, Gurung S, Hansen JM, Bonman JM. Pathogenic and genetic diversity of *Xanthomonas translucens* pv. *undulosa* in North Dakota. Phytopathology. 2012;**102**:390-402

[9] Broadfoot WC, Robertson HT. Pseudo-black chaff of reward wheat. Science in Agriculture. 1933;**13**:512-514

[10] Hagborg WAF. Black chaff, a composite disease. Canadian Journal of Research. 1936;**14**:347-359

[11] McMullen M, Adhikar T. Bacterial leaf streak and black chaff of wheat. In: Plant Disease Management. Fargo, ND, USA: North Dakota State University Extension Publication; 2011. p. 1566

[12] Johnson T, Hagborg WF. Melanism in wheat induced by high temperature and humidity. Canadian Journal of Research. 1944;**22**:7-10

[13] Paul VH, Smith I. Bacterial pathogens of *Gramineae*: Systematic review and assessment of quarantine status for the EPPO region. Bulletin OEPP. 1989;**19**: 33-42. DOI: 10.1111/j.1365-2338.1989.tb00126.x

[14] Sapkota S, Mergoum M, Liu Z. The translucens group of *Xanthomonas translucens*: Complicated and important pathogens causing bacterial leaf streak on cereals. Molecular Plant Pathology. 2020;**21**:291-302. DOI: 10.1111/mpp.12909

[15] Jones LR, Johnson AG, Reddy CS. Bacterial blight of barley. Journal of Agricultural Research. 1917;**11**:625-643

[16] Tubajika KM, Tillman BL, Russin JS, Clark CA, Harrison SA. Relationship between flag leaf symptoms caused by

Xanthomonas translucens pv. *translucens* and subsequent seed transmission in wheat. Plant Disease. 1998;**82**:1341-1344

[17] Curland RD, Gao L, Bull CT, Vinatzer BA, Dill-Macky R, Van Eck L, et al. Genetic diversity and virulence of wheat and barley strains of *Xanthomonas translucens* from the upper midwestern United States. Phytopathology. 2018;**108**:443-453. DOI: 10.1094/PHYTO-08-17-0271-R

[18] Kandel YR, Glover KD, Tande CA, Osborne LE. Evaluation of spring wheat germplasm for resistance to bacterial leaf streak caused by *Xanthomonas campestris* pv. *translucens*. Plant Disease. 2012;**96**:1743-1748

[19] Bamberg RH. Black chaff disease of wheat. Journal of Agricultural Research. 1936;**52**:397-417

[20] Duveiller E. Research on '*Xanthomonas translucens*' of wheat and triticale at CIMMYT. Bulletin. 1989;**19**:97-103

[21] Waldron R. The relationship of black chaff disease of wheat to certain physical and pathological characters. Science. 1929;**70**:268

[22] Forster RL, Schaad NW. Control of black chaff of wheat with seed treatment and a foundation seed health program. Plant Disease. 1988;**72**:935-938

[23] Duveiller E, Maraite H. Study of yield loss due to *Xanthomonas campestris* pv. *undulosa* in wheat under high rainfall temperate conditions. Journal of Plant Diseases Protection. 1993;**100**:453-459

[24] Shane WW, Baumer JS, Teng PS. Crop losses caused by *Xanthomonas* streak on spring wheat and barley. Plant Disease. 1987;**71**:927-930

[25] Tillman BL, Kursell WS, Harrison SA, Russin JS. Yield loss caused by bacterial streak in winter wheat. Plant Disease. 1999;**83**:609-614

[26] Smith EF, Jones LR, Reddy CS. The black chaff of wheat. Science. 1919;**50**:48

[27] Dye DW, Bradbury JF, Goto M, Hayward AC, Lelliott RA, Schroth MN. International standards for naming pathovars of phytopathogenic bacteria and a list of pathovar names and pathotype strains. Review Plant Pathology. 1980;**59**:153-168

[28] Tambong JT, Xu R, Gerdis S, Daniels GC, Chabot D, Hubbard K, et al. Molecular analysis of bacterial isolates from necrotic wheat leaf lesions caused by *Xanthomonas translucens*, and description of three putative novel species, *Sphingomonas albertensis sp*. nov., *Pseudomonas triticumensis sp.* nov. and *Pseudomonas foliumensis sp. nov*. Frontiers in Microbiology. 2021;**12**:666689. DOI: 10.3389/fmicb.2021.666689

[29] Pesce C, Bolot S, Cunnac S, Portier P, Fischer-Le Saux M, Jacques MA, et al. High quality draft genome sequence of the *Xanthomonas translucens* pv. *cerealis* pathotype strain CFBP 2541. Genome Announcements. 2015;**3**:e01574-ee1614

[30] Shah SMA, Haq F, Ma W, Xu X, Wang S, Xu Z, et al. Tal1NXtc01 in *Xanthomonas translucens* pv. *cerealis* contributes to virulence in bacterial leaf streak of wheat. Frontiers in Microbiology. 2019;**10**:2040. DOI: 10.3389/fmicb.2019.02040

[31] Cunfer BM, Scolari BL. *Xanthomonas campestris* pv. *translucens* on triticale and other small grains. Phytopathology. 1982;**72**:683-686

[32] Bragard C, Singer E, Alizadeh A, Vauterin L, Maraite H, Swings J.

Xanthomonas translucens from small grains: Diversity and phytopathological relevance. Phytopathology. 1997;**87**:1111-1117

[33] Kim HK, Sasser M, Sands DC. Selective medium for *Xanthomonas campestris* pv. *yranslucens*. Phytopathology. 1982;**72**:936

[34] Schaad NW, Forster RL. A semi-selective agar medium for isolating *Xanthomonas campestris* pv. *Translucens* from wheat seeds. Phytopathology. 1985;**75**:260-263

[35] Duveiller E. A seed detection method of *Xanthomonas campestris* pv. *undulosa,* using a modification of Wilbrink's agar medium. Parasitica. 1990a;**46**:3-17

[36] Duveiller E. Bacterial leaf streak or black chaff of cereals. Bulletin OEPP/EPPO Bulletin. 1994;**24**:135-157

[37] Mohan SK, Mehta YR. Studies on *Xanthomonas campestris* pv. *undulosa* in wheat and triticale in Paraná state. Fitopatologia Brasileira. 1985;**10**:447-453 (in Portuguese)

[38] Dye DW. The inadequacy of the usual determinative tests for the identification of *Xanthomonas spp*. New Zealand Journal of Science. 1962;**5**:393-416

[39] Miyajima K. Occurrence of bacterial blight *Xanthomonas campestris* pv. *Hordei* (Hagborg) Dye of orchard grass in Hokkaido prefecture. Annual Report Plant Protection North Japan. 1980;**31**:74-77 (in Japanese)

[40] Bradbury JF. Family 1: *Pseudomonaceae*. Genus II: *Xanthomonas* Dowson 1939. In: Krieg NR, Holt JG, editors. Bergey's Manual of Systematic Bacteriology. 1st ed. Baltimore/London: The Williams and Wilkins Co.; 1984. pp. 199-210

[41] Schaad NW. Problems with the pathovar concept. In: Civerolo EL, Collmer A, Davis RE, Gillaspie AG, editors. Proceedings of the 6th International Conference on Plant Pathogenic Bacteria. Maryland. Dordrecht (The Netherlands): Martinus Nijhoff Publishers; 1987. pp. 783-785

[42] Rademaker JLW, Norman DJ, Forster RL, Louws FJ, Schultz MH, De Bruijn FJ. Classification and identification of *Xanthomonas translucens* isolates, including those pathogenic to ornamental asparagus. Phytopathology. 2006;**96**:876-888

[43] Maes M, Garbeva P, Kamoen O. Recognition and detection in seed of the *Xanthomonas* pathogens that cause cereal leaf streak using rDNA spacer sequences and polymerase chain reaction. Phytopathology. 1996;**86**:63-69. DOI: 10.1094/Phyto-86-63

[44] Iqbal MA, Ullah I, Shahbaz MU, Kamran M, Saleem K. Biochemical and molecular identification of *Xanthomonas translucens* pv. *Undulosa* causing bacterial leaf streak of wheat in Punjab, Pakistan. Archives of Phytopathology and Plant Protection. 2013;**47**:417-424. DOI: 10.1080/03235408.2013.811030

[45] Khojasteh M, Taghavi SM, Khodaygan P, Hamzehzarghani H, Chen G, Bragard C, et al. Molecular typing reveals high genetic diversity of *Xanthomonas translucens* strains infecting small-grain cereals in Iran. Applied and Environmental Microbiology. 2019;**85**:e01518-e01519. DOI: 10.1128/AEM.01518-19

[46] Wichmann F, Vorholter FJ, Hersemann L, Widmer F, Blom J, Niehaus K, et al. The noncanonical type III secretion system of *Xanthomonas translucens* pv. *Graminis* is essential for forage grass infection. Molecular Plant Pathology. 2013;**14**:576-588

[47] Gardiner DM, Upadhyaya NM, Stiller J, Ellis JG, Dodds PN, Kazan K, et al. Genomic analysis of *Xanthomonas translucens* pathogenic on wheat and barley reveals cross-kingdom gene transfer events and diverse protein delivery systems. PLoS One. 2014;**9**:e84995

[48] Peng Z, Hu Y, Xie Z, Potnis N, Akhunova A, Jones J, et al. Long read and single molecule DNA sequencing simplifies genome assembly and TAL effector gene analysis of *Xanthomonas translucens*. BMC Genomics. 2016;**17**:21

[49] Langlois PA, Snelling J, Hamilton JP, Bragard C, Koebnik R, Verdier V, et al. Characterization of the *Xanthomonas translucens* complex using draft genomes, comparative genomics, phylogenetic analysis, and diagnostic lamp assays. Phytopathology. 2017;**107**:519-527. DOI: 10.1094/PHYTO-08-16-0286-R

[50] Boosalis MG. The epidemiology of *Xanthomonas translucens* (J.J. and R.) Dowson on cereals and grasses. Phytopathology. 1952;**42**:387-395

[51] Sellam MA, aWilcoxson RD. Bacterial leaf blight of wheat in Minnesota. Plant Disease Report. 1976;**60**:242-245

[52] Otta JD. *Pseudomonas syringae* incites a leaf necrosis on spring and winter wheats in South Dakota. Plant Disease Report. 1974;**58**:1061-1064

[53] Otta JD. Occurrence and characteristics of isolates of *Pseudomonas syringae* on winter wheat. Phytopathology. 1977;**67**:22-26

[54] Scharen AL, Bergman W, Burns EE. Leaf diseases of winter wheat in Montana and losses from them in 1975. Plant Disease Report. 1976;**60**:686-690

[55] Varvaro L. Una batteriosi del Frumento duro (*Triticum durum* Desf.) causata da *Pseudomonas syringae* pv. *atrofaciens* (McCulloch) Young et al. in Italia. Informatore Fitopatologico. 1983;**33**:49-51

[56] Smith J, Hattingh MJ. Fluorescent pseudomonads associated with diseases of wheat in South Africa. Journal of Phytopathology. 1991;**133**:36-48

[57] Akhtar MA. Outbreaks and new records. Pakistan. Bacterial gumming disease of wheat. FAO Plant Protection Bulletin. 1987;**35**:102

[58] Teyssandier E, Sands DC. Wheat leaf blight caused by *Pseudomonas syringae* in Argentina. Phytopathology. 1977;**87**:1766

[59] Elliot C. Manual of Bacterial Plant Pathogens. 2nd ed. Waltham (Massachusetts): Chronica Botanica; 1951. p. 186

[60] Bradbury JF. Guide to Plant Pathogenic Bacteria. CAB International Mycological Institute: Farnham House, Slough, UK; 1986. p. 332

[61] CABI/EPPO. *Pseudomonas syringae pv. syringae*. Distribution Maps of Plant Diseases No. 336. Wallingford, UK: CABI Head Office; 2012

[62] von Kietzell J, Rudolf K. Wheat diseases caused by *Pseudomonas syringae* pathovars. In: Duveiller E, Fucikovsky L, Rudolf K, editors. The Bacterial Diseases of Wheat: Concepts and Methods of Disease Management. Mexico City: CIMMYT; 1997. pp. 49-57

[63] Toben H. Untersuchungen über die basale Spelzenfäule an Weizen und Gerste, hervorgerufen durch *Pseudomonas syringae* pv. *atrofaciens* ((McCull.) Young, Dye, Wilkie) [diploma thesis]. Göttingen, Germany: Universität Göttingen, Institut für Pflanzenpathologie und Pflanzenschutz; 1989

[64] MCculloch L. Basal glume rot of wheat. Journal of Agricultural Research. 1920;**18**:543-551

[65] Matveeva IEV, Pekhtereva ES, Polityko VA, Ignatov AN, Nikolaeva EV, Schaad NW. Distribution and virulence of *Pseudomonas syringae* pv. *atrofaciens*, causal agent of basal glume rot, in Russia. In: Rudolph K, Burr TJ, Mansfield JW, Stead D, Vivian A, and von Kietzell J, editors. *Pseudomonas syringae* and Related Pathogens. The Netherlands: Kluwer Academic Publishers. (2003). P. 97-105. Available from: https://link.springer.com/chapter/10.1007%2F978-94-017-0133-4_11

[66] Duveiller E. Collaborative research network on bacterial diseases of wheat. In: 3rd Annual Report on CIMMYT Project Bacterial Diseases Network of Wheat. Mexico: CIMMYT; 1990. pp. 5-6

[67] Galatchian MJ. Testing the pathogenicity of strains of *Bacterium atrofaciens* (McCull.) under laboratory conditions. C.R. pan. Sov. V.I. Lenin Academy of Agricultural Science, Moscow. 1941;**6**:40-43 (in Russian)

[68] Karov S, Vassilev VI. Studies on wheat bacteriosis in Bulgaria. Scientific Works. 1981;**26**:85-90 (in Bulgarian)

[69] Noble RJ. Basal glume rot. Agricultural Gazette New South Wales. 1933;**44**:107-109

[70] Wilkie JP. Basal glume rot of wheat in New Zealand. New Zealand Journal of Agricultural Research. 1973;**16**:155-160

[71] Dippenaar BJ. 'n Bakteriese wat'n verdorring van die kaffies van koringare veroorsaak. South African Journal of Science. 1931;**28**:280-283 (in Africaans)

[72] Rudolph K, Mavridis A. Über die Isolation von *Pseudomonas syringae* pv. *atrofaciens* aus Winterweizenähren in Schafhausen (Sindelfingen). Phytomedizin, Mitt. Deutsch. Phytomed. Ges. 1987;**17**:20

[73] Toben H, Mavridis A, Rudolph WE. On the occurrence of basal glume rot of wheat and barley caused by *Pseudomonas syringae* pv. *Atrofaciens* in West Germany. Journal of Plant Disease and Protect. 1991;**98**:225-235 (in German)

[74] Vassilev VI, Karov S. Effect of cereal basal bacteriosis on the economic physico-technological and biochemical properties of wheat grain and flour. Plant Science. 1985;**22**:13-20 (in Bulgarian)

[75] Mavridis A, Meyer D, Mielke H, Steinkampf G. Zum Auftreten und zur Schadwirkung der basalen Spelzenfäule beim Sommerweizen. Kali- Briefe (Büntehof). 1991;**20**:469-473

[76] Iacobellis NS, Figliuolo G, Janse J, Scortichini M, Ciuffreda G. Characterization of *Pseudomonas syringae* pv. *Atrofaciens*. In: Rudolph K, Burr TJ, Mansfield JW, Stead D, Vivian A, von Kietzell J, editors. *Pseudomonas syringae* Pathovars and Related Pathogens. Dordrecht (The Netherlands): Kluwer Academic Publishers. 1997. pp. 500-504.

[77] Claflin LE, Ramundo BA. Evaluation of the dot-immunobinding assay for detecting phytopathogenic bacteria in wheat seeds. Journal of Seed Technology. 1987;**11**:52-61

[78] Toben H, Mavridis A, Rudolph KWE. Basal glume rot (*Pseudomonas syringae* pv. *Atrofaciens*) on wheat and barley in FRG and resistance screening of wheat. Bulletin OEPP/EPPO Bulletin. 1989;**19**:119-125

[79] Maraite H, Weyns, J. *Pseudomonas syringae* pv. *aptata* and pv. *atrofaciens*, specific pathovars or members of

pv. *syringae*?. In: Rudolph K, Burr TJ, Mansfield JW, Stead D, Vivian A, von Kietzell J, (eds *Pseudomonas syringae* Pathovars and Related Pathogens.). Dordrecht (The Netherlands):Kluwer Academic Publishers, 1997. pp. 515-520

[80] Mohan SK, Schaad NW. An improved agar plating assay for detecting *Pseudomonas syringae* pv. *syringae* and *P. s.* pv. *phaseolicola* in contaminated bean seed. Phytopathology. 1987;**77**:1390-1395

[81] Carlson RR, Vidaver AK. Bacterial mosaic, a new corynebacterial disease of wheat. Plant Disease. 1982;**66**:76-79

[82] Li X, Tambong J, Yuan KX, Chen W, Xu H, Lévesque CA, et al. Re-classification of *Clavibacter michiganensis* subspecies on the basis of whole-genome and multi-locus sequence analyses. International Journal of Systematic and Evolutionary Microbiology. 2018;**68**:234-240. DOI: 10.1099/ijsem.0.002492

[83] McBeath JH, Adelman M, Jackson L. Screening wheat germplasm for *Corynebacterium michiganense* subsp. *tessellarius*. Phytopathology. 1988;**78**:1566

[84] Scharif G. Corynebacterium iranicum sp. nov. on wheat (*Triticum vulgare* L.) in Iran and a comparative study of it with *C. tritici* and *C. rathayi*. Entomologie et phytopathologie appliquées. 1961;**19**:1-24

[85] Zgurskaya HI, Evtushenko LI, Akimov VN, Kalakovtskii LV. *Rathayibacter gen. nov.*, including the species *Rathayibacter rathayi* comb. *nov.*, *Rathayibacter tritici comb. nov.*, *Rathayibacter iranicus comb. nov.*, and six strains from annual grasses. International Journal of Systematic Bacteriology. 1993;**43**:143-149

[86] Bamdadian A. The importance and situation of wheat diseases in Iran. In: CENTO Panel on Pests and Diseases of Wheat, Tehran University, College of Agriculture; 5-7 February 1973; Karaj, Iran. Tehran Iran: Central Treaty Organization (CENTO); 1973. pp. 57-63

[87] Paruthi IS, Bhatti DS. Estimation of loss in yield and incidence of *Anguina triticina* on wheat in Haryana (India). International Nematology Network Newsletters. 1985;**2**:13-16

[88] Benjama A. Bacterial diseases of cereals in Morocco. Cahiers Agricultural. 1997;**6**:605-610

[89] Fattah FA. Effects of inoculation on the incidence of ear-cockle and 'tundu' on wheat under field conditions. Plant and Soil. 1988;**109**:195-198

[90] Riley IT, Reardon TB. Isolation and characterization of *Clavibacter tritici* associated with *Anguina tritici* in wheat from Western Australia. Plant Pathology. 1995;**44**:805-810

[91] Fucikovsky L, Duvieller E. Other plant pathogenic bacteria reported on wheat. In: Duvieller E, Fucikovsky L, Rudolph K, editors. The Bacterial Diseases of Wheat: Concepts and Methods of Disease Management. Mexico: CIMMYT; 1997. pp. 59-64 https://repository.cimmyt.org/handle/10883/1227

[92] Tanii, Miyajima K, Akita T. The sheath brown rot disease of rice plant and its causal bacterium *Pseudomonas fuscovaginae* A. Tanii, K. Miyajima and T. Akita sp. nov. Annals Phytopathology Socical Japan. 1976;**42**:540-548

[93] Duveiller E, Maraite H. Bacterial sheath rot of wheat caused by *Pseudomonas fuscovaginae* in the highlands of Mexico. Plant Disease. 1990;**74**:932-935

[94] Tohya M, Watanabe S, Tada T, Tin HH, Kirikae T. Genome analysis-based reclassification of *Pseudomonas fuscovaginae* and *Pseudomonas shirazica* as later heterotypic synonyms of *Pseudomonas asplenii* and *Pseudomonas asiatica*, respectively. International Journal of Systematic and Evolutionary Microbiology. 2020;**70**:3547-3552. DOI: 10.1099/ijsem.0.004199

[95] Anonymous. Occurrence of Sheath Rot, a New Disease in Wheat. Vol. 1. Pokhara, Nepal: Lumle Newsletter, Lumle Agricultural Research Centre; 1995. p. 3

[96] Klement Z. Tobacco (HR) test for the quick demonstration of pathogenicity. In: Klement Z, Rudolph K, Sands DC, editors. Methods in Phytobacteriology. Akadémiai Kiadó: Budapest; 1990. pp. 101-102

[97] Cottyn B, Ceres MT, Van outryve MF, Baroga J, Mew TW. Bacterial diseases of rice. I. Pathogenic bacteria associated with sheath rot complex and grain discoloration of rice in the Philippines. Plant Disease. 1996;**80**:429-437

[98] Piening LJ, MacPherson DJ. Stem melanosis, a disease of spring wheat caused by *pseudomonas cichorii*. Canadian Journal of Plant Pathology. 1985;7:168-172

[99] Piening LJ, MacPherson DJ, Mahli SS. The effect of copper in reducing stem melanosis of park wheat. Canadian Journal of Plant Science. 1987;**67**:1089-1091

[100] Piening LJ, MacPherson DJ, Mahli SS. Stem melanosis of some wheat, barley, and oat cultivars on a copper deficient soil. Canadian Journal of Plant Science. 1989;**11**:65-67

[101] Howe ET, Simmonds PM. Bacterial pink blotch of wheat. In: Conners IG, Graigie JH, Vanterpool TC, editors. Proceedings of the Canadian Phytopathological Society; June 28-30, 1937; Ottawa, Canada. 17th ed. 1937. p. 6

[102] Luisetti J, Rapilly F. Sur une altération d'origine bactérienne des graines de blé. Ann. Epiphyt. Phytogenet. 1967;**18**:483-493

[103] Roberts P. *Erwinia* (Millard) Burkholder associated with pink grain of wheat. The Journal of Applied Bacteriology. 1974;**37**:353-358

[104] McMullen MP, Stack RW, Miller JD, Bromel MC, Youngs VL. *Erwinia rhapontici*, a bacterium causing pink wheat kernels. In: Proceedings of the North Dakota Academy of Sciences; Grand Forks, ND. Vol. 38. 1984. p. 78

[105] Dutrecq A, Debras P, Stevaux J, Klaessens D. Estimation de la flore bactérienne sur les épis de froment échaudés. Parasitica. 1990;**46**:69-84

[106] Diekmann M, Putter CAJ. FAO/IPGRI Technical Guidelines for the Safe Movement of Germplasm. No. 14. Small Grain Temperate Cereals. Rome: FAO/IPGRI; 1995. p. 67

[107] Huang HC, Phillippe RC, Phillippe LM. Pink seed of pea: A new disease caused by *Erwinia rhapontici*. Canadian Journal of Plant Pathology;**12**:445-448

[108] Ohuchi A, Ohsawa T, Nishimura J. Two pathogenic bacteria, *Erwinia rhapontici* (Millard 1924) Burkholder 1948 and *Pseudomonas marginalis* pv. *Marginalis* (Brown 1918) Stevens 1925, causing a soft rot of onion. Japanese Annals of the Phytopathological Society. 1983, 1990;**49**:619-626

[109] Choi JE, Han KS. Studies on the bacterial soft rot disease of *Liliaceae*

crops in Korea: 2. Soft rot of garlic caused by *Erwinia*. Korean Journal of Plant Pathology. 1989;**5**:271-276

[110] Huang HC, Erickson RS, Yanke LJ, Mündel H-H. First report of pink seed of common bean caused by Erwinia rhapontici. Plant Disease. 2002;**86**:921

[111] Huang HC, Erickson RS, Yanke LJ, Hsieh TF, Morrall RAA. First report of pink seed of lentil and chickpea caused by *Erwinia rhapontici* in Canada. Plant Disease. 2003;**87**:1398

[112] Stackebrandt E. The richness of prokaryotic diversity: There must be a species somewhere. Food Technology and Biotechnology. 2003;**41**:17-22

[113] Tambong JT. Taxogenomics and systematics of the genus *Pantoea*. Frontiers in Microbiology. 2019;**10**:2463

[114] Thompson CC, Chimetto L, Edwards RA, Swings J, Stackebrandt E, Thompson FL. Microbial genomic taxonomy. BMC Genomics. 2013;**14**:913. DOI: 10.1186/1471-2164-14-913

[115] Meier-Kolthoff JP, Auch AF, Klenk HP, Goker M. Genome sequence-based species delimitation with confidence intervals and improved distance functions. BMC Bioinformatics. 2013;**14**:60. DOI: 10.1186/1471-2105-14-60

[116] Jain C, Rodriguez RL, Phillippy AM, Konstantinidis KT, Aluru S. High throughput ANI analysis of 90K prokaryotic genomes reveals clear species boundaries. Nature Communications. 2018;**9**:5114

[117] Goris J, Konstantinidis KT, Klappenbach JA, Coenye T, Vandamme P, Tiedje JM. DNA-DNA hybridization values and their relationship to whole-genome sequence similarities. International Journal of Systematic and Evolutionary Microbiology. 2007;**57**: 81-91. DOI: 10.1099/ijs.0.64483-0

[118] Teeling H, Meyerdierks A, Bauer M, Amann R, Glockner FO. Application of tetranucleotide frequencies for the assignment of genomic fragments. Environmental Microbiology. 2004;**6**:938-947

[119] Duret L. Evolution of synonymous codon usage in metazoans. Current Opinion in Genetics & Development. 2002;**12**:640-649

[120] Wattam AR, Abraham D, Dalay O, Disz TL, Driscoll T, Gabbard JL, et al. PATRIC, the bacterial bioinformatics database and analysis resource. Nucleic Acids Research. 2014;**42**:581-591

[121] Afgan E, Baker D, Batut B, van den Beek M, Bouvier D, Čech M, et al. The galaxy platform for accessible, reproducible and collaborative biomedical analyses. Nucleic Acids Research. 2018;**46**:W537-W544. DOI: 10.1093/nar/gky379

Section 3

Abiotic Stress

Chapter 8

Effect of Climate Change on Wheat Productivity

Ihsan Elahi, Uzair Saeed, Abdul Wadood, Amjad Abbas, Hira Nawaz and Sohail Jabbar

Abstract

Climate is the average of weather situation in a particular area, which affects all parts of ecosystem. Due to industrialization and urbanization, forests are cutting down and converted into living societies. This change in ecosystem disturbs the balance of ecosystem from decomposers to producers and consumers. Important part of ecosystem is plants (producers) that are energy providers. This alteration affects productivity and sustainability of plants. Wheat is staple food, which is highly affected by temperature and CO_2 elevation. It not only affects wheat yield but also make wheat vulnerable to several diseases. High temperature causes a high rate of transpiration, which causes drought that ultimately leads to low productivity. A model was designed on drought conditions and result showed that global warming causes serious drought in 60% of wheat-growing areas of the world. Currently, drought affects 15% of wheat productivity. It was predicted that every 2°C shift of temperature can cause severe water shortage in the coming 20 to 30 years. Water shortage at milking and grain filling stage will affect yield. This chapter includes factors affecting climate, impact on wheat growth, yield, and elevation of carbon dioxide, impact on disease severity, prediction model for temperature rise, and CO_2 curve in 2050.

Keywords: CO_2 elevation, raised temperature, wheat production, prediction model, global warming, metrological advancement

1. Introduction

Long-term change in the weather pattern is affected by natural and human factors. Climate is changing every day due to several natural processes as well as by human acts. One of the biggest sources of climate change is the accumulation of carbon dioxide in our atmosphere. Carbon dioxide accumulates in the atmosphere by burning fossil fuel, automobile smoke, chlorofluorocarbons released from electric appliances (Air conditions or refrigerators), and volcanic eruptions. Humans release carbon dioxide into the air during respiration. Accumulation of carbon dioxide in an atmosphere enhanced the greenhouse effect because carbon dioxide is considered one of the most important gases in greenhouse gases. It is observed that amount of carbon in the atmosphere is 80% increases today from the time when life on earth started. The main reason for this increased value is humans. In past released carbon was utilized by plants as it is the main element in photosynthesis. But with the passage of time human population increases and agricultural land is being utilized by humans for shelter. With the advancement of colonization deforestation started and agricultural land or cultivated land turned into housing societies. Carbon dioxide released from automobile vehicles accumulates in the air [1]. Other gases include methane, Nitrous oxide, Ozone, Water vapor, Halocarbons. These gases create a sheet around the earth. This sheet is denser in the northern hemisphere because of extreme cold they use more fossil fuels. This sheet of gases causes a rise in temperature on earth also known as global warming. These two terms are inter-related. This temperature rise not only affects humans but also disturbs all the natural habitats and ecosystems on earth. Climate change effect not externally humans, plants, animals, and microbes but also internally by interrupting their genome and causing mutation and cause permanent change on a species level. It causes many animal and plants species endangered. This also interferes with the life cycle of insects and it makes pathogens resistant and cultivars fail to respond better which ultimately leads to food security issues.

Due to global warming agriculture faces serious threats like low crop productivity which leads to global hunger and this low production rise the cost of food commodities and makes it unaffordable for the poor population. Global warming affects the pattern of rainfall which contributes to other disasters. Rise in atmospheric carbon dioxide reduced wheat products as well as nutritional value also down and, in some cases, due to change in the chemical composition some crops start producing toxins [2]. Plant responses to climate change by altering their phenological characteristics. Flowering and fruiting or grain filling in the case of cereals is a very important stage that is particularly affected, it affects pollination, root growth, seed formation, number of seed

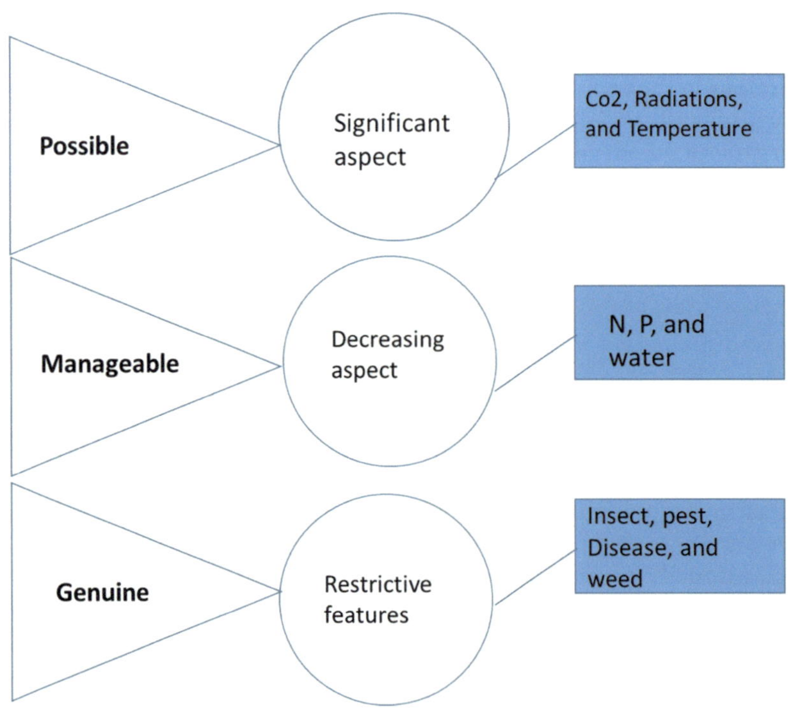

Figure 1.
Crop yield dependency and limiting factors related to climate change [7, 8].

production, leaf expansion, and ripening of the crop. Time for flowering and fruit ripening is affected by the environment, photoperiod, and vernalization [3].

Wheat is a major cereal crop everywhere in the world, it is an important source of energy for the human diet [4]. Wheat is 90% irrigated by arid and semi-arid climates to grow wheat. Wheat in rainfed areas is most affected by climate change. Climate change affects wheat productivity in Australia, Mexico, every year 2.85 billion dollars of wheat loss [5].

Expected food demand will be double by 2050, and production yield losses due to global warming and rise in carbon dioxide concentration. This global warming causes very negative aspects on plants, pathogens, insects, and pests [6] (**Figure 1**).

2. Effect of climate change on wheat growth and production

In this hot climate, every prediction regarding climate shows extreme weather conditions [9]. Climate change has a very different effect on crop productivity. It is estimated that a 1°C increase in temperature can cause a 10–20% decrease in crop yield globally. Similarly, a 1 to 3°C increase in temperature is estimated to reduce 20–30% yield reduction in potato crops [10]. This effect can be even worse till the end of this century it is expected to be 2–4°C even more rise in temperature which affects crop production [11]. Change in weather conditions ultimately made extreme climate shift permanently and affect agriculture in the whole world [12]. These extreme temperature changes during sensitive stages like flowering, anthesis, and milking stage affect wheat yield, grain weight, and grain size at the end of season significantly [13]. Nuttall et al. [14] experimented on wheat production with the combined effect of high temperature and CO_2 enhanced concentration. Results showed that when the temperature was 36 ± 2°C during anthesis it reduced 13 percent and most grains were sterile. Asseng et al. [15] concluded that a 2°C increase in temperature in the Australian core growing area would reduce yield up to 50%. Therefore, heat stress is very critical for future wheat production in Australia, numerous studies are carried out around the globe for risk assessment for yield regarding heat, rainfall, and drought condition along with different cropping patterns [16]. Early maturity rescues the wheat from drought stress in Europe. Drought is linked with low rainfall and high temperature but it is can be managed up to some extent [17].

The wheat crop suffers due to several limiting factors, i.e., biotic (insects, disease, pest, weeds), abiotic stress (heat, cold, drought, and nutrients) effects. At specific wheat growth stages, these factors have decreasing and restricted aspects on wheat crop. The CO_2, radiation, and temperature have positive and significant effects on wheat growth. These factors are directly proportional to the wheat yield [7, 8].

Different studies reported that climate change reported that directly predict crop yield. Every 1-degree rise in temperature decrease the growth attributes and ultimately yield. A comprehensive change in growing season temperature was reported. They predict 100 years crop model for global climate change (variation in temperature & rainfall) and their effect on the wheat yield based on 100 years' data [18].

In the north of Europe, flowering time is very much affected by dry climatic conditions and it causes drastic yield losses [19].

Wilcox and Makowski [20] used 90 articles and made data set of climate change of different regions, i.e., USA, Spain, UK, and Australia. Variability in average yield is high in regions like UK, USA, and Australia, it ranged between −100 and + 90% in the Australian region.

The conclusion of this analysis includes a meta-analysis for wheat production and yield in the future. Analysis was done with high CO_2 concentration, a decline in rainfall along with rising temperature this increased wheat yield but results varied with location. Meta-analysis for wheat explores quick results regarding wheat production [20].

Effect of wheat yield with this climate change scenario till 2050 and impact of this change impart negative effect on wheat production. All studies carried on wheat production were based on global warming and rising temperature with several global climate change models, Hernandez-Ochoa et al. [21] studied the effect of temperature with the rise in carbon dioxide and change in rainfall pattern. The researcher used 5 global climate modeling and 2 ensembles with 2 scaling methods and quantified uncertainty. Spatial and temporal variability on different locations under study showed yield reduction with high temperature and carbon dioxide concentration. The same results were shown with other studies [22].

Under high-temperature spikes, production is reduced and spikes get vulnerable to disease stress. Temperature above 32°C at the time of anthesis, make grain shorter in size, grain filling duration in the spikes is also reduced which ultimately affects the wheat yield [22]. Wheat in rainfed areas is more affected with change in rainfall pattern, rainfall declines and it affects the production of wheat directly, yield decline 5–7 percent with the rise in each degree of temperature [23]. Asseng et al. [24] counted in Sudan, and find a 6 percent yield reduction with a 13°C rise in temperature, which was raised from 27°C.

3. Greenhouse effect and elevation of carbon dioxide on plants

Carbon dioxide in the air is an important source of carbon for plants, unfortunately, this CO_2 level is increasing day by day due to human activities. This elevation not only results in ozone depletion but also affects the growth and yield of field crops. It is observed that an increase in carbon dioxide increases the rate of photosynthesis, it increases water efficiency and high nutrient availability [25]. In C3 plants increase of CO_2 level up to 1 k ppm stimulates the rate of photosynthesis [26] but this does not increase the yield or biomass of the plant. As yield in the wheat crop is depends not only on the rate of photosynthesis but also on the active phase of photosynthesis along with sink capacity of grain [26].

The experiment performed by Fangmeier et al. [27] and concluded that the rise in CO_2 level increases nitrogen sink capacity and also reduces the photosynthetic period which results in poor growth and reduced yield. Another experiment performed by Kimball et al. [28] increase carbon dioxide to 12% with the limited supply of nutrients and yield increase was observed only 7% as compared to control but consume more water. Daepp et al. [29] experimented with wheat by adding nitrogen fertilizer along with elevated CO_2, this nitrogen helps in overcoming carbon sinking especially during the reproductive stage. The researcher concludes that if a crop is grown the plant can be enhanced by using biological nitrogen fixation process, this also favors the yield of legume crops as they already have this natural phenomenon [30].

Carbon dioxide elevation and temperature by a few degrees may disturb the positive aspects. The experiment was done on wheat by doubling CO_2 and increasing 1.5 to 4°C showed a negative effect on wheat yield.

The temperature of the atmosphere is increasing day by day due to global warming and greenhouse gasses. Temperature increases decrease the positive aspects caused by elevated carbon dioxide for plants. The rise in temperature increases the rate of leaf transpiration from the plant [31]. Nevertheless, carbon dioxide elevation can offset the negative effect of high temperature by lowering the stomatal opening and reducing the transpiration rate. Higher temperature can also help in plant production, especially in Mediterranean regions where crop production effects by lower temperature [32]. But elevated CO_2 and temperature change the pattern of rainfall in arid and semi-arid regions which affect plant production very badly. This shift of rainfall has negative as well as positive effects on agriculture. Like in rainfed areas it limits the plant growth while in high rainfall areas it avoids water logging conditions and helps plants to grow well. Wheat is normally grown in the area of less than 550 mm of rainfall and 325 mm is received by the wheat plants in that region. But according to rainfall prediction for 2070, it is expected that by an increase of 10% reduce the winter rain up to 60%, while another research predicted that rainfall will be reduced by 15% till 2030 and 30% till 2070. This prediction is proving right during past years and it is the biggest threat to wheat in rainfed areas of the world [33].

Increased carbon dioxide level is beneficial for C3 plant, as it increases biomass yield and increases metabolism and stomatal conductance as well as an increased rate of photosynthesis. If temperature increases, it changes the uptake of nitrogen, carbon and decreases the nutritional value of the grains [34]. The condition is even worse when drought, rainfall, and less humidity affect plant growth and production [35].

4. Climate change and diseases attack

Fusarium causes different diseases in wheat-like, foot rot, root rot, and head blight in wheat, which causes huge yield losses [36]. *Rhizoctonia solani* is a soil-borne fungus that causes root rot in wheat and causes 50% yield losses in Japan, Europe, and the USA [37]. Climate change has a strong impact on the pathogen population. Temperature and water play important role in the germination and survival of pathogens. In Germany air temperature raised from 0.8–1.1°C from 1900 to 2000, which increase the rainfall during winter, this ultimately suits the pathogen life cycle and helps them for colonization on crop debris and access to particular susceptible hosts [38, 39].

Jacobs et al. [40] experimented on plant decomposition and study on soil temperature and its effect on microbe's survival under natural field conditions. In this experiment amount of bacteria and fungus were counted with the amino sugars and muramic acid. For fungus, ergosterol is used to access fungal biomass as it does not mix with soil organic components plus it is a major part of the fungal cell membrane [41].

Lukas et al. [38] performed an experiment on the survival of three fungal pathogens *F. culmorum, F. graminearum,* and *Rhophitulus solani*, decomposed infected leaf of maize with these disease pathogens, and for temperature heating cables were used. Microbial biomass and fungal colonization were observed after 152 days. Pathogen growth was reported with DNA, saprotrophic biomass with glucosamine and for bacteria, muramic acid was measured and values were compared with control. Moreover, it was also observed that *F. culmorum* produces more DNA so it wasn't affected with soil raised temperature but DNA of *R. solani* decreased significantly. *R. solani* germination and infection varied between soil temperatures ranging 15–25°C this were completely disturbed when temperature fluctuate and at 5°C completely stopped [42]. While Fusarium infected the most in raised temperature which shows strong spatial variability [43].

5. Climate change and insect population

Climate changes where effect every part of agriculture it does not leave insects unaffected. Plant productivity decreases due to the rise in temperature and drought conditions are directed linked with global warming. When plant population decreases it directly affects the insect population which survives on plants. It also contributed to increasing insect outbreaks [44, 45]. This temperature and drought increase causes wildfire and causes plant mortality which ultimately carbon sinks and rising carbon levels in the air [46, 47]. Major insects which threaten wheat yield are wheat stem sawfly, and orange blossom wheat midge, which causes losses up to the economic threshold level [48]. Change in carbon dioxide amount in the atmosphere causes a significant impact on plants, insects, and microorganisms. Insects along with disease pathogens reduce significant yield losses besides all the control strategies [34].

Global warming changes the biochemistry of plants which impacts herbivorous insects and pathogens [49]. Insect populations are disturbed by different abiotic factors due to global warming. Insect population increases in this rising temperature and transmit virus very smoothly from infected to a healthy plant. These climate changes affect badly beneficial insects which cannot survive in dry weather with hot temperatures, it also affects their ability to kill harmful insects [50]. The negative effect of climate change, increase in temperature and CO_2 concentration, an increased photosynthesis rate, and increase productivity but reduce agricultural production due to changing weather patterns [51].

6. Prediction model for disease elevation in correlation to carbon dioxide

The amount of atmospheric carbon dioxide increasing day by day due to the modern standard of living and industrialization. It had been raised to 50% after the industrial revolution [52]. It was 270 μmol/mol which was 408 μmol/mol in 2017 [53]. It was predicted that if carbon dioxide is released at this rate its concentration in the atmosphere will be around 550 μmol/mol till 2050 [54]. Carbon dioxide is the main constituent of photosynthesis so this affects directly to plant growth and metabolism. But this too much carbon dioxide is harmful to C3 plants, this increases the plant biomass and leads to the carbon dioxide fertilization effect. This phenomenon is explained in different perspectives of the agricultural ecosystem but its extent varied from region to region depending upon the environmental condition and temperature and amount of water in soil [55]. Rainfall in arid, semi-arid, and temperate region and its effect on plants are monitored by free air CO_2 enrichment (FACE) [56]. Different regions respond differently to this elevated CO_2 like nutrient uptake, water supply to the plant, water reservoirs in the hot and dry time of year, modeling presents that massive uncertainty in the response of the crop to elevated crop. Wheat is grown everywhere and almost 15% of annual yield is affected by climatic conditions in Mediterranean areas. In Mediterranean regions main water source is rainfall, which is important for the early growth stages of the wheat plant. The grain filling stage in wheat is badly affected due to a shortage of water supply this condition is termed as terminal drought which ultimately affects the yield of the crop [57, 58]. This drought helps wheat to produce a long root system plus reduce stomatal conductance to conserve available water, but this may be harmful at the grain filling stage. On the other hand, if more water is available then it kept on increasing vigor and plant height, and delayed reproductive stages of the plant may be died off before grain formation. The plant also gets vulnerable to disease attacks [59].

7. Prediction model temperature elevation

The reason for global warming and temperature variation is CO_2 elevation and many other greenhouse gases which trap heat and raise the temperature in the atmosphere. Prediction on temperature elevation on this planet till 2100 will be increased from 2, 9.7°C which is 1.1–5.4°C is now. The temperature will fluctuate due to heat-trapping gases. CO_2 is added to the air due to the burning of coal and fossil fuels. So, if humans kept on using these things as an energy source, then exact figures for temperature variation are impossible. Scientists work to develop so much for better understanding and awareness in public, they develop a model named as global climate model for prediction, and this is computerized software. This predicts the number of greenhouse gases and concentrations in the air in different situations. For example, the current concentration of CO_2 in the air is 9 billion metric tons per annum and it would be 12 billion if kept on growing till the end of 2040, but if the situation is controlled it can be reversed to 5 billion which was in 1990. Temperature mainly depends upon the carbon dioxide emission if it increases temperature will go up and vice versa [60, 61].

8. Management strategies

Global warming affects crop productivity throughout the world. Expensive food items are the first sign of a sudden food shortage in the world's crop yield which will be even more shortly if remain uncontrolled. For this reason, scientist needs to develop crop seeds that are resistant to drought, salinity, and major diseases which are the major threatening factors. So that wheat crop production increases to meet the demand of the human population. Adaptation includes agronomic practices like time of sowing, water management, nutrient availability, timely weeding, and resistant cultivars are helpful tools. Genetically modified crops are important tools for production. It is an easy and quick way as compared to a conventional breeding method which is not reliable and time taking. Molecular breeding enhances the wheat productivity to fight with different abiotic and biotic stresses that crops have to face when cultivated in a field. Molecular markers help to identify the insertion and activity of various genes. Advancement in DNA sequencing helps in finding novel resistant genes and their insertion in different crops possible [62]. Government should make management strategies for the minimization of global warming. New projects should be designed for the conservation of water loss, minimizing the use of pesticides in fields. Public awareness campaigns should be initiated at the individual level to stop activities that are changing our ecosystem. Pollution-free water should be used to irrigate agricultural land. There should be instruments for the assessment of Carbon concentration in air, and temperature monetization. Training sessions should be made to practice techniques that are helpful in conservation [63].

9. Conclusion

Climate changes cause an increase in carbon dioxide emission, which causes the greenhouse effect around the globe, it affects all agriculture ecosystems in different ways, sometimes one factor favors plant growth but in combination with other shifts the positive effect into a drastic negative effect. This chapter detail about rise in temperature variation in different regions of the world and its effect on wheat plant growth, biochemistry, grain size and weight, effect on insect pest population, on microbial pathogens. Reviewing literature it has been found that global increase in carbon dioxide helps C3 plants to increase growth, improve plant water uptake capacity and yield of crops. It also favors C3 plants to compete with C4 weed that is grown side by side with the main crop, plants become more resistant to diseases. But these benefits turn negative when the temperature of the area increases, suddenly plants lose the ability to uptake minerals from the soil, and reeducation of grain size, grain weight, and crop resistance towards diseases, increase pest population and water holding capacity of plants. Temperature variation alters the rate of precipitation which ultimately increases drought conditions which is very crucial for wheat growing in rainfed and Mediterranean regions of the world. This condition helps C4 weeds instead of the wheat crop which increases the competition between crop and nutrient for food and water. It is also included as a topic of discussion that if these effects remain uncontrolled it will cause a major food shortage in coming years when food is already expected to be doubled as the world population increases day by day and industrialization increases all these risks. So, it is important to practice the management practice suggested in the chapter to conserve our ecosystem and make this planet a safe place to live.

Author details

Ihsan Elahi[1]*, Uzair Saeed[1], Abdul Wadood[1], Amjad Abbas[2], Hira Nawaz[3] and Sohail Jabbar[1]

1 Department of Computer Science, The University of Faisalabad, Punjab, Pakistan

2 Faculty of Agriculture, Department of Plant Pathology, University of Agriculture, Faisalabad, Punjab, Pakistan

3 Faculty of Agriculture, Department of Plant Protection, Cukurova University, Adana, Turkey

*Address all correspondence to: ahsanelahi18@gmail.com

References

[1] Sadras VO, Calderini DF. Crop physiology: Applications for breeding and agronomy. In: Crop Physiology. Academic Press; 2015. pp. 1-14

[2] Dwivedi S, Sahrawat K, Upadhyaya H, Ortiz R. Food, nutrition and agrobiodiversity under global climate change. In: Advances in Agronomy. Vol. 120. Academic Press; 2013. pp. 1-128

[3] Slafer GA, Savin R, Sadras VO. Coarse and fine regulation of wheat yield components in response to genotype and environment. Field Crops Research. 2014;**157**:71-83

[4] Curtis BC, Rajaram S, Gómez MH. Bread Wheat: Improvement and Production. Food and Agriculture Organization of the United Nations (FAO); 2002

[5] Economics AB. Agricultural commodity statistics. 2018

[6] Alexandratos N, Bruinsma J. World agriculture towards 2030/2050: The 2012 revision

[7] Goudriaan J, Zadoks JC. Global climate change: Modelling the potential responses of agro-ecosystems with special reference to crop protection. Environmental Pollution. 1995;**87**(2):215-224

[8] Fuhrer J. Agroecosystem responses to combinations of elevated CO_2, ozone, and global climate change. Agriculture, Ecosystems & Environment. 2003;**97**(1-3):1-20

[9] Kharin VV, Zwiers FW, Zhang X, Wehner M. Changes in temperature and precipitation extremes in the CMIP5 ensemble. Climatic Change. 2013;**119**(2):345-357

[10] Rose G, Osborne T, Greatrex H, Wheeler T. Impact of progressive global warming on the global-scale yield of maize and soybean. Climatic Change. 2016;**134**(3):417-428

[11] Rogelj J, Den Elzen M, Höhne N, Fransen T, Fekete H, Winkler H, et al. Paris Agreement climate proposals need a boost to keep warming well below 2 C. Nature. 2016;**534**(7609):631-639

[12] Barlow KM, Christy BP, O'leary GJ, Riffkin PA, Nuttall JG. Simulating the impact of extreme heat and frost events on wheat crop production: A review. Field Crops Research. 2015;**171**:109-119

[13] Talukder AS, McDonald GK, Gill GS. Effect of short-term heat stress prior to flowering and early grain set on the grain yield of wheat. Field Crops Research. 2014;**160**:54-63

[14] Nuttall J, Brady S, Brand J, O'Leary G, Fitzgerald GJ. Heat waves and wheat growth under a future climate. In The 16th Australian Agronomy Conference: Climate Change, pp. 14-18. Available from: http://www.regional.org.au/au/asa/2012/climate-change/8085_nuttalljg.htm. [Accessed: October 2012]

[15] Asseng S, Foster IA, Turner NC. The impact of temperature variability on wheat yields. Global Change Biology. 2011;**17**(2):997-1012

[16] Lobell DB, Hammer GL, Chenu K, Zheng B, McLean G, Chapman SC. The shifting influence of drought and heat stress for crops in northeast Australia. Global Change Biology. 2015;**21**(11): 4115-4127

[17] Rezaei EE, Siebert S, Ewert F. Impact of data resolution on heat and drought stress simulated for winter wheat in Germany. European Journal of Agronomy. 2015;**65**:69-82

[18] Wang B, Li Liu D, Asseng S, Macadam I, Yu Q. Modelling wheat yield change under CO_2 increase, heat and water stress in relation to plant available water capacity in eastern Australia. European Journal of Agronomy. 2017;**90**:152-161

[19] Bodner G, Nakhforoosh A, Kaul HP. Management of crop water under drought: A review. Agronomy for Sustainable Development. 2015;**35**(2):401-442

[20] Wilcox J, Makowski D. A meta-analysis of the predicted effects of climate change on wheat yields using simulation studies. Field Crops Research. 2014;**156**:180-190

[21] Hernandez-Ochoa IM, Asseng S, Kassie BT, Xiong W, Robertson R, Pequeno DN, et al. Climate change impact on Mexico wheat production. Agricultural and Forest Meteorology. 2018;**263**:373-387

[22] Martre P, Wallach D, Asseng S, Ewert F, Jones JW, Rötter RP, et al. Multimodel ensembles of wheat growth: Many models are better than one. Global Change Biology. 2015;**21**(2):911-925

[23] Liu B, Asseng S, Müller C, Ewert F, Elliott J, Lobell DB, et al. Similar estimates of temperature impacts on global wheat yield by three independent methods. Nature Climate Change. 2016;**6**(12):1130-1136

[24] Asseng S, Cammarano D, Basso B, Chung U, Alderman PD, Sonder K, et al. Hot spots of wheat yield decline with rising temperatures. Global Change Biology. 2017;**23**(6):2464-2472

[25] Long SP, Ainsworth EA, Rogers A, Ort DR. Rising atmospheric carbon dioxide: Plants FACE the future. Annual Review in Plant Biology. 2004;**55**:591-628

[26] Amthor JS. Effects of atmospheric CO_2 concentration on wheat yield: Review of results from experiments using various approaches to control CO_2 concentration. Field Crops Research. 2001;**73**(1):1-34

[27] Fangmeier A, Chrost B, Högy P, Krupinska K. CO2 enrichment enhances flag leaf senescence in barley due to greater grain nitrogen sink capacity. Environmental and Experimental Botany. 2000;**44**(2):151-164

[28] Kimball BA, Kobayashi K, Bindi M. Responses of agricultural crops to free-air CO_2 enrichment. Advances in Agronomy. 2002;**77**:293-368

[29] Daepp M, Nösberger J, Lüscher A. Nitrogen fertilization and developmental stage alter the response of Lolium perenne to elevated CO2. New Phytologist. 2001;**150**(2):347-358

[30] Lüscher A, Hartwig UA, Suter D, Nösberger J. Direct evidence that symbiotic N_2 fixation in fertile grassland is an important trait for a strong response of plants to elevated atmospheric CO_2. Global Change Biology. 2000;**6**(6):655-662

[31] Peng S, Huang J, Sheehy JE, Laza RC, Visperas RM, Zhong X, et al. Rice yields decline with higher night temperature from global warming. Proceedings of the National Academy of Sciences. 2004;**101**(27):9971-9975

[32] Van Ittersum MK, Howden SM, Asseng S. Sensitivity of productivity

and deep drainage of wheat cropping systems in a Mediterranean environment to changes in CO_2, temperature and precipitation. Agriculture, Ecosystems & Environment. 2003;**97**(1-3):255-273

[33] Ludwig F, Asseng S. Climate change impacts on wheat production in a Mediterranean environment in Western Australia. Agricultural Systems. 2006;**90**(1-3):159-179

[34] Trebicki P. Plant viruses threaten crops as climate warms. Australasian Science. 2016;**37**(4):36-37

[35] Jones RA. Future scenarios for plant virus pathogens as climate change progresses. Advances in Virus Research. 2016;**95**:87-147

[36] Xu X, Nicholson P. Community ecology of fungal pathogens causing wheat head blight. Annual Review of Phytopathology. 2009;**47**:83-103

[37] Kühn J, Rippel R, Schmidhalter U. Abiotic soil properties and the occurrence of Rhizoctonia crown and root rot in sugar beet. Journal of Plant Nutrition and Soil Science. 2009;**172**(5):661-668

[38] Lukas S, Abbas SJ, Kössler P, Karlovsky P, Potthoff M, Joergensen RG. Fungal plant pathogens on inoculated maize leaves in a simulated soil warming experiment. Applied Soil Ecology. 2018;**124**:75-82

[39] Launay M, Caubel J, Bourgeois G, Huard F, de Cortazar-Atauri IG, Bancal MO, et al. Climatic indicators for crop infection risk: Application to climate change impacts on five major foliar fungal diseases in Northern France. Agriculture, Ecosystems & Environment. 2014;**197**:147-158

[40] Jacobs A, Kaiser K, Ludwig B, Rauber R, Joergensen RG. Application of biochemical degradation indices to the microbial decomposition of maize leaves and wheat straw in soils under different tillage systems. Geoderma. 2011;**162**(1-2):207-214

[41] Joergensen RG, Wichern F. Quantitative assessment of the fungal contribution to microbial tissue in soil. Soil Biology and Biochemistry. 2008;**40**(12):2977-2991

[42] Ritchie F, Bain RA, McQuilken MP. Effects of nutrient status, temperature and pH on mycelial growth, sclerotial production and germination of Rhizoctonia solani from potato. Journal of Plant Pathology. 2009;**1**:589-596

[43] Boine B, Renner AC, Zellner M, Nechwatal J. Quantitative methods for assessment of the impact of different crops on the inoculum density of Rhizoctonia solani AG2-2IIIB in soil. European Journal of Plant Pathology. 2014;**140**(4):745-756

[44] Van Mantgem PJ, Stephenson NL, Byrne JC, Daniels LD, Franklin JF, Fulé PZ, et al. Widespread increase of tree mortality rates in the western United States. Science. 2009;**323**(5913):521-524

[45] Westerling AL. Increasing western US forest wildfire activity: Sensitivity to changes in the timing of spring. Philosophical Transactions of the Royal Society B: Biological Sciences. 2016;**371**(1696):20150178

[46] Hicke JA, Johnson MC, Hayes JL, Preisler HK. Effects of bark beetle-caused tree mortality on wildfire. Forest Ecology and Management. 2012;**271**:81-90

[47] Hicke JA, Allen CD, Desai AR, Dietze MC, Hall RJ, Hogg EH, et al.

Effects of biotic disturbances on forest carbon cycling in the United States and Canada. Global Change Biology. 2012;**18**(1):7-34

[48] Beres BL, Rahmani E, Clarke JM, Grassini P, Pozniak CJ, Geddes CM, et al. A systematic review of durum wheat: Enhancing production systems by exploring genotype, environment, and management (G× E× M) synergies. Frontiers in Plant Science. 2020:1665

[49] Ben-Yakir D, Fereres A. The effects of UV radiation on arthropods: A review of recent publications (2010-2015). VIII International Symposium on Light in Horticulture 1134 2016 (pp. 335-342).

[50] Romo CM, Tylianakis JM. Elevated temperature and drought interact to reduce parasitoid effectiveness in suppressing hosts. PLoS One. 2013;**8**(3):e58136

[51] Asseng S, Ewert F, Martre P, Rötter RP, Lobell DB, Cammarano D, et al. Rising temperatures reduce global wheat production. Nature Climate Change. 2015;**5**(2):143-147

[52] Long SP, Ainsworth AE, Rogers AR, Ort D. Rising atmospheric carbon dioxide: Plants FACE the future. Annual Review Plant Biology. 2004;**55**:59

[53] Uddin S, Löw M, Parvin S, Fitzgerald G, Bahrami H, Tausz-Posch S, et al. Water use and growth responses of dryland wheat grown under elevated [CO_2] are associated with root length in deeper, but not upper soil layer. Field Crops Research. 2018;**224**:170-181

[54] Stocker TF, Qin D, Plattner GK, Tignor MM, Allen SK, Boschung J, Nauels A, Xia Y, Bex B, Midgley BM. 2013. IPCC: Climate change 2013: The physical science basis. Contribution of working group I to the Fifth Assessment Report of the Intergovernmental Panel on Climate Change.

[55] Kimball BA. Crop responses to elevated CO_2 and interactions with H_2O, N, and temperature. Current Opinion in Plant Biology. 2016;**31**:36-43

[56] Rosenthal DM, Tomeo NJ. Climate, crops and lacking data underlie regional disparities in the CO_2 fertilization effect. Environmental Research Letters. 2013;**8**(3):031001

[57] McKenzie FC, Williams J. Sustainable food production: Constraints, challenges and choices by 2050. Food Security. 2015;**7**(2):221-233

[58] de Oliveira ED, Bramley H, Siddique KH, Henty S, Berger J, Palta JA. Can elevated CO_2 combined with high temperature ameliorate the effect of terminal drought in wheat? Functional Plant Biology. 2012;**40**(2):160-171

[59] Bahrami H, De Kok LJ, Armstrong R, Fitzgerald GJ, Bourgault M, Henty S, et al. The proportion of nitrate in leaf nitrogen, but not changes in root growth, are associated with decreased grain protein in wheat under elevated [CO_2]. Journal of Plant Physiology. 2017;**216**:44-51

[60] Alley R, Berntsen T, Bindoff NL, Chen Z, Chidthaisong A, Friedlingstein P, et al. Climate change 2007: The physical science basis. Contribution of Working Group I to the Fourth Assessment Report of the Intergovernmental Panel on Climate Change. Summary for Policymakers. Geneva, Switzerland: IPCC Secretariat; 2007. 21p

[61] Friedlingstein P, Houghton RA, Marland G, Hackler J, Boden TA, Conway TJ, et al. Update on CO_2 emissions. Nature Geoscience. 2010;**3**(12):811-812

[62] Kole C, Muthamilarasan M, Henry R, Edwards D, Sharma R, Abberton M, et al. Application of genomics-assisted breeding for generation of climate resilient crops: Progress and prospects. Frontiers in Plant Science. 2015;**6**:563

[63] Ahmed T, Scholz M, Al-Faraj F, Niaz W. Water-related impacts of climate change on agriculture and subsequently on public health: A review for generalists with particular reference to Pakistan. International Journal of Environmental Research and Public Health. 2016; **13**(11):1051

Chapter 9

Nitrogen Use Efficiency in Wheat: Genome to Field

Om Prakash Raigar, Kinjal Mondal, Mehak Sethi, Mohini Prabha Singh, Jasneet Singh, Archana Kumari, Priyanka and Bhallan Singh Sekhon

Abstract

Nitrogen (N), being the most limiting macroelement for optimal plant growth and development needs synthetic N fertilizer usage for uplifting crop yields; nevertheless, an excessive and inefficient use of N fertilizer is a global concern incurring high production costs, environment pollution, and greenhouse gas emissions. Hence, developing crop plants with high nitrogen use efficiency (NUE) is an essential research target to achieve a better agricultural sustainability. NUE being a complex trait depends on our understanding of genetics (G), environment (E), management (M), and their interrelationships (G x E x M). NUE improvement is preceded by key processes such as nitrogen capture, utilization efficiency, nitrogen partitioning, trade-offs between yield and quality aspects, as well as interactions with the capture and utilization of other nutrients. An in-depth knowledge can be attained on NUE mechanisms through the UK Wheat Genetic Improvement Network project (http://www.wgin.org.uk/) using an integrated strategy that look into the physiological, metabolic, molecular, and genetic aspects influencing NUE in wheat. The current book chapter highlights the recent progress in understanding and improving NUE in wheat, focussing on N impact on plant morphology and agronomic performances, using a combination of approaches, including whole-plant physiology and quantitative, forward and reverse genetics.

Keywords: wheat NUE, nitrogen transporters, NUE genes, root traits, N uptake

1. Introduction

Cereal crops are widely farmed across the world in comparison to other crops. Rice (*Oryza sativa* L.), wheat (*Triticum aestivum* L.), and maize (*Zea mays* L.) are the most significant cereals in terms of human nutrition, accounting for 90% of global grain output. Since the Green Revolution, the importance of cereal crops in world agriculture has expanded dramatically. Of them, wheat is well-known to redeem global protein and calorie demands, either directly or indirectly in animals [1]. A number of factors have been found to influence the quality and quantity of cereal crops produced across the world; nitrogen availability is one of them. All plants require an external supply of N as in inorganic form, it functions as a key component of biomolecules, such as proteins, nucleic acids, chlorophyll, and various secondary metabolites. Nitrogen availability is a limiting factor in agricultural activities, and roughly 100 $TgNyr^{-1}$ of reactive nitrogen is administered to crop fields in the form of fertilizers globally [1]. Total N fertilizer usage has increased globally, from 112.5 million tonnes in 2015 to 118.2 million tonnes in 2019. Nitrogenous fertilizer usage has been grown at a faster rate in various nations between 1970 and 2020. In cereals, yield is found to be closely related to nitrogen application [1]. Report says, more than 94 million tonnes of nitrogen fertilizer are applied to cereal crops each year, but unfortunately around only 40% of this is absorbed by the crops, with the rest dispersing in the environment, posing major environmental issues, such as water pollution and greenhouse gas emissions [2, 3]. Among all the cereal crops, barley shows the highest nitrogen recovery (63%), followed by maize (37%), wheat (35–45%), and rice (30–50%) [3]. Nitrogen recovery is affected by various factors, including crop type, ambient conditions, fertilizer type, management technique, and genotype-environment interactions. Fertilizer use will be anticipated to more than double by 2050, rising from 112 Mt in 2015 to 236 Mt in 2050 [4]. Around 50–70% of applied nitrogen is constantly lost in the plant-soil interaction. Overuse of commercially available fertilizers has led to pollution of the air, soil, and water, as well as depletion of natural resources, such as nutrients and water. Nitrogen builds in the soil when nitrogen availability exceeds crop nitrogen requirements, rendering plants sensitive to a variety of loss mechanisms. As a result, enhancing cereal crop resource use efficiency is necessary to mitigate the negative effects of higher output on the environment and natural resources. Improving nitrogen use efficiency (NUE) in cereals must be a goal in breeding efforts to lessen the impact of increased fertilizer usage on climate change and to manage sustainable feeding to the world's rising population. To deal with nitrogen application concerns in fields, it is critical to understand the underlying process of nitrogen usage efficiency.

The utilization of Nin plants requires multiple phases, such as the initial N intake phase, followed by nitrogen reduction to usable forms, amino acid assimilation, translocation, and lastly, nitrogen remobilization to reproductive organs **Figure 1** [5]. The grain yield per unit of nitrogen available in the soil is defined as NUE (nitrogen use efficiency) in the wheat crop **Figure 1** [6]. NUE analysis gives information on plant responses to diverse nitrogen availability conditions. Nitrogen use efficiency may be quantified using a variety of formulas and ideas. Cereal NUE is determined by how efficiently plants gather nitrogen (uptake efficiency, NUpE) and how efficiently plants use the nitrogen that has been taken up (utilization efficiency, NUtE) **Figure 1** [7]. NUpE is calculated by dividing the total amount of above-ground nitrogen content during harvest by the available N in the soil, whereas NUtE is calculated by dividing the nitrogen in grain tissues by the N in above-ground plant biomass at harvest (**Figure 1**). As a result, NUE is determined at harvest, i.e., at the conclusion of the crop cycle. The agronomic efficiency of plants evaluates the efficiency with which they convert applied nitrogen to grain yield, whereas the apparent nitrogen efficiency of plants absorbs nitrogen from the soil [8]. The physiological efficiency of plants is determined by the amount of nitrogen collected and converted to grain production. For major cereal crops, improving resource use efficiency is a must to mitigate the negative effects of greater yield with increased input consumption on the environment and natural resources. The challenge here is to pick the most fertilizer-sensitive stage, to create a plant that maximizes early nitrogen uptake, and to have qualities, such as early vegetative vigor and a large root system for effective fertilizer uptake, all while considering above and below ground components. Later in the growth phase,

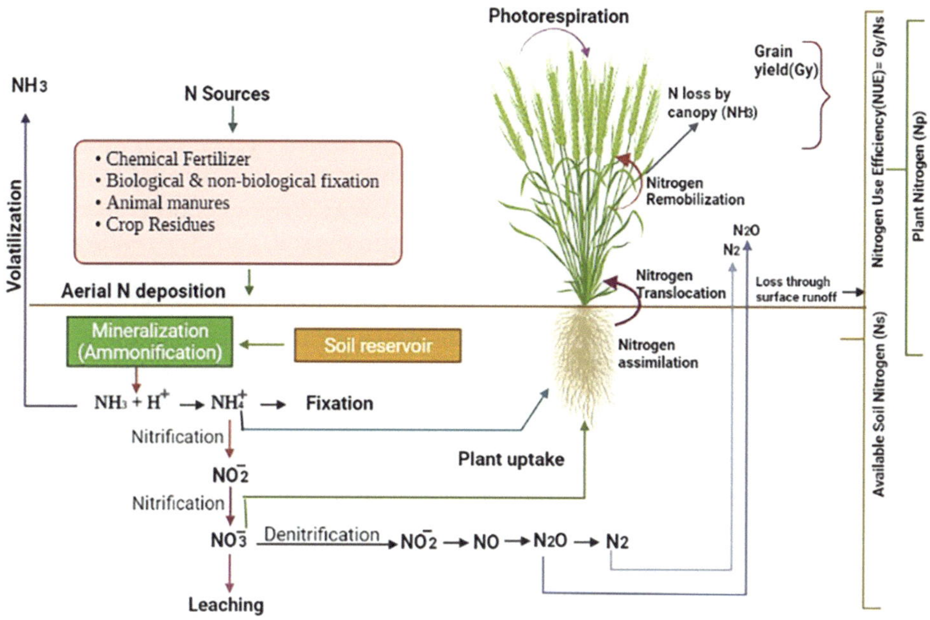

Figure 1.
Schematic representation of the relationship between the nitrogen sources, uptake, utilization, and conversion to the wheat grain yield.

a plant's ability to absorb and remobilize available nitrogen and carbon to the grain is crucial. Major issues include appropriate root phenotyping, genotype x environmental interactions, soil characteristics, water-nutrient management, and nutrient dynamics balance. The primary question is whether it is feasible to improve nutrient absorption while reducing excessive fertilizer input and safeguarding soil health while maintaining optimal production and grain protein content. Nanotechnology, particularly the use of nanofertilizers (1–100 nm in size), is helpful and has been shown to have positive outcomes, while a further study on the impact of nanofertilizers on specific crops is required [9]. Before delving into the biochemistry and genetics of nitrogen use efficiency improvement in cereal crops, it is necessary to comprehend the new potential source of nitrogen fertilizers, the effect of nitrogen at various stages of growth, the nitrogen status of the crop, and development and NUE in the effect of fertilizers [10]. Anhydrous ammonia (82% N), urea (46% N), ammonium nitrate (34% N), ammonium nitrate sulfate (26% N), and aqua ammonia (25% N) are among the fertilizer sources. Organic and inorganic nitrogen fertilizers are the two primary categories of nitrogen fertilizers. In terms of inorganic fertilizers, anhydrous ammonia application contributes the most nitrogen, i.e., greater than 80%. Aqua ammonia, also known as ammonium hydroxide, is the second most significant source of inorganic nitrogen fertilizers and comprises 25–29% ammonia by weight. Another type of nitrogen fertilizer is ammonium nitrate, which is an agronomically relevant mixture of two distinct types of nitrogen (NH_4NO_3). This type of fertilizer is said to improve wheat baking quality [11]. Urea [$CO(NH_2)_2$] is an organic kind of fertilizer [12].

The grain crop goes through numerous stages of development and growth. The rate of nutrient absorption in wheat varies with growth stage, variety, growing conditions, and environment. Detailed research of wheat's nutrient absorption mechanisms is required to determine the optimal time and exact stage of fertilizer applications.

Small amounts of nitrogen are required for seedling viability in the early stages. The mid-tillering stage uses almost half of the nitrogen required [13]. A high nitrogen dose, on the other hand, may damage seedlings and increase vegetative growth early in the season, resulting in poorer yields. Excess nitrogen might cause crop maturity to be pushed back. Nitrogen demand is said to be influenced by a number of factors, and NUE decreases when nitrogen application exceeds demand [14]. NUE is impacted by a number of variables [15], including soil type, the availability of other macro and micronutrients (phosphorus, potassium, etc.) in the soil, and crop rotation, which has been proven to affect nitrogen absorption and utilization [16]. Nitrogen fertilization is influenced by the intensity, timing, and depth of tillage [17, 18]. The most active subject of study to boost N fertilization yield is developing strategies for assessing nitrogen status. Satellite imaging [19], portable hyperspectral sensors [20], drones, chlorophyll meters (SPAD), red edge optical reflectance (R750/R710) [21], NDVI (normalized vegetation index), and RVI (ration vegetation index) [19] all offer the possibility of N estimation in less time.

Wild and primitive cereal crop species are currently undervalued as a source of unique nutrient utilization efficiency differences. Association studies exploiting the best alleles to be assembled in superior varieties, as well as the identification and characterization of candidate genes with non-synonymous and regulatory SNPs, will aid breeders in selecting specific donors to develop resource-efficient high-yielding wheat varieties. Furthermore, because yield and grain protein content, which represent nitrogen use efficiency, are inversely related, it is critical for breeders to design cultivation programs that achieve comparatively successful NUE without sacrificing grain yield [22], and it is critical to understand the details of various genetic, physiological, and biochemical factors affecting NUpE and NUtE to develop such cultivars.

Agronomic practices and field management also had a role in avoiding nitrogen loss to the environment [23]. The present chapter focuses on the myriad biochemical and genetic factors that influence NUE in both direct and indirect ways. The biochemistry of nitrogen absorption and utilization, as well as the genetic system that controls NUE in cereals and the phenotypic results that positively influence NUE, are all covered in this chapter. The associated cereals study will aid in the development of approaches for enhancing NUE while maintaining other desirable characteristics.

2. Traits affecting nitrogen utilization efficiency

2.1 Root architecture

Nutrient availability has a big influence on root growth and root system design. To present, little is known about the root architectural plasticity features, genetic foundation, mechanism, control, and function [24] linked to nutrient absorption. The root architecture is thought to be a key factor in NUE enhancement [25]. In cereal crops (wheat, rice, and maize), root systems can be separated into two types—embryonic (seminal roots) and post-embryonic roots (crown roots). The "steep, inexpensive, and deep" root architecture explains nutrient absorption, especially nitrogen absorption, rather well [26]. It specifies that main roots are responsible for obtaining nitrogen from deeper layers, whilst lateral roots with steep angles are responsible for covering a larger area of soil [27]. Lateral roots are said to be more vulnerable to biotic and abiotic stress, as well as fluctuating nitrogen concentration. Low nitrogen concentration promotes lateral root initiation in the early stages, while severe nitrogen deprivation prevents root emergence and elongation. In the soil, a high nitrate to ammonia ratio had a favorable effect on lateral root length [28].

Understanding the role of root traits in nutrient uptake and dissecting the genetic basis to maximize the potential to breed high-yielding resource-efficient varieties of cereal crops by using modern biotechnological and bioinformatic approaches is required to address the challenge of efficient nutrient uptake. It is critical to uncover the latent potential of root characteristics for enhancing nutrient absorption and identifying important marker correlations that may be used in molecular breeding to develop resource-efficient cultivars. The use of a suitable root prototype as well as strong marker-trait associations/QTL/candidate genes may help to solve the problem of nutrient insufficiency and inadequate nutrient absorption. Efforts to design a robust root system architecture that combines a variety of root traits (nodal root, root hair length, root hair density, root length density, root dry weight, percent lateral root, root branching, root thickness, and root volume) could be a solution to the problem of efficient nutrient uptake, particularly nitrogen (N) (**Figure 2**). The development of root architecture is said to be influenced by a number of elements both above and below ground [25]. At different phases of crop growth and development, different root characteristics are critical for nutrient absorption. Root size and morphological features are directly related to nitrogen uptake efficiency, as it has been observed that among the various forms of nitrogenous compounds present in the soil, particularly nitrate, easily escapes the soil system through leaching, implying the need to improve nitrogen uptake by improving root architecture, including depth, density, and capacity of roots for post-anthesis N uptake [29]. Although primary investigations in Arabidopsis were conducted to determine the molecular regulation of root architecture, multiple homologs in rice and other cereal crops have been found [30]. In wheat, the NAM, ATAF, and CUC transcription factors (TaNAC2-5A) stimulated root growth, whereas the NUCLEAR FACTOR Y (TaNFYA-B1) accelerated root development [31].

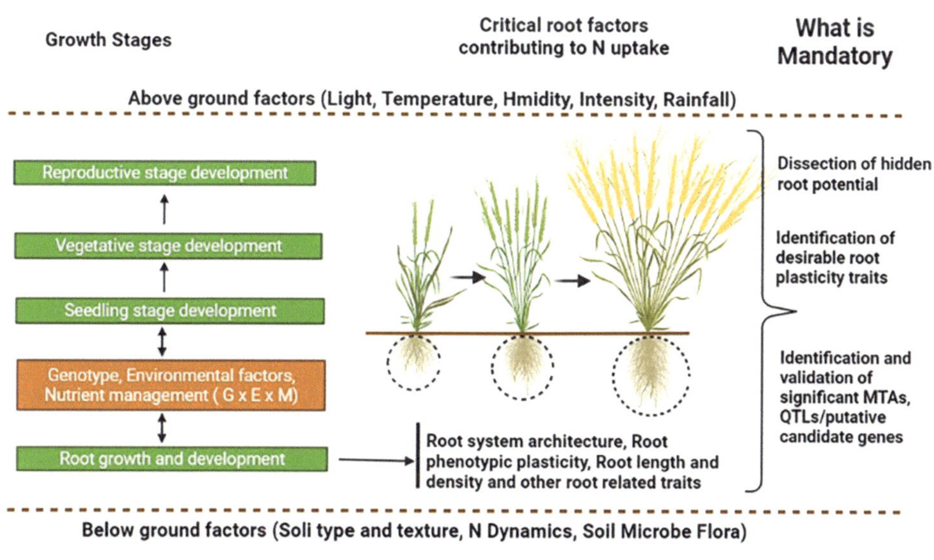

Figure 2.
Role of above and below ground N-affecting factors, G × E × M interactions playing significant roles in the development of root architecture at different stages of plant development.

In wheat, root growth was found to have an important role in increasing nitrogen absorption [32]. As a result, the rooting profile required for nitrate absorption at lower depths was investigated by measuring root length density at a threshold of 1 cm/cm^3 [33], where root length density is a measurement of root length per unit volume of soil [24]. Wheat roots showed a significant level of genetic diversity [24, 34]. Furthermore, a variety of environmental conditions, including soil type and nutrient availability, have a significant impact on root spreading characteristics. Deeper rooting systems have been observed in *Aegilops tauschii* (D genome), a wild cousin of wheat. The found candidate genes might be used in genomics-assisted breeding strategies to create cultivars with reasonably deep root systems. Under low nitrogen circumstances, an increase in the root biomass to total plant biomass ratio (root dry weight ratio; RDWR) was reported to preserve the functional balance between root and shoot development [35]. Even with a limited nitrogen supply, the increase in root-shoot biomass eventually increased crop growth rate (CGR), resulting in better grain production and improved NUE.

Along with root length and density, root hairs are an essential feature to consider for increased nitrogen absorption by active transport. Root hairs play a significant role in increasing the surface area of roots, which may boost nitrogen uptake by active transport. It is challenging to target specific genes for enhanced nitrogen absorption since root structure and function appear to be the result of the cumulative influence of numerous genes [36]. The strategy for increasing nitrogen absorption comprises marker-assisted selection and pyramiding numerous advantageous characteristics. The quantitative trait loci (QTL) for traits, such as root length, root hair number, root density, root angle, and root-to-shoot ratio, are well established in wheat [37, 38], but there is a need to understand the mechanism of orchestrated expression of multiple traits affecting root architecture to positively influence nitrogen uptake.

2.2 N transporter systems in roots

Nitrogen transporters for nitrate (NO^{3-}), ammonium (NH^{4+}), amino acids or peptides, and urea are involved in nitrogen absorption [39, 40]. Nitrogen accumulation by roots is an active process that is mediated by a specific type of nitrogen transport protein. The most common inorganic form of nitrogen in the rhizosphere is NO^-_3, NH^+_4 is also present in the soil, although at much lower concentrations than NO^{3-} [41]. The uptake and transport of nitrate in plants are mediated by five transporter families—the Nitrate Transporter 1/Peptide Transporter (NPF) family [42], the Nitrate Transporter 2 (NRT2) family, the Chloride Channel (CLC) family, the Slow Anion Associated Channel Homolog (SLC/SLAH) family, and aluminum-activated malate transporters (ALMT) [42]. Among the five families described above, NPF and NRT2 have been linked to nitrate absorption and plant localization.

Several kinds of plasma membrane-associated transporter proteins have been identified as being engaged in active transport and have been classed as high- and low-affinity transporters [43, 44]. In higher plants, three types of transport systems are active based on affinity and NO^{3-} content in the rhizosphere—inducible high-affinity transport system (iHATS), constitutively expressed high-affinity transport system (cHATS), and nonsaturable low-affinity transport system (LATS). iHATS is activated at low NO^{3-} concentrations (1–200 lM), and its activity varies depending on plant type and environmental conditions [45]. In wheat, iHATS has a Michaelis constant (Km) of around 27 lM and requires a 10-h induction time before commencing the transport process [46] cHATS, as the name implies, is constitutively produced and exhibited on the plasma membrane even in the absence of NO^{3-}. Both cHATS and iHATS have the trait of becoming saturated once the external NO^{3-} concentration reaches a particular threshold. The third, LATS, has low-affinity transporters and is activated when there is a high concentration of NO^{3-} in the soil (250 lM). Unlike cHATS and iHATS, LATS contains nonsaturable transporters [47]. *NRT1* and *NRT2* are two important gene families involved in NO^{3-} transport in higher plants. NRT1/PTR stands for nitrate transporters, the peptide transporter family (NPF), and the main facilitator superfamily (MFS) of the *NRT2* family [42]. In the absence of NO^{3-}, the plant growth hormone abscisic acid activates the high-affinity transport system in wheat, which is controlled by five genes (*TaNRT 2.1*, TaNRT 2.2, *TaNRT 2.3*, *TaNAR 2.1*, and *TaNAR2.2*) [48]. LATS belongs to the ammonium methylammonium permeases/transporter/Rhesus (MEP/AMT/Rh) family of NH^{4+} permeases and is implicated in NH^{4+} uptake among the three transporter systems examined so far [49]. The activity of these transporters is controlled by post-translational processes, such as phosphorylation, which maintains the quantity of ammonia stored in the plant system under control [25, 50]. Because urea absorption in wheat is so low relative to other inorganic nitrogen sources, determining the kinetics of urea uptake can be problematic [51]. Ammonia, nitrate, and urea are all known to affect the expression of high-affinity urea transporters [52]. However, because urea is mostly utilized as a nitrogen fertilizer in Asian agriculture, further research into the process of urea absorption and metabolic conversion to beneficial physiological components in plant systems is needed.

2.3 Effect of rhizospheric associations

The rhizosphere is the area of the soil that comes into direct contact with the root system, and the organisms that dwell there have a substantial influence on mineral intake, particularly nitrogen uptake by roots [53]. Many higher plants, including wheat, are believed to emit a variety of exudates, including organic acids and sugars, that have a direct influence on the physiological activities of microbes in the root system [54]. Several environmental factors, including climate, water level, soil type, and agricultural practices, also have an influence on these microbial communities [55]. The microbial ecology of the rhizosphere has also been discovered to differ among wheat cultivars [56, 57]. Through the denitrification process, several bacteria minimize nitrogen consumption by converting inorganic nitrates to gaseous nitrogen [58]. As previously stated, denitrification converts nitrogen into an inaccessible form, hence suppressing such processes improves nitrogen absorption; nevertheless, the mechanism in cultivated cereal crops is not well-known [59]. Several attempts have been made to transfer beneficial root-microbial traits from wild relatives of domesticated cereal crops to domesticated cereal crops. A chromosome from Leymusracemosus, a wild wheat relative capable of preventing nitrification in the root rhizosphere, was transferred into cultivated wheat varieties [60, 61].

Improved nitrogen fixation can boost root nitrogen absorption. Although these nitrogen-fixing bacteria are a natural component of the wheat root rhizosphere [62, 63], the artificial introduction of N fixers may increase nitrogen intake, which has a favorable effect on production [64, 65]. The main option for introducing the legume-like system of nitrogen fixation from bacteria to cereal crops is genetic engineering [66]. The non-host-specific endophyte *Pseudomonas stutzeri* and epiphyte

Klebsiella oxytoca, which infiltrate the root systems of rice and wheat, are the most effective strains for nitrogen fixation [67]. Bacterial systems have a wide variety of *nif* gene clusters, ranging in size from 11 to 64 kb operons. The conserved section in these operons comprises nitrogenase (nifHDK) and cofactor (FeMoCo), while the rest of the operon specifies nitrogen fixation under various environmental circumstances [68]. Auxins [69], cytokinins [70], and gibberellins, all of which are regulated by microorganisms in the rhizosphere, were found to impact root architecture by boosting the production of growth hormones. Gibberellins produced by the rhizospheric bacteria and fungi have been reported to boost the primary root elongation and lateral root growth in wheat [71]. Root-associated organisms influenced nitrogen uptake as well as the activation of plant defense systems against pathogenic infections [72, 73]. The pathogenic defense-related transcriptional accumulates in wheat when infected with the bacterium *Pseudomonas fluorescens* Q8r1-96 [74]. Overall, the microbial association with nitrogen absorption is a broad issue that must be investigated and explored to improve nitrogen uptake efficiency in wheat and other cereal crops.

3. Traits associated with nitrogen utilization efficiency

3.1 Nitrate assimilation

Nitrate is an essential component of the nitrogen cycle and a major player in inorganic nitrogen assimilation in cereals [75]. The nitrate assimilation is primarily driven by its reduction and incorporation of nitrogen into the carbon skeleton to generate biologically active, organic nitrogen form. Nitrate uptake in plants is root transporter-mediated, inside root cells nitrate is targeted by nitrate reductase (NR) enzyme along with NAD(P)H cofactor. NR is the key enzyme that is involved in the very first step of nitrogen utilization and its conversion into biologically active molecules. It is reported that in hexaploid wheat two genes encode the NADH-dependent nitrate reductase [76]. NR leads to the conversion of nitrate into nitrite. Nitrite is further reduced to ammonia by the action of enzyme nitrite reductase (NiR) which is usually present in plastids of the plant cell [77]. In the case of NiR, ferredoxin is associated with NiR and the electrons for reduction are provided by ferredoxin [78]. Ammonia released by the action of NiR is used for amino acid formation. The primary amino acid involved in ammonia incorporation is glutamate. Glutamine synthetase (GS) and glutamate synthase (GOGAT) are the two enzymes that act in conjugation for amino acid formation [79]. GS is present in two isoforms in different cellular organelles. GS1 is prevalent in the cytosol of plant cells and GS2 works in plastids of roots and etiolated tissues [79]. It is reported that in wheat, the expression of GS2 is uniform throughout the plant development and comes to a halt toward maturity, and the expression of GS1 isoenzyme is consistently observed in senescing tissues and phloem [80]. Second enzyme GOGAt or glutamate synthase works with the primary enzyme in the formation of two amino acids glutamate and glutamine. These two amino acids are further involved in amino acid, nucleic acid formation by acting as donors of the amino group for nitrogen-containing compounds [79]. Two isomeric forms of GOGAT are present in the plant system. Both isoforms vary in terms of cofactors that they use and the process they are involved in. One is FD-GOGAT; this form is ferredoxin dependent; it is involved in the reassimilation of ammonia released from the cycle of photorespiration. The second isoform of GOGAT is NADH dependent which is primarily involved in amino acid synthesis which is channelized for protein formation involved in the growth and development of photosynthetic and accessory organs [79]. Almost 95% of ammonia availed by plants is dependent on GS and GOGAT as reported from several mutational studies [79]. These amino acids lead to increased protein formation which ultimately enhances productivity [81].

3.2 Carbon metabolism in N assimilation

Multiple factors are reported to affect nitrogen assimilation but carbon metabolism is the major player having direct interaction with nitrogen metabolic pathways. The role of photosynthesis on nitrogen accumulation was analyzed in different target plants to dissect the interaction between carbon and nitrogen metabolic pathways. It was observed that nitrogen assimilation was changed when the photosynthetic rate was changed and vice versa. This is so because carbon fixation requires enzymes, such as RuBISCO, and as enzymes are protein a continuous flow of amino acid is needed for enzyme formation which further depends upon nitrogen availability [82]. So, nitrogen is critically important as it affects the photosynthetic activity which further regulates crop yield. Along with it, nitrogen metabolism is dependent on carbon metabolism as most of the enzymes involved in nitrogen metabolism need electron donors for their activity which is provided by carbon metabolism. Along with it, the GS/GOGAT pathway requires a carbon skeleton (Ketoglutarate) for ammonia assimilation which is the product of the TCA (tri carboxylic acid) cycle, an important regulator of carbon metabolism. So, carbon skeleton and other accessory elements needed for nitrate assimilation are provided by the carbon cycle [83]. So, overall nitrate assimilation is an interlinked metabolic pathway where several factors of carbon metabolism are critically related. Therefore, NUE is affected directly by components of nitrogen metabolic pathways and indirectly by players of carbon metabolism [75]. So, while targeting breeding programs for enhanced NUE enzymes and proteins associated with nitrogen and carbon metabolism can be targeted.

3.3 Photosynthesis and canopy traits

As discussed earlier, carbon fixation is an important process of plant growth and development. Rubisco is the major enzyme regulating the most critical step of Calvin cycle. Rubisco is the most abundant protein in the biosphere. The nitrogen accumulated by the plant is directly related to the amount of Rubisco formed which further defines the photosynthetic activity of the mesophyll cells. Almost 75% of N in wheat leaves is driven toward Rubisco enzyme formation which is important for photosynthesis [84, 85]. It is reported that in nitrogen-limited conditions, Rubisco content decreases which lead to reduced photosynthetic activity and reduced organic matter production. It is observed that photosynthetic activity is associated with leaf morphogenesis as it is the main region for carbon fixation. Leaf structure and canopy directly affect the yield output in crop plants [86]. High NUE increases the nitrogen uptake and utilization which enhances source and sink abilities and increases dry matter output and crop yield. The theory of optimization for canopy photosynthesis indicates that the coefficient of both light gradient (KL) and nitrogen (KN) positively contributes to photosynthesis [86]. Although the gradients for nitrogen observed in wheat were less steep than optimization theory [86]. Nitrogen utilization is majorly affected by the photosynthetic rate per unit of nitrogen. In light-saturated conditions, the photosynthetic rate was increased by 20–30 lmol $CO_2/m^2/s$ for around 2 g N/m^2 in C_3 crops, such as wheat. The important aspect to target nitrogen utilization efficiency is to identify wheat cultivars with the capacity of accumulating around 2.0 g N/m^2 under favorable conditions. A wide range of genetic variability was observed among various wheat lines specific leaf nitrogen (SLN) which is an indicator of leaf nitrogen content per unit leaf area. In earlier, Araus et al. [87] were grown a panel of 144 durum wheat genotypes in two rain-fed conditions and 125 of these were grown under supplementary irrigation before heading stage, and revealed that the SLN in these genotypes varied from 1.4 to 2.6 g/m^2. Another study by Giunta et al. [88] reported that SLN varied from 2.1 to 2.4 g/m^2 for the 17 durum wheat cultivars. A study in 16 bread wheat cultivars SLN varied from 1.4 to 2.2 g/m^2 [86]. The nitrogen content in different tissues, including stem, leaf lamina, and leaf sheath, at anthesis show heritability of >0.60 under low nitrogen in winter wheat. So, these traits can be used in targeted breeding programs [89]. The genetic diversity associated with nitrogen utilization efficiency in wheat germplasm can be used to achieve the desired modification in photosynthetic components. It was reported earlier that around 30% improvement in photosynthesis can be attained by targeting photo-respiration, along with its other mechanisms contributing to 15–22% increase in photosynthetic activity [90]. There is a need to understand the intricacy of the molecular mechanisms affecting the pathways for leaf development, photorespiration, and majorly photosynthesis. The recent advancement in technologies for gene editings, such as CRISPR-Cas9 or specific promoter expression can be used in regulating pathways for leave development. This can generate diverse germplasm with high NUE and ultimately high yield potential [91].

3.4 N remobilization and senescence

Nitrogen distribution in the plant is source-sink relation dependent. Initially nitrogen uptake by roots acting as source and transpiration of absorbed nitrogen from roots to leaves and buds acting as major sink organ. This source-sink relation changes with the plant's developmental stage, as it is observed that toward maturity the capability of the plant for nitrogen uptake decreases so the root does not act as a major source of nitrogen for the rest of the plant. During maturity, the leaf acts as a source, as toward senescence the old leaves die off and their protein components are degraded to release nitrogen which is remobilized to the younger leaves [92]. Leaf lamina is a major storage house of nitrogen in above-ground tissue during anthesis in wheat under optimal N supply. Other tissues, such as true stem, ear, and leaf sheath, also retain nitrogen [93], whereas the trend of nitrogen accumulation changes under nitrogen-limiting conditions, with more nitrogen in ears as compared to other parts of the plant [93]. Although, the NUE is majorly determined by nitrogen remobilization from leaves to its developing parts during the grain-filling stage which further defines the crop yield. So, during the grain-filling stage, the photosynthates and proteins stored in the older leaves act as a major source of nutrients for developing seeds. Autophagy is the basic mechanism that affects remobilization during the grain-filling stage. Autophagy is programmed cell death for the regulated release of stored compounds which is regulated by senescence-associated genes (ATG and metacaspases) [94]. Specific tissue-specific transporters are activated during the reproductive stage which is important for nitrogen remobilization. NRT1.7 is an important nitrogen transporter and its gene is reported to be controlled by nitrogen limitation adaptation regulators which are further under the control of miRNA827 [95]. This double-level control over tissue-specific nitrogen transporters suggests that the remobilization of nitrogen is tightly regulated. The remobilization process is under multiple regulatory controls along with transporters the enzymes, such as GOGAT, are reported to be involved in ammonia recycling during remobilization [96]. Along with its certain transcription factors, such as NAM-B1, efficiently increase nitrogen remobilization toward grains from mature leaves in wheat [97]. As in the case of cereals grain nitrogen, almost 50–90% is contributed by nitrogen from leaves [5]. The stage of nitrogen remobilization in grains from flag leaves can be used as a phenotypic marker [97]. As it is established that an inverse relation exists between grain yield and grain protein content, so higher grain yield is associated with delayed senescence of flag leaf in cereals. Among multiple proteins present in the leaf during senescence, the Rubisco (the most abundant protein in the biosphere) acts as a major contributor to remobilized nitrogen. In older leaves, chloroplast is degraded first as compared to other cellular components because of upregulation of proteases enzymes [98]. The tissue breakdown in older tissue is programmed by autophagy (chloroplast and Rubisco degradation) by the action of exopeptidases and endopeptidases present in cell vacuoles during senescence [98].

3.5 Stay-green phenotype

The stay-green phenotype is a marker for the tendency of a genotype to remain green during the grain-filling stage. The plants with stay-green phenotype remain photosynthetically active after anthesis [99]. Stay green-phenotype is a trait of interest to enhance NUE in plants and a wide range of genetic diversity is reported for this trait in hexaploid wheat [100]. Along with stay-green phenotypes traits, such as Rubisco degradation, and stem nitrogen assimilation are important targets for efficient nitrogen remobilization to the grains post-anthesis. The target of high yield with balanced protein content in wheat depends on an in-depth understanding of the mechanisms affecting post-anthesis nitrogen accumulation and remobilization toward developing grains.

3.6 Grain yield and grain protein content

In cereals, endosperm contributes to the maximum nutritive value of the grain due to its size ratio as compared to germ. The metabolic composition of endosperm is very essential for grain with high nutritive value. In cereals, starch is the prevalent biomolecule, along with its protein is also present with starch. Among different storage forms, Gluten is the major storage fraction of endosperm. Glutens have two components polymeric glutenins and monomeric gliadins. This storage protein contributes to 60–70% of the nitrogen in seed endosperm. Glutens provide the dough-making properties to wheat. Gliadin is responsible for dough viscosity and glutelins ensure dough elasticity. This dough-making capacity is important for consumable products of wheat, including pasta, bread, and noodles. The gluten synthesis is dependent on the protein accumulation which depends on the nitrogen utilization efficiency. Grain protein quality changes under different genetic backgrounds in wheat [101, 102]. Grain protein content and grain yield are both affected by NUtE although they are inversely related to each other [22, 103] which creates a barrier in attaining both simultaneously. The inverse relation between grain yield and grain protein content is due to metabolic competition between carbon and nitrogen fluxes for biomolecule accumulation [104], so dilution in NUtE depends on the accumulation of carbon-based compounds [105]. The efficient nitrogen in grain can be calculated by calculating grain protein deviation (GPD). GPD is a measure of deviation from the regression between grain protein concentration (GPC) and grain yield. Identification of genotypes with higher GPC from an expected GY can be estimated by calculating GPD [106]. In cereals, grain yield is dependent on coordinated regulation between several factors, majorly competition between photosynthesis and photorespiration [107]. The correlation between yield and nitrogen uptake and utilization is important for high wheat yields. There is a need to completely understand the mechanisms and regulatory pathways for nutrient uptake and its transport to stems, sheaths, leaves, and finally to developing grains. Along with this, it is important to understand the mechanisms for improvement of slow and ineffective filling of grains [108].

4. Genetic factors

The number of genetic factors is associated with controlling NUE traits for cereal crops that include majorly six categories—transporters, signal molecules, amino acid biosynthesis, nitrate assimilation, transcription factors, and other genes. The upregulation and downregulation of these genes depend on nitrogen levels in the environment and thus are controlled by mechanisms as discussed in the following text.

4.1 QTLs related to NUE

One of the complex quantitative traits is nitrogen use efficiency (NUE) which is controlled by multiple genes and dissected using a powerful tool called QTL mapping [109, 110]. A successful QTL mapping for such a complicated trait relies on various factors, such as the selection of suitable parents, appropriate population size, multi-location testing, and genome coverage. QTL is conventionally affected by environmental variation where constitutive QTL is consistent over environments, while adaptive QTL shows an expression in a specific environment, or modulates its effect with a change in an environment. QTL analysis provides ample opportunities to identify correlations among different traits. A genetically and functionally linked trait is evident through co-localized QTL linked to phenotypically different traits.

Nitrogen use efficiency of cereal crops can be improved by employing classical genetics involving both conventional breeding and QTL mapping in combination with marker-assisted selection (MAS). To develop genomic knowledge for complex genomes of cereal crops, such as wheat, advances in next-generation sequencing and agronomically relevant traits can now be identified [111]. Wheat improvement could be heightened with the identification of cheap, easy-to-use, widely distributed, codominant, trait-associated, and regulatory SNPs, candidate genes, and regulatory pathways. Association mapping studies assist in accessing allelic diversity and identifying the best alleles to be assembled in superior varieties. Accuracy for identifying QTL for nitrogen uptake and utilization-related traits can be improved by using high-throughput genotyping techniques. In this regard, several promising means have also been proposed, such as focusing on root architecture [112] or senescence and remobilization [113].

Previous case studies reported various QTLs for NUE in the model crop plant, i.e., *Arabidopsis*as well as in other cereals, such as maize, rice, and wheat [25, 114]. Significant QTLs were detected in the wheat RIL population (TN18 × LM6) for grain yield; root NUE and shoot dry [115]. A major QTL was observed on the short arm of chromosome 6B controlling grain protein content in wheat accounting for 66% of the phenotypic variation where the cloning of functional gene named Gpc-B1 was carried out [97]. Various novel NUE-related traits and alleles in adapted breeding materials

[116], landraces [117, 118], and wheat wild relatives [119] were identified. One such report is on winter wheat where the QTL associated with NUE on chr 1D, 6A, 7A, and 7D with LOD scores ranging from 2.63 to 8.33 and phenotypic variation up to 18.1% were instigated [120].

The identification of genomic regions (QTL) associated with nitrogen response would enable more efficient cultivar selection [121]. This approach allows breeders to proficiently develop high nitrogen use efficient cultivars by screening germplasm and studying the genetic markers associated with nitrogen response. As per previous work on rice and wheat, identification of the novel traits, alleles, genes/QTL, adapted breeding lines, landraces, and wild relatives improving NUE differences in cereal crops were well established. Using bi-parental populations, genes/QTL influencing nitrogen uptake have been mapped in wheat under different doses of fertilizer application [122, 123]. Genome-wide association studies for nitrogen uptake and use efficiency associated with variability and marker-trait selection have been reported [95, 124]. The development of synthetic wheat introgression libraries through Genome-wide association studies (GWAS) was made possible at Punjab Agricultural University, Ludhiana (India) to exploit their phenotypic variability. Several marker-trait associations related to root and plant morphological traits, grain yield, and yield-related traits have been well documented. Other than wheat, rice also shows highly conserved sequences, new genes, and regulatory elements to link genomes, genes, proteins, and traits controlling traits of interest across different species and genera through comparative mapping. These inter-genome relational patterns can lead to new hypotheses, knowledge, and predictions about the related species and can pave the way for genetic gain for future cereal crops.

5. Genes related to NUE

Regulation of nitrogen utilization efficiencies, such as nitrogen absorption, accumulation, and remobilization, is controlled by multiple sets of genes in crop plants (wheat, rice, etc.). These genes are majorly classified into six categories, including transporters, signal molecules, amino acid biosynthesis, nitrate assimilation, transcription factors, and other genes. The detailed description of genes regulating nitrogen use efficiency in wheat crops is presented in **Table 1**. Transporters and nitrate assimilation genes actively participate in nitrogen uptake, while amino acid biosynthesis genes are involved in nitrogen utilization. On the contrary, signaling molecules, transcription factors, and other genes have a passive role in both nitrogen uptake and nitrogen utilization [126, 127]. Nitrate, being the most common form of nitrogen present in soil needs to be transported in a plant which is done by nitrate transporters that encode for NRT families. The first reported case was studied in *Arabidopsis* where NRT families were categorized into two subfamilies, i.e., the NRT1 family (low-affinity transporters) and the NRT2/NRT family (high-affinity transporters) [128]. Using the reciprocal best hit (RBH) approach, orthologs of NRT transporter genes were found in cereal crops. It was observed that around 16 low-affinity nitrate transporter NPFs showed their expression in wheat which was homologous to *Arabidopsis* NPFs [125]. For an expression of transporter in wheat, information on the nitrogen status of the plant and soil is a prerequisite, thereby indicating its role in the regulation of NPF genes in wheat.

Nitrate transporters, although are the main players in nitrogen uptake in most plants, in certain cases, such as rice, ammonia is the predominant form in the soil. Nitrogen uptake is followed by nitrogen assimilation. A crucial metabolic step

Category	Gene	Chr	Location	IWGSC Gene ID	References
Nitrogen transporters	*TaNPF6.2*	1A	373,766,258–373,768,702	*TraesCS1A02G210900*	[125]
	TaNPF6.5	1A	14,519,757–14,525,659	*TraesCS1A02G031300*	[125]
	TaNPF7.7	1A	355,624,056–355,628,073	*TraesCS1A02G197600*	[5]
	TaNPF7.7	1B	385,644,930–385,648,470	*TraesCS1B02G212200*	[5]
	TaNPF7.7	1D	284,040,636–284,044,349	*TraesCS1D02G201100*	[5]
	TaNPF2.3	2A	17,869,278–17,871,731	*TraesCS2A02G045500*	[125]
	TaNPF1.1	3A	540,654,271–540,656,804	*TraesCS3A02G304400*	[125]
	TaNPF2.4	3A	660,436,466–660,444,074	*TraesCS3A02G418700*	[125]
	TaNRT2.3	3B	457,633,984–457,635,782	*TraesCS3B02G285900*	[5]
	TaNRT2.3	3D	356,623,041–356,624,585	*TraesCS3D02G254900*	[5]
	TaNAR2.1	4A	640,232,228–640,233,158	*TraesCS4A02G367300*	[5]
	TaNRT1	4B	483,508,916–483,514,108	*TraesCS4B02G231500*	[5]
	TaNPF2.1	5A	3,085,412–3,088,853	*TraesCS5A02G004400*	[125]
	TaNPF2.2	5A	34,980,804–34,986,700	*TraesCS5A02G037900*	[125]
	TaNPF6.6	5A	599,204,895–599,208,619	*TraesCS5A02G409600*	[125]
	TaNPF6.1	6A	486,547,388–486,550,355	*TraesCS6A02G263500*	[125]
	TaNPF7.1	6AL/BL/DL	486,547,388–486,550,355	*TraesCS6A02G263500*	[125]
	TaNAR2.2	6B	415,788,848–415,790,024	*TraesCS6B02G238700*	[5]
	TaAMT1.2/1.3	6B	458,486,050–458,487,918	*TraesCS6B02G254800*	[5]
	TaNAR2.2	6D	267,236,634–267,237,837	*TraesCS6D02G193100*	[5]
	TaAMT1.2/1.3	6D	293,801,873–293,803,683	*TraesCS6D02G208200*	[5]
	NRT1 PTR	7A	169,020,411–169,025,550	*TraesCS7A02G206400*	[42]

Category	Gene	Chr	Location	IWGSC Gene ID	References
	TaLHT1	7A	109,262,804–109,265,004	*TraesCS7A02G156600*	[5]
	TaNRT2.4	7B	583,923,053–583,926,829	*TraesCS7B02G328700*	[5]
N assimilation	*TaNiR1*	6B	636,392,631–636,397,024	*TraesCS6B02G364600*	[5]
	TaNiR1	6D	422,078,484–422,081,985	*TraesCS6D02G313100*	[5]
Amino acid biosynthesis (glutamine synthase)	*TaAlaAT10–1/TaAlaAT-4*	1A	71,689,760–71,695,155	*TraesCS1A02G085600*	[5]
	TaASN2	1A	553,535,726–553,542,082	*TraesCS1A02G382800*	[5]
	TaASP6	1A	287,681,550–287,684,692	*TraesCS1A02G160200*	[5]
	TaAlaAT10–1/TaAlaAT-4	1B	112,748,629–112,753,960	*TraesCS1B02G102700*	[5]
	TaASN2	1B	635,920,024–635,926,285	*TraesCS1B02G408200*	[5]
	TaASP6	1B	317,791,804–317,795,107	*TraesCS1B02G176400*	[5]
	TaASP6	1D	221,915,283–221,918,343	*TraesCS1D02G157400*	[5]
	TaGOX4	2D	301,816,850–301,819,891	*TraesCS2D02G251800*	[5]
	TaASP4	3A	541,257,235–541,261,301	*TraesCS3A02G305400*	[5]
	TaASP4	3B	536,074,881–536,079,450	*TraesCS3B02G331100*	[5]
	TaGOGAT1/3	3B	481,595,302–481,606,660	*TraesCS3B02G299800*	[5]
	TaGOGAT1/3	3D	369,790,549–369,802,074	*TraesCS3D02G266400*	[5]
	TaASN1	4B	417,737,785–417,741,607	*TraesCS4B02G194400*	[5]
	TaGS1	4B	499,898,695–499,901,767	*TraesCS4B02G240900*	[5]
	TaGGT2	4B	573,273,107–573,276,702	*TraesCS4B02G288100*	[5]
	TaGGT3	4B	363,644,060–363,647,074	*TraesCS4B02G167100*	[5]
	TaAlaAT10–2	5B	74,659,823–74,670,378	*TraesCS5B02G066600*	[5]
	TaAS	5B	107,190,378–107,196,256	*TraesCS5B02G084600*	[5]

Category	Gene	Chr	Location	IWGSC Gene ID	References
	TaGDH1	5D	494,216,160–494,219,691	*TraesCS5D02G442000*	[5]
	TaASP1	6B	668,432,728–668,437,537	*TraesCS6B02G393600*	[5]
	TaGS1	6B	577,183,711–577,187,787	*TraesCS6B02G327500*	[5]
	TaGS1	6AL/BL/DL	531,394,366–531,398,363	*DQ124209;DQ124210; DQ124211*	[125]
Transcription factors	*TaNF-YB2.1*	1A	572,334,701–572,336,969	*TraesCS1A02G411700*	[5]
	TaNF-YB2.1	1B	662,783,949–662,786,278	*TraesCS1B02G442000*	[5]
	TaNF-YB2.2	3B	605,665,548–605,668,470	*TraesCS3B02G385600*	[5]
	TaNF-YB2.2	3D	458,624,044–458,626,934	*TraesCS3D02G347000*	[5]
	TaHLHm1	4B	639,452,139–639,453,299	*TraesCS4B02G345800*	[5]
	TaFBX94	5B	133,417,326–133,419,111	*TraesCS5B02G100300*	[5]
	TaHLHm4	5B	13,081,769–13,086,120	*TraesCS5B02G013000*	[5]
	TaHLHm4	5D	13,313,304–13,318,505	*TraesCS5D02G020600*	[5]
	TaNAC9/NAM	6B/1B	51,579,298–51,580,659	*TraesCS6B02G075200*	[5]
Other genes (kinases)	*TaSAPK6*	1A	381,819,326–381,822,599	*TraesCS1A02G215900*	[5]
	TaSAPK6	1B	411,987,863–411,990,884	*TraesCS1B02G229400*	[5]
	TaSAPK6	1D	304,838,300–304,841,343	*TraesCS1D02G218200*	[5]
(Rubisco)	*Rbcs*	2A	171,076,784–171,079,172	*TraesCS2A02G198700*	[89]

Table 1.
Genes associated with nitrogen use efficiency in wheat.

regulating the grain yield and NUE is the nitrogen uptake followed by nitrogen assimilation in the form of amino acids which is usually carried out by glutamine synthetase (GS)/glutamate synthase (GOGAT) cycle. Increased GS1 activity is observed in the leaves of wheat crop directing an accumulation of nitrogen in grains and also enhanced dry grain matter. At high N content, the GS1 gene gets overexpressed thereby enhancing the nitrogen harvest index and NUE while at low N content, NUE does not change. Nitrogen remobilization is the last step in nitrogen use efficiency (NUE) for seeds during maturity. Generally monocots, dicots, C_3, and C_4 plants share a common mechanism for the nitrogen remobilization [5]. Asparagine and glutamine are common amino acid transport forms for nitrogen remobilization from leaves to reproductive tissues catalyzed by enzymes GS and GOGAT, respectively [129]. In durum wheat, asparagine synthetase encoding genes (AsnS1) are prominent for nitrogen remobilization from flag leaf to developing grains where their concentration increase in phloem sap during senescence of leaves [130]. Leaf senescence affects high yield in cereal crops as even though delayed leaf promotes prolonged photosynthesis for improving grain yield, it however decreases nitrogen remobilization efficiency and grain protein content [5].

6. Transcription factors concerned with NUE

Plant regulatory network is governed by transcription factors and like several other metabolic processes, NUE imperatively relies on coordinated transcription factors presented in **Table 1** [131]. Transcription factors for lateral root growth in response to nitrate belong to the MADS-box family analogous to ANR1, a transcription factor reported in *Arabidopsis* [132]. It is reported that DOF1.3 (DNA-binding with one finger) gene gets overexpressed in wheat under stress conditions, such as nitrogen starvation [132]. Differential expression studies between nitrogen-stressed and control durum wheat tissues are controlled by a total of 170 unique genes encoding transcription factors belonging to different families, including bHLH (helix loop helix), MYB, bZIP, C2C2-Dof, TERF, WRKY, NF-Y, NAC, AUX/IAA, and the auxin-modulated ARF, etc.

7. miRNA involved in NUE

miRNAs have been reported to play a significant role in NUE along with several transcription factors. The miRNA169 family is instigated to regulate the expression of genes for nitrogen transport in durum wheat under the nitrogen starvation stage in both roots and leaves [133]. In a recent study on the durum wheat plant, ttu-miR169h and ttu-miR169c at the seedling and grain-filling stages and ttu-novel-61 belonging to the miR169family showed down-regulation under nitrogen-deficient conditions in both roots and leaves. These miRNAs negatively regulate the CCAAT box-binding transcription factors in several tissues influencing NUE-related genes in durum wheat plants [133]. Another report indicated the role of the NAM-B1 gene in bread wheat as a NAC transcription factor that affects the grain nutrient concentration as well as increases the remobilization of nutrients from leaves to developing grains in wild wheat [134].

At low nitrogen levels, upregulation of TaMIR1129, TaMIR1118, and TaMIR1136 and downregulation of TaMIR1133 in roots of wheat were reported. The miRNA expression was inversely proportional to the concentration and duration of nitrogen

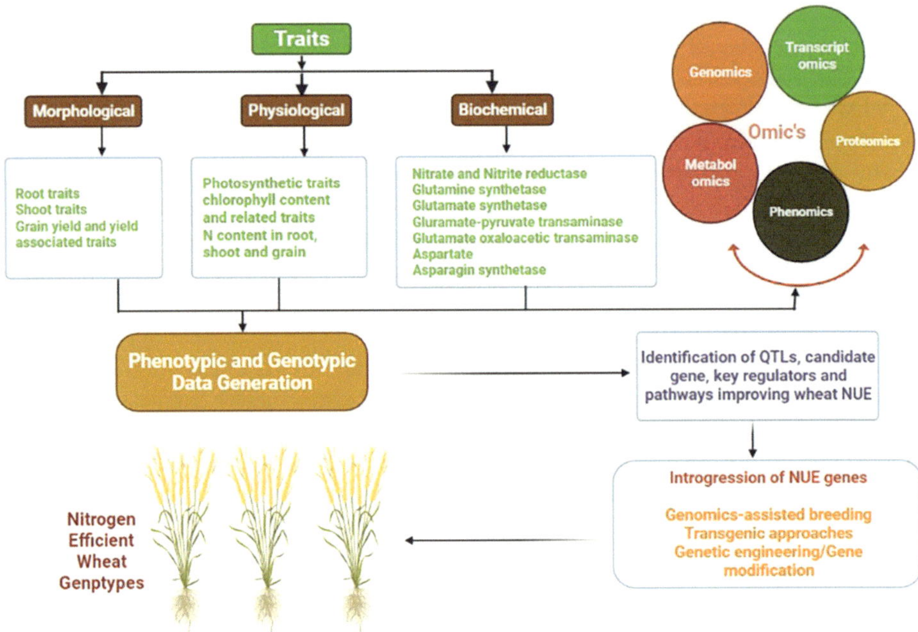

Figure 3.
Schematic representation of flow work to the development of nitrogen-efficient wheat genotypes.

application [135]. A gradual uprise in the expression of TaMIR2275 during nitrogen starvation was observed which was restored progressively once nitrogen level is recovered. Overexpression of produced plants with increased nitrogen accumulation and biomass is obtained from overexpression of TaMIR2275, while knockdown mutants showed the reverse. Inevitably, several classes of miRNAs are involved in nitrogen metabolism by altering multiple processes associated directly or indirectly with NUE. To comprehend, it is crucial to have a deep understanding of the precise network of miRNA expression and interaction for channelizing the mechanism underlying NUE.

The development of nutrient efficient varieties calls for the identification of suitable traits, and candidate genes underlying QTL that may provide new opportunities for the introgression of these QTL and genes into elite genetic backgrounds (**Figure 3**).

8. Conclusion

Immense use of nitrogen fertilizers even though uplift grain yields of cereal crops, negatively affect the environment by causing water, soil, air pollution, and greenhouse gas emissions. It thus poses an economic impact globally due to the high production costs of nitrogen fertilizer. To combat this, the challenge to improve NUE in cereal crops lies in achieving both high yield and high nitrogen use efficiency (NUE) simultaneously. Crop improvement can be achieved by improving our knowledge of agronomic management, suitable traits, QTL, genes, and the mechanisms and functions of genes associated with nitrogen use efficiency. Selection of diverse genotypes, exploitation of natural variation, exploring root architecture, high-throughput precise phenotyping, standardized field trials, new techniques for efficient fertilizer application, appropriate field management practices, and identification of new QTL/genes/nitrogen transporters, as well as signaling molecules, could contribute in reducing fertilizer consumption in the near future. Thus, an improvement in basic research in combination with agronomical, marker-aided molecular breeding and biotechnological strategies will help to achieve higher nitrogen use efficiency in cereal crops.

Author details

Om Prakash Raigar[1*], Kinjal Mondal[2], Mehak Sethi[1], Mohini Prabha Singh[1], Jasneet Singh[1], Archana Kumari[1], Priyanka[3] and Bhallan Singh Sekhon[4]

1 School of Agricultural Biotechnology, Punjab Agricultural University, Ludhiana, Punjab, India

2 Department of Molecular Biology and Biotechnology, MPUAT, Udaipur, Rajasthan, India

3 Department of Genetics and Plant Breeding, CSK-HPKV, Palampur, Himachal Pradesh, India

4 Department of Vegetable Science, KVK Mansa, Punjab Agricultural University, Ludhiana, Punjab, India

*Address all correspondence to: ompingoliya96@gmail.com

References

[1] Ladha JK, Tirol-Padre A, Reddy CK, Cassman KG, Verma S, Powlson DS. Global nitrogen budgets in cereals: A 50-year assessment for maize, rice and wheat production systems. Scientific Reports. 2016;**6**:19355

[2] Plett DC, Holtham LR, Okamoto M, Garnett TP. Nitrate uptake and its regulation in relation to improving nitrogen use efficiency in cereals. Seminars in Cell and Developmental Biology. 2018;**74**:97-104

[3] Sandhu N, Sethi M, Kumar A, Dang D, Singh J, Chhuneja P. Biochemical and genetic approaches improving nitrogen use efficiency in cereal crops: A review. Frontiers in Plant Science. 2021;**12**:657629

[4] Tilman D, Balzer C, Hill J, Befort BL. Global food demand and the sustainable intensification of agriculture. Proceedings. National Academy of Sciences. United States of America. 2011;**108**:20260-20264

[5] Masclaux-Daubresse C, Daniel-Vedele F, Dechorgnat J, Chardon F, Gaufichon L, Suzuki A. Nitrogen uptake, assimilation and remobilization in plants: Challenges for sustainable and productive agriculture. Annals of Botany. 2010;**105**:1141-1157

[6] Moll RH, Kamprath EJ, Jackson WA. Analysis and interpretation of factors which contribute to efficiency of nitrogen utilization 1. Agronomy Journal. 1982;**74**:562-564

[7] Hansen NJS, Plett D, Berger B, Garnett T. Tackling nitrogen use Efficiency in cereal crops using high-throughput phenotyping. In: Shrawat A, Zayed A, Lightfoot D, editors. Engineering Nitrogen Utilization in Crop Plants. Cham: Springer Nature; 2018. pp. 121-139

[8] Craswell ET, Godwin DC. The efficiency of nitrogen fertilizers applied to cereals grown in different climates. In: Tinker PB, Lauchli A, editors. Advances in Plant Nutrition. Vol. 1. Westport, CT: Praeger Publishers; 1984. pp. 1-55

[9] Cowling SA, Field CB. Environmental control of leaf area production: Implications for vegetation and land-surface modelling. Global Biogeochemical Cycles. 2003;**17**:7-14

[10] Cameron K, Di H, Moir J. Nitrogen losses from the soil/plant system: A review. Annals of Applied Biology. 2013;**162**:145-173

[11] Dobermann A, Cassman KG. Plant nutrient management for enhanced productivity in intensive grain production systems of the United States and Asia. In: Progress in Plant Nutrition: Plenary Lectures of the XIV International Plant Nutrition Colloquium. Netherlands; 2002. pp. 153-175

[12] Franzen DW, Hopkins DH, Sweeney MD, Ulmer MK, Halvorson AD. Evaluation of soil survey scale for zone development of site-specific nitrogen management. Agronomy Journal. 2002;**94**:381

[13] Miller RO, Jacobsen JS, Skogley EO. Aerial accumulation and partitioning of nutrients by hard red spring wheat. Communications in Soil Science and Plant Analysis. 1993;**24**:2389-2407

[14] Mullen RW, Freeman KW, Raun WR, Johnson GV, Stone ML, Solie JB.

Identifying an in-season response index and the potential to increase wheat yield with nitrogen. Agronomy Journal. 2003;**95**:347-351

[15] Inman D, Khosla R, Reich RM, Westfall DG. Active remote sensing and grain yield in irrigated maize. Precision Agriculture. 2007;**8**:241-252

[16] Hatfield JL, Gitelson AA, Schepers JS, Walthall CL. Application of spectral remote sensing for agronomic decisions. Agronomy Journal. 2008;**100**:117-131

[17] Cassman KG, Dobermann A, Walters DT. Agroecosystems, nitrogen-use efficiency, and nitrogen management. AMBIO: A Journal of the Human Environment. 2002;**31**:132-140

[18] Osborne SL, Schepers JS, Francis DD, Schlemmer MR. Use of spectral radiance to estimate in-season biomass and grain yield in nitrogen- and water-stressed corn. Crop Science. 2002;**42**:165

[19] Sharma L, Bu H, Franzen D, Denton A. Use of corn height measured with an acoustic sensor improves yield estimation with ground based active optical sensors. Computers and Electronics in Agriculture. 2016;**124**:254-262

[20] Shaver T, Khosla R, Westfall D. Evaluation of two crop canopy sensors for nitrogen recommendations in irrigated maize. Journal of Plant Nutrition. 2014;**37**:406-419

[21] Sharma LK, Bu H, Franzen DW. Comparison of two ground-based active-optical sensors for in-season estimation of corn (*Zea mays* L.) yield. Journal of Plant Nutrition. 2015;**39**:957-966

[22] Oury F, Godin C. Yield and grain protein concentration in bread wheat: How to use the negative relationship between the two characters to identify favourable genotypes? Euphytica. 2007;**157**:45-57

[23] Karamanos RE, Hanson K, Stevenson FC. Nitrogen form, time and rate of application, and nitrification inhibitor effects on crop production. Canadian Journal of Plant Science. 2014;**94**:425-432

[24] Ford KE, Gregory PJ, Gooding MJ, Pepler S. Genotype and fungicide effects on late-season root growth of winter wheat. Plant and Soil. 2006;**284**:33-44

[25] Li H, Hu B, Chu C. Nitrogen use efficiency in crops: Lessons from Arabidopsis and rice. Journal of Experimental Botany. 2017;**68**: 2477-2488

[26] Lynch JP. Steep, cheap and deep: An ideotype to optimize water and N acquisition by maize root systems. Annals of Botany. 2013;**112**:347-357

[27] Mandal VK, Sharma N, Raghuram N. Molecular targets for improvement of crop nitrogen use efficiency: Current and emerging options. In: Shrawat A, Zayed A, Lightfoot D, editors. Engineering Nitrogen Utilization in Crop Plants. Cham: Springer; 2018. DOI: 10.1007/978-3-319-92958-3_577

[28] Qin S, Sun X, Hu C, Tan Q, Zhao X, Xin J. Effect of NO^{3-}:NH^{4+} ratios on growth, root morphology and leaf metabolism of oilseed rape (*Brassica napus* L.) seedlings. Acta Physiologiae Plantarum. 2017;**39**:198

[29] Ju C, Buresh RJ, Wang Z, Zhang H, Liu L, Yang J. Root and shoot traits for rice varieties with higher grain yield and higher nitrogen use efficiency at lower nitrogen rates application. Field Crops Research. 2015;**175**:47-55

[30] Shahzad Z, Amtmann A. Food for thought: How nutrients regulate root system architecture. Current Opinion in Plant Biology. 2017;**39**:80-87

[31] He X, Qu B, Li W, Zhao X, Teng W, Ma W. The nitrate-inducible NAC transcription factor TaNAC2-5A controls nitrate response and increases wheat yield. Plant Physiology. 2015;**169**:1991-2005

[32] Carvalho P, Foulkes MJ. Chapter 195: Roots and the uptake of water and nutrients. In: Meyers RA, editor. Encyclopedia of Sustainability Science and Technology. Heidelberg: Springer; 2011. pp. 1390-1404

[33] Gregory PJ, Brown SC. Root growth, water use and yield of crops in dry environments: What characteristics are desirable? Aspects of Applied Biology. 1989;**22**:234-243

[34] Ehdaie B, Whitkus RW, Waines JG. Root biomass, water-use efficiency, and performance of wheat–rye translocations of chromosomes 1 and 2 in spring bread wheat 'Pavon'. Crop Science. 2003;**43**:710

[35] Robinson D. Root proliferation, nitrate inflow and their carbon costs during nitrogen capture by competing plants in patchy soil. Plant and Soil. 2001;**232**:41-50

[36] Hall AJ, Richards RA. Prognosis for genetic improvement of yield potential and water-limited yield of major grain crops. Field Crops Research. 2013;**143**:18-33

[37] Bai C, Liang Y, Hawkesford MJ. Identification of QTLs associated with seedling root traits and their correlation with plant height in wheat. Journal of Experimental Botany. 2013;**64**:1745-1753

[38] Atkinson JA, Wingen LU, Griffiths M, Pound MP, Gaju O, Foulkes MJ. Phenotyping pipeline reveals major seedling root growth QTL in hexaploid wheat. Journal of Experimental Botany. 2015;**66**:2283-2292

[39] Crawford NM, Glass AD. Molecular and physiological aspects of nitrate uptake in plants. Trends in Plant Science. 1998;**3**:389-395

[40] Kant S. Understanding nitrate uptake, signaling and remobilisation for improving plant nitrogen use efficiency. Seminars in Cell & Developmental Biology. 2017;**74**:89-96

[41] Nieder R, Benbi DK, Scherer HW. Fixation and defixation of ammonium in soils: A review. Biology and Fertility of Soils. 2010;**47**:1-14

[42] Léran S, Varala K, Boyer JC, Chiurazzi M, Crawford N, Daniel-Vedele F. A unified nomenclature of NITRATE TRANSPORTER 1/PEPTIDE TRANSPORTER family members in plants. Trends in Plant Science. 2014;**19**:5-9

[43] Loqué D, Von Wirén N. Regulatory levels for the transport of ammonium in plant roots. Journal of Experimental Botany. 2004;**55**:1293-1305

[44] Dechorgnat J, Nguyen CT, Armengaud P, Jossier M, Diatloff E, Filleur S. From the soil to the seeds: The long journey of nitrate in plants. Journal of Experimental Botany. 2010;**62**:1349-1359

[45] Feng H, Yan M, Fan X, Li B, Shen Q, Miller AJ. Spatial expression and regulation of rice high-affinity nitrate transporters by nitrogen and carbon status. Journal of Experimental Botany. 2011;**62**:2319-2332

[46] Goyal SS, Huffaker RC. The uptake of NO3-, NO2-, and NH4+ by intact wheat (*Triticum aestivum*) seedlings. Plant Physiology. 1986;**82**:1051-1056

[47] Von Wirén N, Gazzarrini S, Frommer WB. Regulation of mineral uptake in plants. Plant and Soil. 1997;**196**:191-199

[48] Cai C, Zhao X, Zhu Y, Li B, Tong Y, Li Z. Regulation of the high-affinity nitrate transport system in wheat roots by exogenous Abscisic acid and glutamine. International Journal of Plant Biology. 2007;**49**:1719-1725

[49] Wiren NV, Merrick M. Regulation and function of ammonium carriers in bacteria, fungi, and plants. In: Molecular mechanisms controlling transmembrane transport. Berlin, Heidelberg: Springer; 2004. pp. 95-120

[50] Xuan W, Beeckman T, Xu G. Plant nitrogen nutrition: Sensing and signaling. Current Opinion in Plant Biology. 2017;**39**:57-65

[51] Criddle RS, Ward MR, Huffaker RC. Nitrogen uptake by wheat seedlings, interactive effects of four nitrogen sources: $NO3^-$, $NO2^-$, $NH4^+$, and urea. Plant Physiology. 1988;**86**:166-175

[52] Reddy MM, Ulaganathan K. Nitrogen nutrition, its regulation and biotechnological approaches to improve crop productivity. American Journal of Plant Sciences. 2015;**6**:2745-2798

[53] Richardson AE, Barea J, McNeill AM, Prigent-Combaret C. Acquisition of phosphorus and nitrogen in the rhizosphere and plant growth promotion by microorganisms. Plant and Soil. 2009;**321**:305-339

[54] Nguyen C. Rhizodeposition of organic C by plants: Mechanisms and controls. Agronomy. 2003;**23**:375-396

[55] Mazzola M, Funnell D, Raaijmakers J. Wheat cultivar-specific selection of 2,4-diacetylphloroglucinol-producing fluorescent Pseudomonas species from resident soil populations. Microbial Ecology. 2004;**48**:338-348

[56] Germida J, Siciliano S. Taxonomic diversity of bacteria associated with the roots of modern, recent and ancient wheat cultivars. Biology and Fertility of Soils. 2001;**33**:410-415

[57] Wu H, Haig T, Pratley J, Lemerle D, An M. Allelochemicals in wheat (*Triticum aestivum* L.): Cultivar difference in the exudation of phenolic acids. Journal of Agricultural and Food Chemistry. 2001;**49**:3742-3745

[58] Herold MB, Baggs EM, Daniell TJ. Fungal and bacterial denitrification are differently affected by long-term pH amendment and cultivation of arable soil. Soil Biology and Biochemistry. 2012;**54**:25-35

[59] Bardon C, Piola F, Bellvert F, Haichar FE, Comte G, Meiffren G. Evidence for biological denitrification inhibition (BDI) by plant secondary metabolites. New Phytologist. 2014;**204**:620-630

[60] Subbarao GV, Tomohiro B, Masahiro K, Osamu I, Samejima H, Wang HY. Can biological nitrification inhibition (BNI) genes from perennial Leymus racemosus (*Triticeae*) combat nitrification in wheat farming? Plant and Soil. 2007;**299**:55-64

[61] Ortiz R, Braun H, Crossa J, Crouch JH, Davenport G, Dixon J. Wheat genetic resources enhancement by the international maize and wheat improvement center (CIMMYT). Genetic Resources and Crop Evolution. 2008;**55**:1095-1140

[62] Nelson DR, Mele PM. The impact of crop residue amendments and lime on microbial community structure and

nitrogen-fixing bacteria in the wheat rhizosphere. Soil Research. 2006;**44**:319

[63] Venieraki A, Dimou M, Pergalis P, Kefalogianni I, Chatzipavlidis I, Katinakis P. The genetic diversity of culturable nitrogen-fixing bacteria in the rhizosphere of wheat. Microeconomics. 2010;**61**:277-285

[64] Behl RK, Ruppel S, Kothe E, Narula N. Wheat x Azotobacter x VA Mycorrhiza interactions towards plant nutrition and growth–a review. Journal of Applied Botany and Food Quality. 2012;**81**:95-109

[65] Neiverth A, Delai S, Garcia DM, Saatkamp K, de Souza EM, de Oliveira PF. Performance of different wheat genotypes inoculated with the plant growth promoting bacterium Herbaspirillum seropedicae. European Journal of Soil Biology. 2014;**64**:1-5

[66] Geddes BA, Ryu MH, Mus F, Costas AG, Peteres JW, Voigt CA. Use of plant colonizing bacteria as chassis for transfer of N2-fixation to cereals. Current Opinion in Biotechnology. 2015;**32**:216-222

[67] Triplett EW, Kaeppler SM, Chelius MK. *Klebsiella pneumoniae* inoculants for Enhancing Plant Growth. US patent 7,393,678. Madison, 2008; WI: Wisconsin Alumni Research.

[68] Boyd ES, Costas AMG, Hamilton TL, Mus F, Peters JW. Evolution of molybdenum nitrogenase during the transition from anaerobic to aerobic metabolism. Journal of Bacteriology. 2015;**197**:1690-1699

[69] Ortíz-Castro R, Contreras-Cornejo HA, Macías-Rodríguez L, López-Bucio J. The role of microbial signals in plant growth and development. Plant Signaling & Behavior. 2009;**4**:701-712

[70] Cassan F, Perrig D, Sgroy V, Masciarelli O, Penna C, Luna V. Azospirillum brasilense Az39 and Bradyrhizobium japonicum E109, inoculated singly or in combination, promote seed germination and early seedling growth in corn (*Zea mays* L.) and soybean (*Glycine max* L.). European Journal of Soil Biology. 2009;**45**:28-35

[71] Bottini R, Cassian F, Piccoli P. Gibberellin production by bacteria and its involvement in plant growth promotion and yield increase. Applied Microbiology and Biotechnology. 2004;**65**:497-503

[72] Couillerot O, Combes-Meynet E, Pothier JF, Bellvert F, Challita E, Poirier M. The role of the antimicrobial compound 2,4-diacetylphloroglucinol in the impact of biocontrol Pseudomonas fluorescens F113 on Azospirillum brasilense phytostimulators. Microbiology. 2011;**157**:1694-1705

[73] Almario J, Moënne-Loccoz Y, Muller D. Monitoring of the relation between 2,4-diacetylphloroglucinol-producing pseudomonas and Thielaviopsis basicola populations by real-time PCR in tobacco Black root-rot suppressive and conducive soils. Soil Biology and Biochemistry. 2013;**57**:144-155

[74] Maketon C, Fortuna A, Okubara PA. Cultivar-dependent transcript accumulation in wheat roots colonized by Pseudomonas fluorescens q8r1-96 wild type and mutant strains. Biological Control. 2012;**60**:216-224

[75] Ali A. Nitrate assimilation pathway in higher plants: Critical role in nitrogen signalling and utilization. Plant Science Today. 2020;**7**:182-192

[76] Boisson M, Mondon K, Torney V, Nicot N, Laine A, Bahrman N. Partial

sequences of nitrogen metabolism genes in hexaploid wheat. Theoretical and Applied Genetics. 2005;**110**:932-940

[77] Sétif P, Hirasawa M, Cassan N, Lagoutte B, Tripathy JN, Knaff DB. New insights into the catalytic cycle of plant nitrite reductase. Electron transfer kinetics and charge storage. Biochemistry. 2009;**48**:2828-2838

[78] Sakakibara Y, Kimura H, Iwamura A, Saitoh T, Ikegami T, Kurisu G. A new structural insight into differential interaction of cyanobacterial and plant ferredoxins with nitrite reductase as revealed by NMR and X-ray crystallographic studies. The Journal of Biochemistry. 2012;**151**:483-492

[79] Lea PJ, Miflin BJ. Nitrogen assimilation and its relevance to crop improvement. In: Zhang H, editor. Annual Plant Reviews, Nitrogen Metabolism in Plants in the Post-Genomic Era. Chichester: Wiley-Blackwell; 2011. pp. 1-40

[80] Yamaya T, Kusano M. Evidence supporting distinct functions of three cytosolic glutamine synthetases and two NADH-glutamate synthases in rice. Journal of Experimental Botany. 2014;**65**:5519-5525

[81] Howarth JR, Parmar S, Jones J, Shepherd CE, Corol D, Galster AM. Co-ordinated expression of amino acid metabolism in response to N and S deficiency during wheat grain filling. Journal of Experimental Botany. 2008;**59**:3675-3689

[82] Kant S, Seneweera S, Rodin J, Materne M, Burch D, Rothstein SJ. Improving yield potential in crops under elevated CO2: Integrating the photosynthetic and nitrogen utilization efficiencies. Frontiers in Plant Science. 2012;**3**:162

[83] Xu G, Fan X, Miller AJ. Plant nitrogen assimilation and use efficiency. Annual Review of Plant Biology. 2012;**63**:153-182

[84] Evans JR. Nitrogen and photosynthesis in the flag leaf of wheat (*Triticum aestivum* L.). Plant Physiology. 1983;**72**:297-302

[85] Guo TT, Wang DF, Fang J, Zhao JF, Yuan SJ, Xiao LT. Mutations in the rice OsCHR4 gene, encoding a CHD3 family chromatin remodeler, induce narrow and rolled leaves with increased cuticular wax. International Journal of Molecular Sciences. 2019;**20**:2567

[86] Moreau D, Allard V, Gaju O, Le Gouis J, Foulkes MJ, Martre P. Acclimation of leaf nitrogen to vertical light gradient at anthesis in wheat is a whole-plant process that scales with the size of the canopy. Plant Physiology. 2012;**160**:1479-1490

[87] Araus JL, Amaro T, ZuhairY NMM. Effect of leaf structure and water status on carbon isotope discrimination in field-grown durum wheat. Plant, Cell & Environment. 1997;**20**:1484-1494

[88] Giunta F, Motzo R, Deidda M. SPAD readings and associated leaf traits in durum wheat, barley and triticale cultivars. Euphytica. 2002;**125**: 197-205

[89] Laperche A, Devienne-Barret F, Maury O, Le Gouis J, Ney B. A simplified conceptual model of carbon/nitrogen functioning for QTL analysis of winter wheat adaptation to nitrogen deficiency. Theoretical and Applied Genetics. 2006;**112**:797-807

[90] Long SP, Zhu X, Naidu SL, Ort DR. Can improvement in photosynthesis increase crop yields? Plant, Cell & Environment. 2006;**29**:315-330

[91] Zhu H, Li C, Gao C. Author correction: Applications of CRISPR-Cas in agriculture and plant biotechnology. Nature Reviews Molecular Cell Biology. 2020;**21**:782

[92] Lemaire G, van Oosterom E, Sheehy J, Jeuffroy MH, Massignam A, Rossato L. Is crop N demand more closely related to dry matter accumulation or leaf area expansion during vegetative growth? Field Crops Research. 2007;**100**:91-106

[93] Gaju O, Allard V, Martre P, Le Gouis J, Moreau D, Bogard M. Nitrogen partitioning and remobilization in relation to leaf senescence, grain yield and grain nitrogen concentration in wheat cultivars. Field Crops Research. 2014;**155**:213-223

[94] Havé M, Marmagne A, Chardon F, Masclaux-Daubresse C. Nitrogen remobilisation during leaf senescence: Lessons from Arabidopsis to crops. Journal of Experimental Botany. 2016;**68**:2513-2529

[95] Liu W, Sun Q, Wang K, Du Q, Li WX. Nitrogen limitation adaptation (NLA) is involved in source-to-sink remobilization of nitrate by mediating the degradation of NRT1.7 in Arabidopsis. New Phytologist. 2016;**214**:734-744

[96] Zeng DD, Qin R, Li M, Alamin M, Jin XL, Liu Y. The ferredoxin-dependent glutamate synthase (OsFd-GOGAT) participates in leaf senescence and the nitrogen remobilization in rice. Molecular Genetics and Genomics. 2016;**292**:385-395

[97] Uauy C, Distelfeld A, Fahima T, Blechl A, Dubcovsky J. A NAC gene regulating senescence improves grain protein, zinc, and iron content in wheat. Science. 2006;**314**:1298-1301

[98] Ishida H, Izumi M, Wada S, Makino A. Roles of autophagy in chloroplastrecycling. Biochimica et Biophysica Acta. 2014;**1837**:512-521

[99] Thomas H, Smart CM. Crops that stay green. Annals of Applied Biology. 1993;**123**:193-219

[100] Naruoka Y, Sherman JD, Lanning SP, Blake NK, Martin JM, Talbert LE. Genetic analysis of green leaf duration in spring wheat. Crop Science. 2012;**52**:99-109

[101] Shewry PR, Halford NG. Cereal seed storage proteins: Structures, properties and role in grain utilization. Journal of Experimental Botany. 2002;**53**:947-958

[102] Ravel C, Martre P, Romeuf I, Dardevet M, El-Malki R, Bordes J. Nucleotide polymorphism in the wheat transcriptional activator spa influences its pattern of expression and has pleiotropic effects on grain protein composition, dough viscoelasticity, and grain hardness. Plant Physiology. 2009;**151**:2133-2144

[103] Bogard M, Allard V, Brancourt-Hulmel M, Heumez E, Machet J, Jeuffroy M. Deviation from the grain protein concentration–grain yield negative relationship is highly correlated to post-anthesis N uptake in winter wheat. Journal of Experimental Botany. 2010;**61**:4303-4312

[104] Munier-Jolain NG, Salon C. Are the carbon costs of seed production related to the quantitative and qualitative performance? An appraisal for legumes and other crops. Plant, Cell & Environment. 2005;**28**:1388-1395

[105] Acreche MM, Slafer GA. Variation of grain nitrogen content in relation with grain yield in old and modern Spanish wheats grown under a wide range of agronomic conditions in a Mediterranean

region. The Journal of Agricultural Science. 2009;**147**:657-667

[106] Monaghan JM, Snape JW, Chojecki AJ, Kettlewell PS. The use of grain protein deviation for identifying wheat cultivars with high grain protein concentration and yield. Euphytica. 2001;**122**:309-317

[107] Sinclair TR, Rufty TW, Lewis RS. Increasing photosynthesis: Unlikely solution for world food problem. Trends in Plant Science. 2019;**24**:1032-1039

[108] Li G, Cui K. Sucrose translocation and its relationship with grain yield formation in rice. Journal of Plant Physiology. 2014;**50**:735-740

[109] Yang X, Xia X, Zhang Z, Nong B, Zeng Y, Xiong F. QTL mapping by whole genome re-sequencing and analysis of candidate genes for nitrogen use efficiency in rice. Frontiers in Plant Science. 2017;**8**:1634

[110] Sun J, Guo Y, Zhang G, Gao M, Zhang G, Kong F. QTL mapping for seedling traits under different nitrogen forms in wheat. Euphytica. 2012;**191**:317-331

[111] Guo Y, Kong F, Xu Y, Zhao Y, Liang X, Wang Y. QTL mapping for seedling traits in wheat grown under varying concentrations of N. P and K nutrients. Theoretical and Applied Genetics. 2011;**124**:851-865

[112] Foulkes M, Hawkesford M, Barraclough P, Holdsworth M, Kerr S, Kightley S. Identifying traits to improve the nitrogen economy of wheat: Recent advances and future prospects. Field Crops Research. 2009;**114**:329-342

[113] Distelfeld A, Avni R, Fischer AM. Senescence, nutrient remobilization, and yield in wheat and Barley. Journal of Experimental Botany. 2014;**65**:3783-3798

[114] Gallais A, Hirel B. An approach to the genetics of nitrogen use efficiency in maize. Journal of Experimental Botany. 2004;**55**:295-306

[115] Zhang M, Gao M, Zheng H, Yuan Y, Zhou X, Guo Y. QTL mapping for nitrogen use efficiency and agronomic traits at the seedling and maturity stages in wheat. Molecular Breeding. 2019;**39**:71

[116] Fontaine JX, Ravel C, Pageau K, Heumez E, Dubois F, Hirel B. A quantitative genetic study for elucidating the contribution of glutamine synthetase, glutamate dehydrogenase and other nitrogen-related physiological traits to the agronomic performance of common wheat. Theoretical and Applied Genetics. 2009;**119**:645-662

[117] Pozzo T, Higdon SM, Pattathil S, Hahn MG, Bennett AB. Characterization of novel glycosyl hydrolases discovered by cell wall glycan directed monoclonal antibody screening and metagenome analysis of maize aerial root mucilage. PLoS One. 2018;**13**:e0204525

[118] Van Deynze A, Zamora P, Delaux PM, Heitmann C, Jayaraman D, Rajasekar S. Nitrogen fixation in a landrace of maize is supported by a mucilage-associated diazotrophic microbiota. PLoS Boilogy. 2018;**16**:e2006352

[119] Hu B, Wang W, Ou S, Tang J, Li H, Che R. Variation in NRT1.1B contributes to nitrate-use divergence between rice subspecies. Nature Genetics. 2015;**47**:834

[120] Brasier K, Ward B, Smith J, Seago J, Oakes J, Balota M. Identification of quantitative trait loci associated with nitrogen use efficiency in winter wheat. PLoS One. 2020;**15**:e0228775

[121] Cormier F, Foulkes J, Hirel B, Gouache D, Moenne-Loccoz Y,

Le Gouis J. Breeding for increased nitrogen-use efficiency: A review for wheat (*T. aestivum* L.). Plant Breeding. 2016;**135**:255-278

[122] An D, Su J, Liu Q, Zhu Y, Tong Y, Li J. Mapping QTLs for nitrogen uptake in relation to the early growth of wheat (*Triticum aestivum* L.). Plant and Soil. 2006;**284**:73-84

[123] Mahjourimajd S, Kuchel H, Langridge P, Okamoto M. Evaluation of Australian wheat genotypes for response to variable nitrogen application. Plant and Soil. 2016;**399**:247-255

[124] Monostori I, Szira F, Tondelli A, Arendas T, Gierczik K, Cattivelli L. Genome-wide association study and genetic diversity analysis on nitrogen use efficiency in a central European winter wheat (*Triticum aestivum* L.) collection. PLoS One. 2017;**12**:e0189265

[125] Buchner P, Hawkesford MJ. Complex phylogeny and gene expression patterns of members of the nitrate transporter 1/peptide transporter family (NPF) in wheat. Journal of Experimental Botany. 2014;**65**:5697-5710

[126] Zhou Y, Cai H, Xiao J, Li H, Zhang Q, Lian X. Over-expression of aspartate amino transferase genes in rice resulted in altered nitrogen metabolism and increased amino acid content in seeds. Theoretical and Applied Genetics. 2009;**118**:1381-1390

[127] Saini DK, Chopra Y, Pal N, Chahal A, Srivastava P, Gupta PK. Meta-QTLs, ortho-MQTLs and candidate genes for nitrogen use efficiency and root system architecture in bread wheat (*Triticum aestivum* L.). Physiology and Molecular Biology of Plants. 2021;**27**:2245-2267

[128] Plett D, Toubia J, Garnett T, Tester M, Kaiser BN, Baumann U. Dichotomy in the NRT gene families of dicots and grass species. PLoS One. 2010;**5**:e15289

[129] Tabuchi M, Abiko T, Yamaya T. Assimilation of ammonium ions and reutilization of nitrogen in rice (*Oryza sativa* L.). Journal of Experimental Botany. 2007;**58**:2319-2327

[130] Curci PL, Bergès H, Marande W, Maccaferri M, Tuberosa R, Sonnante G. Asparagine synthetase genes (AsnS1 and AsnS2) in durum wheat: Structural analysis and expression under nitrogen stress. Euphytica. 2018;**214**:36

[131] Spitz F, Furlong EE. Transcription factors: From enhancer binding to developmental control. Nature reviews genetics. 2012;**13**:613-626

[132] Zhang H. An Arabidopsis MADS box gene that controls nutrient-induced changes in root architecture. Science. 1998;**279**:407-409

[133] Zuluaga DL, De Paola D, Janni M, Curci PL, Sonnante G. Durum wheat miRNAs in response to nitrogen starvation at the grain filling stage. PLoS One. 2017;**12**:e0183253

[134] Waters BM, Uauy C, Dubcovsky J, Grusak MA. Wheat (*Triticum aestivum*) NAM proteins regulate the translocation of iron, zinc, and nitrogen compounds from vegetative tissues to grain. Journal of Experimental Botany. 2009;**60**:4263-4274

[135] Zhao Y, Guo L, Lu W, Li X, Chen H, Guo C. Expression pattern analysis of microRNAs in root tissue of wheat (*Triticum aestivum* L.) under normal nitrogen and low nitrogen conditions. Journal of Plant Biochemistry and Biotechnology. 2013;**24**:143-153

Chapter 10

Insights into Physiological, Biochemical and Molecular Responses in Wheat under Salt Stress

Hari Kesh, Sunita Devi, Naresh Kumar, Ashwani Kumar, Arvind Kumar, Pooja Dhansu, Parvender Sheoran and Anita Mann

Abstract

Globally, wheat is a major staple food crop that provides 20% of food calories for 30% of the human population. Wheat growth and production are significantly affected by salt stress at various stages and adversely affect germination, vegetative growth, stomatal conductance, photosynthesis, reproductive behavior, protein synthesis, enzymatic activity and finally hampered grain yield. Maintenance of low Na^+/K^+ ratio, antioxidants and hormonal regulation, and accumulation of compatible osmolytes such as glycine betaine, proline and trehalose help the wheat genotypes to mitigate the negative effects of salt stress. Recent studies have reported various mechanisms at the physiological, biochemical and molecular levels to adapt the salinity stress in various ecologies. Salt tolerant genotypes can be developed by conventional breeding approaches and through biotechnological approaches. This chapter reviews the updates on mechanisms and recent approaches to structure the salt-tolerant and high-yielding genotypes.

Keywords: wheat, salt stress, ion homeostasis, conventional approaches, molecular breeding

1. Introduction

Bread wheat is a major staple food cultivated throughout the world with a global yield of 8.8 million tons [1]. Global demand for wheat is increasing day by day due to its unique features such as bio-fortified and processed products like biscuits, cookies, doughnuts, porridge and pastries [2]. However, the production and productivity of wheat have decreased due to various biotic and abiotic stresses. Different climate models estimated that wheat production could reduce by 6% due to adverse climatic conditions [3]. Among the abiotic stresses, salinity stress significantly affects the growth and production of wheat crops. Up to 40 percent yield loss worth $ 27 billion

US annually have reported in salt-affected regions [4, 5]. Soil salinity negatively affects the morphological traits such as germination percentage, grain per spike, plant height, grain yield and harvest index [6–8]; physiological traits like relative water content, membrane stability, chlorophyll fluorescence and mineral uptake [9, 10]; biochemical traits like proline content, gluten content, protein synthesis and enzymatic activity involved in various metabolic processes [11–14]. Salinity stress disturbs the ionic balance due to the accumulation of Na^+ which reduces the mineral uptake and their translocation to grains [15, 16]. Salt stress also causes the production of reactive oxygen species which hampers plant growth and development [13, 17]. Wheat grain yield reduces greatly when soil pH or electrical conductivity reaches 8.5 or 4 dS m^{-1}. Salt stress creates a water deficit which makes it difficult for roots to draw water from their surroundings [18, 19]. Early-stage exposure to salt causes osmotic stress, which adversely affects the normal cell metabolism, stomatal opening and transpiration process. Long-term stress leads to ionic stress due to a high concentration of NaCl. Ionic stress causes chlorosis and necrosis of leaves and reduces photosynthesis and protein synthesis [20]. Irrigated water with the salt content of 2–3 g L^{-1} or 3–5 g L^{-1} reduces the grain yield of wheat by 7–13% or 13–24%, respectively [21]. To cope with these adverse effects of salt stress, plants use different mechanisms such as the exclusion of sodium ions and increase in potassium concentration, maintenance of high K^+/Na^+ ratio, increased stomatal conductance and transpiration efficiency, osmotic adjustment and antioxidant defence [22–24]. Therefore, knowledge and understanding of the physiological and biochemical mechanisms are very essential for selecting and developing salt-tolerant wheat genotypes. Moreover, an integrated approach of conventional and molecular breeding can be used to improve wheat productivity and under salinity stress. Therefore, the present chapter summarizes the negative effect of salt stress, tolerance mechanism and potential breeding methods to improve the resilience in wheat.

2. Effect of salt stress in wheat

2.1 Germination

Germination is the basic and dynamic process that determines the further growth and development of plants. The seed germination process may be divided into three distinct phases. Phase one initiates with imbibition of water by dry seed, phase two causes activation of enzymatic activity and metabolic processes and phase three is a post-germination phase that includes rupturing of endosperm and radicle elongation followed by seedling establishment [25–27]. Salt stress lowers the osmotic potential of the germination medium which disrupts the normal functioning of the enzyme responsible for protein metabolism, deteriorates seed food reserves and ultimately grain yield [28]. These consequences together cause inhibition of cell expansion and cell division. Besides enzymatic imbalance, seed dormancy, hard seed coat, seed vigor and viability, temperature, moisture content and light intensity also affects the seed germination [29]. Previous studies reported that the accumulation of mucilage, callose, lignin and suberin increase the seed dormancy by limiting the permeability of water and diffusion of oxygen through the seed coat, delaying the germination process [30, 31]. Delayed and decreased germination of the wheat seed was reported at 12.5 dS m^{-1} salinity level [32]. Germination percentage in wheat also depends on the type of wheat i.e. spring or winter or differences in the cultivar. For example, wheat variety Kharchia 65 was found more salt-tolerant than KRL 1–9 due to its high chlorophyll content, membrane stability and relative water content. Significant variation in wheat cultivars was observed for percent germination, rate of germination and germination index [33, 34]. Cultivar Shakha 93 and Shakah 94 were found positive while Masr 1 was negative for most of the germination traits under salt stress condition [35]. Similarly, Charushahi et al. [36] observed complete inhibition of germination at a high salinity rate due to limited uptake of water. High salt tolerance of Al-Hussein variety at germination stage was due to high tolerance index and chlorophyll stability at different salt concentration [37].

2.2 Plant growth

Salt stress severely affects wheat growth at both the vegetative and reproductive stages. Further, salt stress at the seedling stage may cause seedling chlorosis, necrosis or even death [38]. Early maturity under salt stress reduces the plumule length, leaf area and plant height [39]. Moreover, reduction in leaf size, number of leaves, root colonization, leaf expansion and dry matter of shoot were also noticed in wheat [40, 41]. The root is the first and most important organ of plants which is essential for the uptake of water and nutrient from the soil to maintain the growth and various developmental processes. Salinity inhibited growth of root and shoot dry weight, root length and diameter and root volume in wheat. Salinity reduces the root length and coleoptiles length and seedling establishment [42]. Otu et al. [43] reported a significant effect of increasing salinity level on root and shoot length, root fresh weight and elongation rate (**Figure 1**). A serious injuries effect in growth parameters of wheat like the relative growth rate of roots and leaves was seen under salt stress in

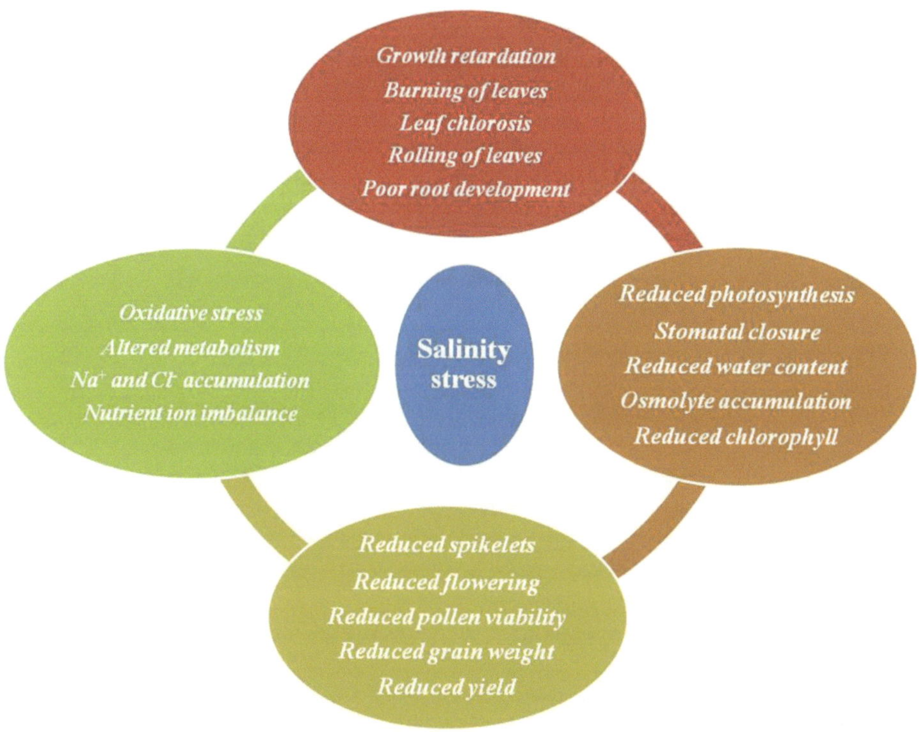

Figure 1.
Effects of salinity stress on wheat morphological, physiological, biochemical traits and yield attributes.

comparison to normal condition [44]. Many earlier studies have reported a reduction in growth parameters like root and shoot length, seedling length, leaf area, the relative growth rate of root and shoot, fresh and dry weight of root and shoot, plant height and tillering capacity at different salinity levels [45–48] (**Table 1**).

Salinity level	Effects	References
5.40 and 10.60 dS m^{-1}	Decreased chlorophyll, carotenoids and relative water content, reduced grain yield, increased hydrogen peroxides and thiobarbituric acid reactive substances	[33]
16 dS m^{-1}	Reduction in grain filling duration and harvest index	[49]
6.85 and 12.3 dS m^{-1}	Decreased relative water content, chlorophyll content, membrane stability index and increased hydrogen peroxide, SOD, ascorbate peroxidase (APOX) and GR	[50]
150 mM NaCl	Reduced stomatal conductance, potassium content and photosynthetic rate	[51]
0, 25, 50, 75, 100, 125, 150 mM NaCl	Decrease in root and shoot length, fresh and dry weight of roots and shoots, protein content and increase in proline content	[47]
150 and 300 mM NaCl	Increased hydrogen peroxide and lipid peroxidation, reduced glutathione and glutathione disulfide, glutathione S-transferase, decreased ascorbate content	[52]
100 and 200 mM NaCl	Declined leaf area, chlorophyll content, relative water content, grain yield, N, Fe, Mn and Mg content and increased Cu and Zn content.	[53]
150 mM NaCl	Reduced plant height, root length, shoot dry weight, glutathione (GSH), chlorophyll and carotenoid content, increased MDA, H2O2 and superoxide radical.	[54]
10 mM NaCl	Reduced water potential, osmotic potential, relative water content, decreased N, P and K uptake, reduced grain yield	[55]
200 mmol L^{-1} NaCl	Decreased net photosynthetic rate, stomatal conductance, maximum and actual photochemical efficiency of PSII and increase in intercellular CO_2 concentration.	[56]
10 dS m^{-1}	Reduced membrane stability, increased MDA and hydrogen peroxide content	[57]
6.25 dS m^{-1}	Increase in lipid peroxidation, lipoxygenase enzyme activity, H_2O_2 content, decrease in fresh and dry weight of shoots	[58]
100 mM NaCl	increase in NADPH oxidase activity, H_2O_2 and proline content in roots	[59]
250 mM NaCl	Increase lipid peroxidation, hydrogen peroxide and proline content, decreased relative water and chlorophyll content	[60]
0.23, 3.0, 6.0 and 12.0 dS m^{-1}	Reduced carbon fixation, chlorophyll content, leaf area, plant height, number of vascular bundles, phloem tissue thickness and pith cavity	[61]
50, 100, and 200 mM NaCl	Decrease in level of catechin hydrate, quercetin, and benzoic acid, reduced shoot and root length, increase in epicatechin levels	[62]
50, 100, and 200 mM NaCl	Reduction in germination %, root and shoot length, total chlorophyll content and increase in MDA content	[7]
6.0 and 10.0 dS m^{-1}	Reduced K content in roots and shoots, relative water and chlorophyll content	[8]
0, 100, and 200 mol/L NaCl	Increased Na + concentration, decreased K+ concentration, decreased chlorophyll fluorescence, chlorophyll content, shoot and root length	[48]

Table 1.
Effect of salt stress on physiological, biochemical and yield attributes.

2.3 Photosynthesis

Salinity stress has severed effect of various physiological processes such as respiration, membrane stability, ion toxicity and photosynthesis. The photosynthesis process involves photosynthetic apparatus, PS-I and PS-II, electron transport chain, carbon dioxide reduction pathways. Any damage at any stage leads to a reduction in the photosynthetic efficiency of a crop plant [63]. Salinity stress greatly reduces the amount of photosynthetic pigments at different salt concentrations and it was found more in salt-sensitive genotypes than tolerant. This decreased pigment content may be a due accumulation of ions in chloroplast and the high activity of chlorophyllase enzyme [64, 65]. At the vegetative stage, salt stress affects the carbohydrate synthesis while its translocation to grains during the grain filling stage [66]. Sodium chloride treatment decreases the stomatal conductance, CO_2 uptake required for carboxylation reaction and activity of RUBISCO (**Figure 1**) which ultimately reduces the photosynthetic efficiency [67]. Kafi [68] observed varied responses of wheat genotypes depending upon the growth stage, the concentration of salt and period of salt exposure. Stomatal conductance and reduced variable to maximum fluorescence were found major limiting factors affecting photosynthesis under salt stress. The toxic concentration of Na^+ and Cl^- in leaves, decreases the photosynthetic rate by disrupting the chlorophyll structure and PS-II [69]. Furthermore, reduced stomatal conductance decreases the electron transport chain efficiency, which results a decline in adenosine triphosphate (ATP) and nicotinamide adenine dinucleotide phosphate (NADPH) consumption in the photosynthesis process and ultimately in quantum yield of PS-II (**Table 1**). The reduction in PS-II quantum yield was more in salt-sensitive genotypes under salinity condition [70]. All the physiological and biochemical processes are depending upon the accessibility of water. High salt concentration osmotic and ionic stress which lowers the water potential of wheat plants [71]. Relative water content reduced 3.5% intolerance while 6.7% in sensitive genotypes of wheat after six days of salt stress resulted a drastic decline in water use efficiency [72, 73]. In general, water stress at heading and after the anthesis stage significantly affects the productivity of wheat [74]. Salt sensitive wheat variety HD 2687 showed a higher decrease in chlorophyll content, membrane stability and relative water content under stress compared to Kharchia 65 indicating their salt-tolerant nature [50].

2.4 Mineral uptake

One of the most severe effects of salinity stress is the accumulation of Na^+ content in leaves over the control condition. A high concentration of Na^+ and Cl^- ions in root zone reduces the uptake of essential cationic and anionic nutrients like calcium (Ca^{2+}), potassium (K^+) and nitrate (No_3^-) and decreases the amount of calcium, potassium, phosphorus and magnesium (**Figure 1**) in different plant parts [75, 76]. However, the differential response was seen for Na^+ concentration in winter and spring wheat. Winter-type wheat cultivars accumulate high Na^+ than spring type. Slat tolerant genotypes have the better ability to maintain more K^+ and K^+/Na^+ ratios and accumulate less Na^+ in their leaves [77, 78]. The findings of Hussain et al. [55] showed that the grain yield and tolerance power of wheat can be increased by enhancing Na^+/H^+ type antiporter. These antiporters are responsible for the transition of Na + from the cytoplasm to apoplast [79]. Moreover, tolerant genotypes have two pore K+ channels and one selective cation channel for K permeability [80]. Reduced plant growth under salt stress may be due to the high plasma membrane injury due to Na + toxicity [48, 81]. Poor membrane stability due to the replacement of Ca^{2+} by Na + causes the influx of heavy metals [82] like Zn^{2+} and Cu^{2+}. Iron and manganese content were drastically reduced under salinity stress; however, this reduction was lesser in tolerant genotypes (**Table 1**). Salt sensitive cultivars like HD 2687 and WL 711 showed a significant reduction in magnesium, nitrogen, iron, manganese and an increase in zinc and copper. Kharchia 65 gave good performance due to its better nutrient uptake capacity and ion partitioning [53]. Shaaban and El-Nour [83] also reported a significant reduction in nitrogen, potassium, phosphorus, calcium, magnesium, iron, manganese concentration and uptake; this may be due to the increase in osmotic pressure of root when irrigated with saline water.

2.5 Grain yield

Grain yield in wheat depends on several agronomic and physiological traits such as tillers number, earhead length, size and number of grains, root and shoot length, chlorophyll content, membrane stability and stomatal conductance. Reduction in any of the above-mentioned traits in salt condition directly affect the grain yield of wheat. However, the percent reduction in grain yield depends on the salt concentration and tolerance power of genotypes. Hussain et al. [84] reported significant differences among the 40 genotypes of wheat under salt stress. The sensitive genotypes had fewer yields than tolerant genotypes mainly due to decreased size and number of grains and reduced tillering capacity. The tolerant genotypes produce more productive tillers, the high number of fertile spikelets and have a better capacity of photo-assimilates translocation to developing grains. Less availability of photosynthates and their translocation from source to sink (**Figure 1**) is the main reason for lower grain yield in sensitive genotypes [85, 86]. Salt stress reduced the thousand kernel weight by 20% and starch content of grains by 6% in wheat compared to control condition [87, 88]. Wheat plants grown at high salinity level 10 dSm^{-1} significantly reduced the spike length by 24%, the number of spikelets by 21%, thousand-grain weight by 70%, straw yield by 20% and grain yield by 67% [89] (**Table 1**). A number of previous studies also reported a significant decline in wheat grain yield with increasing levels of salinity [90–93]. Reduced grain yield under salt stress may be due to low germination percentage and small size and number of medium and small veins in leaves of wheat [94, 95]. As far as the wheat quality is concerned, carbohydrates, proteins, fibers and gluten index in grains declined significantly under salt stress [96]. Salt stress at the grain maturation stage promotes leaf senescence due to which protein deposition takes place in grains over starch [66]. A high reduction in protein content of wheat was observed in wheat than triticale [97].

3. Mechanism of salt stress tolerance

3.1 Ion homeostasis

Salinity stress causes an ionic imbalance in wheat by affecting the Na^+ and K^+ concentrations in different plant tissues. A high concentration of Na^+ disturbs the uptake of nutrients like K^+ and Ca^{2+} causing lesions on different plant parts with declined leaf dry weight and shoot growth. Furthermore, high concentrations of Cl^- disturb the nutrient uptake by impairing anion uptake. Antagonistic effect of Cl^- has been observed with nitrate and phosphate [98, 99] causing a reduction in wheat growth and yield. Under salt stress, Na^+ is the major cause of both ionic and osmotic stress. Thus, maintaining ionic homeostasis is very essential for plant growth and development under salinity stress. Plants maintain the ionic balance of Na^+ and Cl^- inside the cell by removing excess salts via primary and secondary transport systems and their compartmentalization into vacuole [100]. Na^+ exclusion in plants can be achieved by different ion channels and transporters present in the cell membrane. Apart from being an essential micronutrient, K plays an important role in maintaining a low Na^+ to K^+ ratio. Previous studies in wheat reported a positive association of low Na^+ concentration in leaves with salt tolerance. Yadav et al. [93] reported that salt stress tolerance in wheat was associated with a high K^+ to Na^+ ratio in roots and shoots. Low K^+ to Na^+ ratio in the upper leaves of wheat reduces the plant growth. The high affinity potassium transported (HKT) gene family plays a major role in Na^+ exclusion via minimizing the entry of Na^+ into the roots from the soil [101]. The Nax 1 and Nax 2 genes belonging to the HKT gene gamily were initially identified in durum wheat. These genes exclude the Na^+ from xylem tissues and maintain its low concentration in the leaves. The Nax genes have already been utilized in the breeding program for developing the salt-tolerant genotypes in durum and hexaploid wheat. The presence of the Nax 2 gene in the durum wheat variety produced 25% more grain yield under salt stress conditions. While the presence of both Nax 1and Nax 2 in bread wheat reduces Na^+ concentration by 60% in leaves [102]. Salt tolerance in wheat genotypes can be achieved by down-regulation of TaHKT 2 gene [103].

3.2 Osmotic protection

Osmotic stress due to high salt concentration reduces the water uptake, cell expansion in roots, growth and development of plants. Likewise, the accumulation of high Na in leaves affects the photosynthesis process which results in leaf chlorosis and necrosis [104]. Osmoprotectants such as sugars e.g. trehalose, sucrose and fructose, amino acids e.g. proline and pipecolic acid, quaternary ammonium compounds e.g. glycine betaine, pipecolate betaine, alanine betaine and hydroxyl pro betaine, polyols e.g. mannitol, sorbitol and inositol and polyamines e.g. spermidine, putrescine and spermine [105] acts as a defensive mechanism in plants by lowering the cell water potential, detoxifying reactive oxygen species, activating anti-oxidants activity and stabilizing normal structures of proteins and enzymes [106–108]. Production of compatible osmolytes in wheat plays an important role in providing tolerance against salt injury. Accumulation of glycine betaine in transgenic lines of wheat improves the salt tolerance by protecting the photosystem II reaction centers and oxygen-evolving complex thus enhancing the photosynthetic activity [109]. Salinity stress disrupts the function of the thylakoid membrane which affects the photosynthesis process and ultimately grain yield of wheat. Glycine betaine improves salt tolerance by maintaining ionic balance, increasing osmotic adjustment and neutralizing ROS [110]. Wheat seedlings treated with showed a diminished level of malondialdehyde and an enhanced level of glutathione under salt stress [111]. Exogenous application of glycine betaine increased the activities of antioxidants such as CAT and POD to neutralize ROS damage in wheat [112]. Similarly, proline is the well-known osmolytes produced under salt stress condition [113]. Proline accumulation in wheat generally found in the cytoplasm where it acts as a shield against salt injury. Exogenous application of proline significantly enhanced the root length, seedling fresh and dry weight, photosynthetic pigments and K^+/Na^+ ratio and thus efficiently sustains the wheat growth under salt stress [114, 115]. Proline acts as defending agents for electron transport chain and RIBISCO enzyme from salt stress damage and increases the CO_2 assimilation rate, chlorophyll content and photosynthetic rate [116]. The sugars help in the regulation and stabilization of the native structure of proteins and enzymes which enables their normal functioning. These sugars may contribute up to 50% of osmoregulation in leaves of glycophytes. It is reported that galactose plays a major role in ascorbic acid pathways and enhances salt tolerance in wheat [117].

3.3 Antioxidants

Salinity stress disrupts the availability of CO_2 in leaves and electron transport chain in mitochondria and chloroplast due to which reactive oxygen species like singlet of oxygen ($^1O_2$), superoxide radicle (O_2^-), hydroxyl radicle (OH^-) and hydrogen peroxide (H_2O_2) are produced [118, 119]. Accumulation of these ROS at high concentrations is extremely harmful to plants. Chloroplast, mitochondria and peroxisomes are the primary site of ROS production. Photosystem I and II in chloroplasts, respiratory complex I and III in mitochondria and glycolate oxidase in peroxisomes are the major source of ROS generation [120–123]. ROS cause protein oxidation, lipid peroxidation, damage to nucleic acid, inhibition of enzyme activity and programmed cell death [124]. Lipid peroxidation is caused by the oxidative burst of the cell membrane which can be estimated by the content of malondialdehyde (MDA). Lipid peroxidation increases electrolyte leakage, disturbs membrane permeability and activates the oxidation of protein and DNA. Up to 73% increase in MDA content at 300 mM and 35% increase at 100 mM have been observed when wheat plants exposed to salinity stress [52]. Plants have the natural defense system antioxidant to detoxify the harmful effect of ROS. The enzymatic antioxidants are catalase (CAT), superoxide dismutase (SOD), peroxidase (POX), ascorbate peroxidase (APX), glutathione reductase (GR), glutathione peroxidases (GPX), dehydro-ascorbate reductase (DHAR) and mono-dehydro-ascorbate reductase (MDHAR) while non-enzymic antioxidants are glutathione (GSH), ascorbate (AsA), tocopherol and carotenoids [124, 125]. Sairam et al. [33] reported an increased concentration of catalase in both salt-tolerant and sensitive cultivars of wheat. Mandhania et al. [126] observed enhanced activity of SOD and CAT in wheat which detoxify H_2O_2 and break it down as H_2O and O_2 under salt stress. Tolerant wheat genotypes produced a high concentration of AsA and catalase to counter the effect of salinity in comparison to sensitive genotypes [45]. Likewise, exogenous applications of ferulic acid, caffeic acid and sinapic acid up-regulate the CAT and POX activity in stresses plant of tolerant genotypes. These phenolic acids decrease the H_2O_2 and MDA content in roots and shoots of both sensitive (cv. HD 2329) and tolerant (cv. Kharchia local) cultivars [57].

4. Approaches for salt stress tolerance

4.1 Conventional breeding

Genetic improvement for grain yield, quality traits, biotic stress and abiotic stress including salinity stress are the major breeding objective in wheat. Different methods such as screening of genotypes, pedigree method, hybridization, genetic transformation and marker-assisted breeding have been used in for increasing salt tolerance in wheat. Target breeding for salt stress is mainly done in India and Pakistan. The salt-tolerant genotypes in India are KRL 19, KRL 1–4, KRL 210, KRL 213 and KRL 283 developed by Central Soil Salinity Research Institute, Karnal (India). Almost all the tolerant genotypes in India are developed using Kharchia 65 as donor parent.

Kharchia 65 is collected from Kharchi in Pali district of Rajasthan possessing very high salinity and sodicity tolerance. KRL 1–4 developed in 1990 using the pedigree method from a cross between Kharchia 65 and WL 711 [127]. KRL 19 (PBW 255/KRL 1–4) which can tolerate salinity up to EC 5–7 dSm^{-1} was released in 2000. It has yield potential in saline soil is 2.5–3.5 ton ha^{-1}. KRL 210 (PBW 65/2*Pastor) and KRL 213 (Cndo/r143//Ente/Mexi-2/3 *Aegilops squarrosa* (taus)/4/Weaver/5/2*Kauz) were released in 2010 with yield potential 3.0–5.0 ton ha^{-1}. KRL 283 (CPAN 3004/Kharchia 65//PBW 343) was released in 2018 using bulk selection method [128] with yield potential up to 41 q ha^{-1}. Apart from the released variety, genetic stock of salt-tolerant wheat lines i.e. KRL 35, KRL 99 and KRL 3–4 have been registered with NBPGR. Similarly, two varieties LU26S and SARC-1 were developed in Pakistan by Saline Agriculture Research Cell (SARC) at Faisalabad and one variety Sakha 8 was developed in Egypt by Agricultural Research Centre at Giza [128]. KTDH a double haploid line with good Na + exclusion ability was a product of a cross between Kharchia 65 with TW161. This line matured early and performed average under saline conditions of Spain [129].

4.2 Molecular breeding

Salt stress is a major constraint in wheat production and productivity worldwide. Salt stress causes the accumulation of Na+, Cl^- ions and reactive oxygen species which disrupts the nutrient uptake, hormonal balance leads to a reduction in growth and development of wheat plants. Salt stress tolerance is a polygenic trait governed by multiple QTLs and interaction effects. Understanding the inheritance pattern of slat tolerance is the major step in developing the improved genotypes for salinity stress. Identification of QTLs with major effects helps in marker-assisted selection of salt-tolerant wheat genotypes. Several QTLs associated with salt tolerance-related traits have been mapped in wheat. A major QTL for salt tolerance was identified on linkage group 4DL controlling K/Na ratio in wheat [130]. To enhance the salt tolerance capacity, two major Na + exclusion genes Nax 1 and Nax 2 have been introgressed into durum wheat from *Triticum monococcum* [131, 132]. Genetic analysis mapped Nax 1 and Nax 2 locus on the long arm of linkage group 2A and 5A, respectively. Both of the genes were also introgressed into *Triticum aestivum* cv. Westonia from durum wheat and showed reduced Na + concentration in leaves [133]. In a RIL mapping population between Pasban 90 x Frontana, a total 60 QTLs for various physiological traits related to salinity tolerance has been identified on linkage group 1B, 1D, 2A, 2B, 2D, 3A, 3B, 3D, 4A, 4B, 4D, 5A, 5B, 5D, 6A, 6B, 6D, 7A, 7B and 7D. Out of these, one for chlorophyll a, three for proline content, four each for osmotic potential, superoxide dismutase, chloride content, five each for relative water content and water potential, six for membrane stability index, seven for total chlorophyll and eight for chlorophyll b [134]. Low Na^+ and high K^+ content in leaves is an important cellular mechanism that help the plant to withstand under salt stress condition [73, 135]. For Na content, 3 QTLs were identified by Amin and Diab [136], one by Asif et al. [137], eight by Devi et al. [138], six by Hussain et al. [104], five by Ilyas et al. [134], one by Lindsay et al. [139], four by Masoudi et al. [140] and six by Xu et al. [141]; for K content, four QTLs were detected by Amin and Diab [136], two by Devi et al. [138], five by Hussain et al. [104], four by Ilyas et al. [134], ten by Masoudi et al. [140], and eight by Xu et al. [141]; for K/Na ratio six QTLs were mapped by Amin and Diab [136], two by Asif et al. [137], four by Ilyas et al. [134], twelve by Masoudi et al. [140] and three by Xu et al. [141] (**Table 2**).

Traits	Gen	No. of QTLs	Linkage group	References
Na conc.	DH	3	2B, 4B, 5D	[136]
K conc.		4	2B, 2D, 3D, 5D	
K/Na ratio		6	2B, 2D, 3B, 4A, 5B, 7A	
Growth rate		5	1A, 2A, 2B, 3A, 4A	
Leaf fresh weight		5	3A, 3B, 4B, 5B, 5D	
Leaf dry weight		3	1D, 4B, 5D	
Water content		2	3D, 5B	
No. of spikes/plant		5	2B, 4B, 5B, 7D	
No. of spikelets/spike		6	1D, 2B, 3B, 4B, 5B, 7D	
No. of grains/plant		5	1B, 1D, 2B, 3B, 5D	
Grain weight/plant		5	1D, 2B, 3B, 3D, 5B	
Total dry weight		6	1D, 2A, 3A, 4B, 5D	
Shoot growth	RILs	3	5A, 7B	[137]
Na accumulation		1	2A	
Chloride accumulation		3	1A, 2A, 3A	
K/Na ratio		2	2B, 2D	
Germination %	RILs	5	2A, 2B, 4A, 6D, 7B	[142]
Germination index		5	2A, 2B, 4A, 6D, 7B	
Seedling vigor index		5	2A, 2B, 4A, 6D, 7B	
Root length		12	1B, 2A, 2D, 3B, 3D, 4D, 5A, 5B, 6A, 6B, 6D	
Shoot length		5	2D, 3D, 5D, 6D, 7B	
Seedling fresh weight		7	1D, 2A, 2D, 3B, 3D, 6A, 6B	
Seedling dry weight		5	1B, 2B, 5B, 5D, 6A	
Sodium content	RILs	8	1B, 2D, 5D, 6A, 7D	[138]
Potassium content		2	1B, 2D	
Proline content		3	2D	
Plant height		6	2D, 6A	
Length of ear head		3	5D, 6A, 6B	
Thousand-grain weight		3	2D	
Grain yield		4	1A, 2D, 6A, 7D	
Tiller number per plant		3	2D, 4D, 6A	
Number of earhead		1	4D	
Days to heading		2	2D	
Days to anthesis		1	2D	

Traits	Gen	No. of QTLs	Linkage group	References
Shoot height	RILs	8	1D, 2B, 3A, 3B, 5B	[143]
Shoot fresh weight		11	1A, 1D, 2A, 3B, 4B, 5B, 6B	
Shoot dry weight		5	1A, 3B, 6A, 6B	
Chlorophyll content		7	2B, 5A, 6B, 7B, 7D	
Root boron conc.	F2	3	2A, 2B, 3D	[104]
Root calcium conc.		3	3B, 6B	
Root copper conc.		2	1D, 7B	
Root iron conc.		3	2A, 6A, 6B	
Root potassium conc.		3	2A, 4B, 3D	
Root magnesium conc.		1	5A	
Root manganese conc.		3	2A, 6B	
Root sodium conc.		3	2A, 6A, 7A	
Root phosphorus conc.		1	7B	
Root sulfur conc.		5	2A, 3B, 6B, 7B	
Root zinc conc.		3	2A, 6A, 7A	
Shoot boron conc.		3	3B	
Shoot calcium conc.		2	6A, 6B	
Shoot potassium conc.		2	2A, 6A	
Shoot magnesium conc.		2	2A, 6B	
Shoot manganese conc.		1	4B	
Shoot sodium conc.		3	2A, 7A	
Shoot phosphorus conc.		2	4B, 1D	
Shoot sulfur conc.		3	1A, 2A, 4B	
Shoot zinc conc.		1	7B	
Relative water content	RIL	5	2A, 4A, 7A, 7B	[134]
Membrane stability index		6	3A, 3D, 4A 5B, 7B, 7D	
Water potential		5	2A, 5B, 5D, 6A, 6B	
Osmotic potential		4	2B, 5D, 7A, 7B	
Total chlorophyll		7	1B, 3D, 5B, 6A, 6B, 6D, 7D	
Chlorophyll a		1	7D	
Chlorophyll b		8	1D, 3A, 3B, 4A, 6B, 7A, 7B	
Proline content		3	1B, 4B, 7A	
Superoxide dismutase		4	1B, 1D, 2A, 6D	
Sodium content		5	1D, 2A, 2B, 3B, 6B	
Potassium content		4	2B, 4A, 5A, 6A	
Chloride content		4	1D, 2B, 3B, 7A	
Na/K		4	1D, 2D, 3A, 4D	
Sodium exclusion	F2	1	2A	[139]

Traits	Gen	No. of QTLs	Linkage group	References
Salt tolerance index	RIL	3	3A, 4D, 5A	[144]
Fresh weight of radicle		1	4D	
Dry weight of radicle		3	3A, 3B, 7A	
Fresh weight of plumule		2	3A, 3B	
Dry weight of plumule		1	4D	
Salt injury index		5	3A, 5B, 6B, 6D	
Root fresh weight index		2	4A, 6D	
Shoot fresh weight		6	2A, 2B, 3B, 3D, 4A	
Plant fresh weight index		5	3B, 3D, 4A, 7B	
Root dry weight index		2	3D, 6D	
Shoot dry weight index		2	1A, 3B	
Plant dry weight index		5	2A, 3B	
Root/shoot length ratio		8	1A, 2A, 2D, 3A, 3D, 6A, 6D	
Chlorophyll content		2	3D, 7A	
Shoot height	RIL	7	1A, 1B, 4A, 4B, 6B, 7B	[140]
Shoot fresh weight		9	1A, 1B, 3A, 3B, 6B	
Shoot dry weight		7	1A, 1B, 3A, 3B, 7B	
Chlorophyll content		6	1B, 1D, 3B, 6A, 6B, 7B	
Salt injury index		5	1B, 3B, 6A, 6B	
Shoot Na^+ conc		3	1B, 3B, 5A	
Shoot K^+ conc		3	2A, 2B	
Shoot Na^+/K^+		6	1B, 2D, 3A, 3B, 5A	
Root Na^+ conc		1	7A	
Root K^+ conc		7	1A, 2B, 3B, 3A, 4A	
Root Na^+/K^+		6	1A, 2B, 3A, 3B, 7D	
Na^+ translocation from roots to shoots		3	1B, 2B, 3B	
K^+ translocation from roots to shoots		2	4A	
Maximal root length	RIL	4	1B, 5A, 6A, 7B	[141]
Shoot height		3	4A, 4B, 5A	
Root dry weight		3	2D, 3B, 5A	
Shoot dry weight		1	2A	
Total dry weight		2	2A, 2D	
Chlorophyll content		1	5B	
Root K^+ conc		4	1D, 5A, 5B	
Root Na^+ conc		2	2B, 3B	
Root K^+/Na^+ concentration		3	4B, 5B, 7D	
Shoot K^+ conc		4	2B, 3B, 4B, 6A	
Shoot Na^+ conc		2	5A, 7A	

Table 2.
QTLs associated with salt tolerance related traits in wheat.

DOI: http://dx.doi.org/10.5772/intechopen.102740

4.3 Salt-tolerant gens

One of the approaches to improve salinity tolerance is the identification of genes playing a significant role in the tolerance mechanism. Till now massive information about tolerant gens, transcription factors that are either up-regulated or down-regulated have been identified using genomic or trancriptomics approaches. There is increasing evidence for the involvement of dehydrin and expansion proteins, transcription factors like TaSRG, TaMYB2A, TaNAC29, TdERF1 and Sodium antiporter, transporters and vacuolar pyrophosphatase in the salt stress response in wheat [145–147]. Some of the examples of salt-responsive genes are listed in **Table 3**.

Gene	Protein	Function	References
DHN-5	Dehydrin	Higher seed germination and growth rate, high proline contents, and lower water loss	[148]
TNHX1 and TVP1	Sodium antiporter and vacuolar pyrophosphatase	High relative water and K^+ content,	[149, 150]
TaSTRG		Higher fresh weight, chlorophyll content, proline and soluble sugar contents	[151]
TaMYB2A	Transcription factor	High membrane stability, water retention capacity, high photosynthetic efficiency	[145]
TmHKT1;4-A2 and TmHKT1;5-A	Na + transporters	TmHKT1;4-A2 excludes Na + from the xylem in roots and leaf sheaths; mHKT1;5-A excludes Na + from the xylem only in the roots	[152]
TaSC	Calcium-dependent protein kinase	High germination rate, seedling length, K+/ Na + ratio and proline content	[153]
TaEXPB23	Expansins	Root elongation, improved water retention ability reduced osmotic potential	[154]
TaERF3	Ethylene-response factor	Accumulation of proline and chlorophyll increased while and H_2O_2 content decreased	[146]
TaNAC29		TaNAC29 enhance salt tolerance by reducing H_2O_2 production, membrane damage and by enhancing the antioxidant (SOD, POD, APX and CAT) activity	[155]
TdERF1	Ethylene-response factor	Maintain high membrane stability, and soluble sugar content	[147]
TNHXS1 and TVP1	Sodium antiporter and vacuolar pyrophosphatase	Higher biomass, chlorophyll content and catalase (CAT) activity, more K^+ and less Na^+ content	[155]
TMKP1	Mitogen-activated protein kinase phosphatases	Increased antioxidants activities, namely SOD, CAT and peroxidase, reduced MDA and H_2O_2	[156]
TaPUB1	Ubiquitin-protein ligase	higher germination rate, less reduction in chlorophyll, higher photosynthetic capacity and antioxidants activity and lower Na+/ K+ ratio	[88]
TaEXPA2	Expansin proteins	higher germination rates, longer root length, more lateral roots, higher survival rates and more green leaves, lower Na + but higher K+	[157]

Table 3.
Genes conferring salt tolerance in wheat.

These transcription factors can change the gene expression by specific binding in the promoter region of a large number of genes. Rong et al. [146] characterized the function of ethylene response factor TaERF3 and observed that overexpression of TaERF3 improved the salt and drought tolerance in wheat. Electrophoretic mobility shift assay showed that TaERF3 protein interacted with the GCC-box present in the promoters of seven TaERF3-activated stress-related genes, suggesting that TaERF3 positively regulated wheat adaptation responses to salt and drought stress through the activation of stress-related genes. Similarly, Up-regulation of bZIP genes was found insensitive and down-regulation in tolerant cultivar of wheat under salinity stress [158]. Overexpression of TaEXPA2 an α-expansin gene of wheat provides salt tolerance in transgenic lines of tobacco. The enhanced salt tolerance was associated with improved relative water content, selective ion absorption and increased antioxidant activity. Moreover, ABA signaling positively participated in regulating theTaEXPA2-enhanced salt stress tolerance but how ABA participates in regulating salt stress tolerance needs to be studied further [157].

Author details

Hari Kesh[1], Sunita Devi[1], Naresh Kumar[1], Ashwani Kumar[1]*, Arvind Kumar[1], Pooja Dhansu[2], Parvender Sheoran[1] and Anita Mann[1]

1 ICAR – Central Soil Salinity Research Institute, Karnal, India

2 ICAR – Sugarcane Breeding Research Institute, Regional Center, Karnal, India

*Address all correspondence to: ashwani.kumar1@icar.gov.in

References

[1] FAO. Food Outlook—Biannual Report on Global Food Markets: June 2020. Food Outlook. 2020;**1**. DOI: 10.4060/ca9509en

[2] Nawaz F, Majeed S, Ahmad KS, Hamid A, Shabbir RN, Aqib M, et al. Use of osmolytes in improving abiotic stress tolerance to wheat (*Triticum aestivum* L.). In: Wheat Production in Changing Environments. Singapore: Springer; 2019. pp. 497-519

[3] Asseng S, Ewert F, Martre P, Rötter RP, Lobell DB, Cammarano D, et al. Rising temperatures reduce global wheat production. Nature Climate Change. 2015;**5**:143-147

[4] Qadir M, Quillérou E, Nangia V, Murtaza G, Singh M, Thomas RJ, et al. Economics of salt-induced land degradation and restoration. Natural Resources Forum. 2014;**38**:282-295

[5] Munns R, Gilliham M. Salinity tolerance of crops—What is the cost? The New Phytologist. 2015;**208**:668-673

[6] Kumar A, Sharma SK, Lata C, Devi R, Kulshrestha N, Krishnamurthy SL, et al. Impact of water deficit (salt and drought) stress on physiological, biochemical and yield attributes on wheat (*Triticum aestivum*) varieties. Indian Journal of Agricultural Sciences. 2018;**88**:1624-1632

[7] Datir S, Singh N, Joshi I. Effect of NaCl-induced salinity stress on growth, osmolytes and enzyme activities in wheat genotypes. Bulletin of Environmental Contamination and Toxicology. 2020;**104**:351-357

[8] Soni S, Kumar A, Sehrawat N, Kumar A, Kumar N, Mann A. Effect of saline irrigation on plant water traits, photosynthesis and ionic balance in durum wheat genotypes. Saudi Journal of Biological Sciences. 2021;**28**:2510-2517

[9] Shoresh M, Spivak M, Bernstein N. Involvement of calcium-mediated effects on ROS metabolism in the regulation of growth improvement under salinity. Free Radical Biology & Medicine. 2011;**51**:1221-1234

[10] Jamshidi GK, Riahi-Madvar A, Rezaee F, Pakzad R, Jadid Bonyad F, Ghazizadeh AM. Effect of salinity stress on enzymes activity, ions concentration, oxidative stress parameters, biochemical traits, content of sulforaphane, and CYP79F1 gene expression level in *Lepidium draba* plant. Journal of Plant Growth Regulation. 2019;**6**:1-20

[11] Szabados L, Savouré A. Proline: A multifunctional amino acid. Trends in Plant Science. 2010;**15**:89-97

[12] Hasanuzzaman M, Alam M, Rahman A, Hasanuzzaman M, Nahar K, Fujita M. Exogenous proline and glycine betaine mediated upregulation of antioxidant defense and glyoxalase systems provides better protection against salt-induced oxidative stress in two rice (*Oryza sativa* L.) varieties. BioMed Research International. 2014:757219

[13] Soni S, Kumar A, Sehrawat N, Lata C, Kumar A, Kumar N, et al. Variability of durum wheat genotypes in terms of physio-biochemical traits against salinity stress. Cereal Research Communications. 2021;**49**:45-54

[14] Sheoran P, Kumar A, Sharma R, Prajapat K, Kumar A, Barman A, et al. Quantitative dissection of salt tolerance for sustainable wheat production

in sodic agro-ecosystems through farmers' participatory approach: An Indian experience. Sustainability. 2021;**13**(6):3378

[15] Arif Y, Singh P, Siddiqui H, Bajguz A, Hayat S. Salinity induced physiological and biochemical changes in plants: An omic approach towards salt stress tolerance. Plant Physiology and Biochemistry. 2020;**156**:64-77. DOI: 10.1016/j.plaphy.2020.08.042

[16] Sharma PC, Kumar A, Mann A. Physiology of salinity tolerance in Crop plants. In: Minhas PS, Yadav RK, Sharma PC, editors. Managing Soil Salinity. Karnal: ICAR; 2021. pp. 331-356

[17] Luo MB, Liu F. Salinity-induced oxidative stress and regulation of antioxidant defense system in the marine macroalga *Ulva Prolifera*. Journal of Experimental Marine Biology and Ecology. 2011;**409**:223-228

[18] Sairam RK, Srivastava GC. Changes in antioxidant activity in sub-cellular fractions of tolerant and susceptible wheat genotypes in response to long term salt stress. Plant Science. 2002;**162**:897-904

[19] Goudarzi M, Pakniyat H. Salinity causes increase in proline and protein contents and peroxidase activity in wheat cultivars. Journal of Applied Sciences. 2009;**9**(2):348-353

[20] Woodrow P, Ciarmiello LF, Annunziata MG, Pacifico S, Iannuzzi F, Mirto A, et al. Durum wheat seedling responses to simultaneous high light and salinity involve a fine reconfiguration of amino acids and carbohydrate metabolism. Physiologia Plantarum. 2017;**159**:290-312

[21] Alzahrani O, Abouseadaa H, Abdelmoneim TK, Alshehri MA, Mohamed EM, El-Beltagi HS, et al. Agronomical, physiological and molecular evaluation reveals superior salt-tolerance in bread wheat through salt-induced priming approach. Notulae Botanicae Horti Agro botanici Cluj-Napoca. 2021;**49**:12310-12310

[22] Rahman MA, Chikushi J, Yoshida S, Yahata H, Yasunaga E. Effect of high air temperature on grain growth and yields of wheat genotypes differing in heat tolerance. Journal of Agricultural Meteorology. 2005;**60**:605-608

[23] Chahine K, Sourour A, Youssef T, Hajer S. Salinity effect on plant growth at the seedling stage of durum wheat (*Triticum durum Desf*.). Journal of Plant Breeding and Crop Science. 2013;5:20-25

[24] Rahman A, Nahar K, Hasanuzzaman M, Fujita M. Calcium supplementation improves Na^+/K^+ ratio, antioxidant defense and glyoxalase systems in salt-stressed rice seedlings. Frontiers in Plant Science. 2016;7:609. DOI: 10.3389/fpls.2016.00609

[25] Weitbrecht K, Müller K, Leubner MG. First off the mark: Early seed germination. Journal of Experimental Botany. 2011;**62**(10):3289-3309

[26] Chamorro D, Luna B, Ourcival JM, Kavgac A, Sirca C, Mouillot F, et al. Germination sensitivity to water stress in four shrubby species across the Mediterranean Basin. Plant Biology. 2017;**19**:23-31. DOI: 10.1111/plb.12450

[27] Kumari A, Kaur R. Evaluation of benzyl-butyl phthalate induced germination and early growth vulnerability to barley seedlings (Hordeumvulgare L.). Indian Journal of Ecology. 2018;**45**:174-177

[28] Hasanuzzaman M, Nahar K, Fujita M, Ahmad P, Chandna R, et al.

Enhancing plant productivity under salt stress: Relevance of poly-omics. In: Ahmad P, Azooz MM, Prasad MNV, editors. Salt Stress in Plants: Omics, Signaling and Responses. Berlin, Germany: Springer; 2013. pp. 113-156. DOI: 10.1007/978-1-4614-6108-1_6

[29] Wahid A, Farooq M, Basra SMA, Rasul E, Siddique KHM. Germination of seeds and propagules under salt stress. In: Pessarakli M, editor. Handbook of Plant and Crop Stress. 3rd ed. Boca Raton: CRC Press; 2011. pp. 321-337

[30] Beisson F, Li Y, Bonaventure G, Mike P, Ohlrogge JB. The acyltransferase GPAT5 is required for the synthesis of suberin in seed coat and root of Arabidopsis. Plant Cell. 2007;**19**:351-368. DOI: 10.1105/tpc.106.048033

[31] Yang XJ, Baskin J, Baskin CC, Huang ZY. More than just a coating: Ecological importance, taxonomic occurrence and phylogenetic relationships of seed coat mucilage. Perspectives in Plant Ecology. 2012;**14**:434-442. DOI: 10.1016/j.ppees.2012.09.002

[32] Akbarimoghaddam H, Galavi M, Ghanbari A, Panjehkeh N. Salinity effects on seed germination and seedling growth of bread wheat cultivars. Trakia Journal of Sciences. 2011;**9**:43-50

[33] Sairam RK, Rao KV, Srivastava GC. Differential response of wheat genotypes to long term salinity stress in relation to oxidative stress, antioxidant activity and osmolyte concentration. Plant Science. 2002;**163**:1037-1046. DOI: 10.1016/S0168-9452(02)00278-9

[34] Badridze G, Weidner A, Asch F, Börner A. Variation in salt tolerance within a Georgian wheat germplasm collection. Genetic Resources and Crops Evolution. 2009;**56**:1125-1130

[35] Kandil A, Sharief A, Elokda M. Germination and seedling characters of different wheat cultivars under salinity stress. Journal of Basic & Applied Sciences. 2012;**8**:585-596

[36] Charushahi V, Bargali K, Bargali SS. Influence of seed size and salt stress on seed germination and seedling growth of wheat (*Triticum aestivum*). Indian Journal of Agricultural Sciences. 2015;**85**(9):1134-1137

[37] Alwan AA, Hussein KA, Jaddoa KA. Effect of sodium chloride on response of two wheat cultivars (*Triticum aestivum* L.) at germination and early seedling stages. International Journal of Applied Agricultural Sciences. 2015;**1**:60-65

[38] Goyal E, Amit SK, Singh RS, Mahato AK, Chand S. Transcriptome profiling of the salt-stress response in *Triticum aestivum* cv. Kharchia Local. Scientific Reports. 2016;**6**:1-14

[39] Ahmed IM, Cao F, Zhang M, Chen X, Zhang G, Wu F. Difference in yield and physiological features in response to drought and salinity combined stress during anthesis in Tibetan wild and cultivated barleys. PLoS One. 2013;**8**:e77869. DOI: 10.1371/journal.pone.0077869

[40] Bacilio M, Rodriguez H, Moreno M, Hernandez JP, Bashan Y. Mitigation of salt stress in wheat seedlings by a gfp-tagged *Azospirillum lipoferum*. Biology and Fertility of Soils. 2004;**40**:188-193

[41] El-Hendawy SE, Ruan Y, Hu Y, Schmidhalter U. A comparison of screening criteria for salt tolerance in wheat under field and environment controlled conditions. Journal of Agronomy and Crop Science. 2009;**49**:1-9

[42] Bera AK, Pati MK, Bera A. Bassionolide ameliorates adverse effect

on salt stress on germination and seedling growth of rice. Indian Journal of Plant Physiology. 2006;**11**:182-189

[43] Otu H, Celiktas V, Duzenli S, Hossain A, El Sabagh A. Germination and early seedling growth of five durum wheat cultivars (*Triticum durum* Desf.) is affected by different levels of salinity. Fresenius Environmental Bulletin. 2018;**27**:7746-7757

[44] Guo R, Yang Z, Li F, Yan C, Zhong X, et al. Comparative metabolic responses and adaptive strategies of wheat (*Triticum aestivum* L.) to salt and alkali stress. BMC Plant Biology. 2015;**15**:170. DOI: 10.1186/s12870-015-0546-x

[45] Athar HUR, Khan A, Ashraf M. Exogenously applied ascorbic acid alleviates salt-induced oxidative stress in wheat. Environmental and Experimental Botany. 2007;**63**:224-231. DOI: 10.1016/j.envexpbot.2007.10.018

[46] Afzal I, Rauf S, Basra SMA, Murtaza G. Halopriming improves vigor, metabolism of reserves and ionic contents in wheat seedlings under salt stress. Plant, Soil and Environment. 2008;**54**:382-388

[47] Datta JK, Nag S, Banerjee A, Mondal NK. Impact of salt stress on five varieties of Wheat (*Triticum aestivum* L.) cultivars under laboratory condition. Journal of Applied Sciences and Environmental Management. 2009;**13**:93-97

[48] Saddiq MS, Iqbal S, Hafeez MB, Ibrahim AM, Raza A, et al. Effect of salinity stress on physiological changes in winter and spring wheat. Agronomy. 2021;**11**:4. DOI: 10.3390/agronomy11061193

[49] Poustini K, Siosemardeh A. Ion distribution in wheat cultivars in response to salinity stress. Field Crops Research. 2004;**85**:25-133

[50] Sairam RK, Srivastava GC, Agarwal S, Meena RC. Differences in antioxidant activity in response to salinity stress in tolerant and susceptible wheat genotypes. Biologia Plantarum. 2005;**49**:85-91

[51] Wahid A, Perveen M, Gelani S, Basra SMA. Pretreatment of seed with H_2O_2 improves salt tolerance of wheat seedlings by alleviation of oxidative damage and expression of stress proteins. Journal of Plant Physiology. 2007;**164**:283-294. DOI: 10.1016/j.jplph.2006.01.005

[52] Hasanuzzaman M, Hossain MA, Fujita M. Nitric oxide modulates antioxidant defense and the methylglyoxal detoxification system and reduces salinity-induced damage of wheat seedlings. Plant Biotechnology Reports. 2011;**5**:353-365. DOI: 10.1007/s11816-011-0189-9

[53] Lekshmy S, Sairam RK, Kushwaha SR. Effect of long-term salinity stress on growth and nutrient uptake in contrasting wheat genotypes. Indian Journal of Plant Physiology. 2013;**18**:344-353

[54] Qiu Z, Guo J, Zhu A, Zhang L, Zhang M. Exogenous jasmonic acid can enhance tolerance of wheat seedlings to salt stress. Ecotoxicology and Environmental Safety. 2014;**104**:202-208

[55] Hussain B, Khan AS, Ali Z. Genetic variation in wheat germplasm for salinity tolerance at seedling stage: Improved statistical inference. Turkish Journal of Agriculture and Forestry. 2015;**39**:182-192

[56] Tian F, Wang W, Liang C, Wang X, Wang G, Wang W. Overaccumulation

of glycine betaine makes the function of the thylakoid membrane better in wheat under salt stress. The Crop Journal. 2016;**174**:1-10

[57] Kaur H, Bhardwaj RD, Grewal SK. Mitigation of salinity-induced oxidative damage in wheat (*Triticum aestivum* L.) seedlings by exogenous application of phenolic acids. Acta Physiologiae Plantarum. 2017;**39**:1-15

[58] Sadak MS. Physiological role of trehalose on enhancing salinity tolerance of wheat plant. Bulletin of the National Research Centre. 2019;**43**:1-10. DOI: 10.1186/s42269-019-0098

[59] Liu X, Chen D, Yang T, Huang F, Fu S, Li L. Changes in soil labile and recalcitrant carbon pools after land-use change in a semi-arid agro-pastoral ecotone in Central Asia. Ecological Indicators. 2020;**110**:105925. DOI: 10.1016/j.ecolind.2019.105925

[60] Mohsin SM, Hasanuzzaman M, Parvin K, Fujita M. Pretreatment of wheat (*Triticum aestivum* L.) seedlings with 2, 4-D improves tolerance to salinity-induced oxidative stress and methylglyoxal toxicity by modulating ion homeostasis, antioxidant defenses, and glyoxalase systems. Plant Physiology and Biochemistry. 2020;**152**:221-231

[61] Nassar R, Kamel HA, Ghoniem AE, Alarcón JJ, Sekara A, et al. Physiological and anatomical mechanisms in wheat to cope with salt stress induced by seawater. Plants. 2020;**9**:237. DOI: 10.3390/plants9020237

[62] Cuong DM, Kwon SJ, Nguyen BV, Chun SW, Kim JK, Park SU. Effect of salinity stress on phenyl propanoid genes expression and related gene expression in wheat sprout. Agronomy. 2020;**10**:390

[63] Ashraf M, Harris PJC. Photosynthesis under stressful environments: An overview. Photosynthetica. 2013;**51**:163-190. DOI: 10.1007/s11099-013-0021-6(2013)

[64] Chookhampaeng S. The effect of salt stress on growth, chlorophyll content proline content and antioxidative enzymes of pepper (*Capsicum annuum* L.) seedling. European Journal of Scientific Research. 2011;**49**:103-109

[65] Abedini M. Physiological responses of wheat plant to salinity under different concentrations of Zn. Acta Biologica Szegediensis. 2016;**60**:9-16

[66] Fernandez-Figares I, Marinetto J, Royo C, Ramos JM, Del Moral LG. Amino-acid composition and protein and carbohydrate accumulation in the grain of triticale grown under terminal water stress simulated by a senescing agent. Journal of Cereal Science. 2000;**32**:249-258. DOI: 10.1006/jcrs.2000.0329

[67] Shafi M, Zhang G, Bakht J, Khan MA, Islam UE, Khan MD, et al. Effect of cadmium and salinity stresses on root morphology of wheat. Pakistan Journal of Botany. 2010;**42**:2747-2754

[68] Kafi M. The effects of salinity and light on photosynthesis, respiration and chlorophyll fluorescence in salt-tolerant and salt sensitive wheat (*Triticum aestivum* L.) cultivars. Journal of Agricultural Science and Technology. 2009;**11**:535-547

[69] Chen TH, Murata N. Glycine betaine protects plants against abiotic stress: Mechanisms and biotechnological applications. Plant, Cell and Environment. 2011;**34**:1-20

[70] El-Hendawy SE, Hassan WM, Al-Suhaibani NA, et al. Comparative performance of multivariable agro-physiological parameters for detecting

salt tolerance of wheat cultivars under simulated saline field growing conditions. Frontiers in Plant Science. 2017;**8**:435

[71] Qamar H, Ilyas M, Jan SA, Mustafa HSB, Arshad A, Yar MS, et al. Recent trends in molecular breeding and biotechnology for the genetic improvement of Brassica species against drought stress. Fresenius Environmental Bulletin. 2020;**29**:19-25

[72] Zhang X, Shi Z, Tian Y, Zhou Q, Cai J, Dai T, et al. Salt stress increases content and size of gluten in macropolymers in wheat grain. Food Chemistry. 2016;**197**:516-521. DOI: 10.1016/j.foodchem.2015.11.008

[73] Gil-Muñoz F, Pérez-Pérez JG, Quiñones A, Primo-Capella A, Cebolla J, Ángeles Forner-Giner M, et al. A cross population between D. Kaki and D. Virginiana shows high variability for saline tolerance and improved salt stress tolerance. Plos One. 2020;**15**:e0229023. DOI: 10.1371/journal.pone.0229023

[74] Barutcular C, Yildirim M, Koc M, Akinci C, Topta I, Albayrak O, et al. Evaluation of SPAD chlorophyll in spring wheat genotypes under different environments. Fresenius Environmental Bulletin. 2016;**25**:1258-1266

[75] Munns R. Comparative physiology of salt and water stress. Plant, Cell and Environment. 2002;**25**:239-250

[76] Esechie HA, Al-Saidi A, Al-Khanjari S. Effect of sodium chloride salinity on seedling emergence in chickpea. Journal of Agronomy and Crop Science. 2002;**188**:155-160

[77] Saddiq MS, Afzal I, Basra SM, Ali Z, Ibrahim AM. Sodium exclusion is a reliable trait for the improvement of salinity tolerance in bread wheat. Archives of Agronomy and Soil Science. 2018;**64**:272-284

[78] Elkelish AA, Soliman MH, Alhaithloul HA, El-Esawi MA. Selenium protects wheat seedlings against salt stress-mediated oxidative damage by up-regulating antioxidants and osmolytes metabolism. Plant Physiology and Biochemistry. 2019;**137**:144-153

[79] Munns R, Tester M. Mechanism of salinity tolerance. Annual Review of Plant Biology. 2008;**59**:651-681. DOI: 10.1146/annurev.arplant.59.032607.092911

[80] Shabala S, Pottosin I. Regulation of potassium transport in plants under hostile conditions: Implications for abiotic and biotic stress tolerance. Physiologia Plantarum. 2014;**151**:257-279

[81] Loutfy N, Sakuma Y, Gupta DK. Modifications of water status, growth rate and antioxidant system in two wheat cultivars as affected by salinity stress and salicylic acid. Journal of Plant Research. 2020;**133**:549-570

[82] Di Donato RJ, Roberts LA, Sanderson T, Eisley RB, Walker EL. Arabidopsis yellow stripe-like 2 (YSL2): A metal-regulated gene encoding a plasma membrane transporter of nicotianamine–metal complexes. The Plant Journal. 2004;**39**:403-414

[83] Shaaban Mahmoud M, Abou El-Nour EAA. Nutritional status and growth of wheat plants grown under salinity stress conditions as responded to different nitrogen sources. American Journal of Plant Physiology. 2013;**8**:32-40

[84] Hussain S, Hussain S, Ali B, Ren X, Chen X, et al. Recent progress in understanding salinity tolerance in plants: Story of Na^+/K^+ balance and beyond. Plant Physiology and

Biochemistry. 2021;**160**:250-256. DOI: 10.1016/j.plaphy.2021.01.029

[85] Arfan M, Athar HR, Ashraf M. Does exogenous application of salicylic acid through the rooting medium modulate growth and photosynthetic capacity in differently adapted spring wheat cultivated under salt stress. Journal of Plant Physiology. 2007;**6**:685-694

[86] Dugasa T, Abebie B, Tomer RPS, Barnabas J. Tolerance of *Triticum aestivum* L. (Bread wheat) varieties for growth yield in high salinity soils of Ethiopia. International Journal of Science and Research. 2016;**5**:139-153

[87] Dadshani S, Sharma RC, Baum M, Ogbonnaya FC, Léon J, Ballvora A. Multi-dimensional evaluation of response to salt stress in wheat. PLoS One. 2019;**14**:e0222659. DOI: 10.1371/journal.pone.0222659

[88] Jamal Y, Shafi M, Bakht J, Arif M. Seed priming improves salinity tolerance of wheat varieties. Pakistan Journal of Botany. 2011;**43**:2683-2686

[89] Kalhoro NA, Rajpar I, Kalhoro SA, Ali A, Raza S, Ahmed M, et al. Effect of salts stress on the growth and yield of wheat (Triticumaestivum L.). American Journal of Plant Sciences. 2016;**7**:2257-2271

[90] Chinnusamy V, Jagendorf A, Zhu JK. Understanding and improving salt tolerance in plants. Crop Science. 2005;**45**:437-448. DOI: 10.2135/cropsci2005.0437

[91] Asgari HR, Cornelis W, Damme PV. Salt stress effect on wheat (*Triticum aestivum* L.). growth and leaf ion concentrations. International Journal of Plant Production. 2012;**6**:195-208

[92] Afzal I, Basra SMA, Cheema MA, Farooq M, Jafar MZ, Shahid M, et al. Seed priming: A shotgun approach for alleviation of salt stress in wheat. International Journal of Agriculture and Biology. 2013;**15**:1199-1203

[93] Yadav S, Solanki YPS, Singh V, Yashveer S, Kesh H. Assessment of diversity for yield and its component traits in F2 population of wheat (*Triticum aestivum* L.) under salinity conditions. Applied Biological Research. 2018;**20**:256-261

[94] Singh KN, Chatrath R. Salinity tolerance. In: Reynolds MP, Ortiz-Monasterio JJ, McNab A, editors. Application of Physiology in Wheat Breeding. Mexico: CIMMYT; 2001. pp. 101-110

[95] Hu YC, Schmidhalter U. Drought and salinity. A comparison of their effects on mineral nutrition of plants. Journal of Plant Nutrition and Soil Science. 2007;**168**:541-549

[96] Katerji N, Van Hoorn JW, Fares C, Hamdy A, Mastrorilli M, Oweis T. Salinity effect on grain quality of two durum wheat varieties differing in salt tolerance. Agricultural Water Management. 2005;**75**:85-91

[97] Darvey NL, Naeem H, Gustafson JP. Triticale: Production and utilization. In: Klup K, Ponte J, editors. Handbook of Cereal Science and Technology. 2nd ed. New York: Marcel Dekker; 2000

[98] Abdelgadir EM, Oka M, Fujiyama H. Characteristics of nitrate uptake by plants under salinity. Journal of Plant Nutrition. 2005;**28**:33-46. DOI: 10.1081/PLN-200042156

[99] Massa D, Mattson NS, Lieth HJ. Effects of saline root environment (NaCl) on nitrate and potassium uptake kinetics for rose plants: A Michaelis–Menten modelling approach. Plant and

Soil. 2009;**318**:101-115. DOI: 10.1007/s11104-008-9821-z

[100] Li J, Li X, Guo L, Lu F, Feng X, He K, et al. A subgroup of MYB transcription factor genes undergoes highly conserved alternative splicing in Arabidopsis and rice. Journal of Experimental Botany. 2006;**57**:1263-1273

[101] Tester M, Davenport R. Na^+ tolerance and Na^+ transport in higher plants. Annals of Botany. 2003;**91**:503e527

[102] James RA, Davenport RJ, Munns R. Physiological characterization of two genes for Na^+ exclusion in durum wheat, Nax1 and Nax21. Plant Physiology. 2006;**142**:1537-1547. DOI: 10.1104/pp.106.086538

[103] Singh A, Bhushan B, Gaikwad K, Yadav OP, Kumar S, Rai RD. Induced defence responses of contrasting bread wheat genotypes under differential salt stress imposition. Indian Journal of Biochemistry & Biophysics. 2015;**52**:75-85

[104] Hussain B, Lucas SJ, Ozturk L, Budak H. Mapping QTLs conferring salt tolerance and micronutrient concentrations at seedling stage in wheat. Scientific Reports. 2017;**7**:15662

[105] Wang Y, Nii N. Changes in chlorophyll, ribulosebisphosphate carboxylase-oxygenase, glycine betaine content, photosynthesis and transpiration in *Amaranthus tricolor* leaves during salt stress. The Journal of Horticultural Science and Biotechnology. 2000;**75**:623-627. DOI: 10.1080/14620316.2000.11511297

[106] Bose J, Rodrigo-Moreno A, Shabala S. ROS homeostasis in halophytes in the context of salinity stress tolerance. Journal of Experimental Botany. 2014;**65**:1241-1257. DOI: 10.1093/jxb/ert430

[107] Blum A. Osmotic adjustment is a prime drought stress adaptive engine in support of plant production. Plant, Cell and Environment. 2017;**40**:4e10

[108] Farooq M, Nawaz A, Chaudhry MAM, Indrasti R, Rehman A. Improving resistance against terminal drought in bread wheat by exogenous application of proline and gamma-aminobutyric acid. Journal of Agronomy and Crop Science. 2017;**203**:464-472

[109] Tian F, Wang W, Liang C, Wang X, Wang G, Wang W. Over accumulation of glycine betaine makes the function of the thylakoid membrane better in wheat under salt stress. The Crop Journal. 2017;**5**(1):73-82

[110] Liang C, Zhang XY, Luo Y, Wang GP, Zou Q, Wang W. Over accumulation of glycine betaine alleviates the negative effects of salt stress in wheat. Russian Journal of Plant Physiology. 2009;**56**(3):370-376

[111] Salama KH, Mansour MMF, Al-Malawi HA. Glycinebetaine priming improves salt tolerance of wheat. Biologia. 2015;**70**:1334-1339

[112] Raza SH, Athar HR, Ashraf M, Hameed A. Glycine betaine-induced modulation of antioxidant enzymes activities and ion accumulation in two wheat cultivars differing in salt tolerance. Environmental and Experimental Botany. 2007;**60**:368-376

[113] Rady MM, Kuşvuran A, Alharby HF, Alzahrani Y, Kuşvuran S. Pretreatment with proline or an organic bio-stimulant induces salt tolerance in wheat plants by improving antioxidant redox state and enzymatic activities and reducing the oxidative stress. Journal of Plant Growth Regulation. 2019;**38**:449-462

[114] Ashraf M, Foolad MR. Roles of glycine betaine and proline in improving plant abiotic stress resistance. Environmental and Experimental Botany. 2007;**59**:206-216. DOI: 10.1016/j.envexpbot.2005.12.006

[115] Mahboob W, Khan MA, Shirazi MU. Induction of salt tolerance in wheat (*Triticum aestivum* L.) seedlings through exogenous application of proline. Pakistan Journal of Botany. 2016;**48**:861-867

[116] Hamilton EW, Heckathorn SA. Mitochondrial adaptations to NaCl. Complex I is protected by anti-oxidants and small heat shock proteins, whereas complex II is protected by proline and betaine. Plant Physiology. 2001;**126**:1266-1274

[117] Darko E, Gierczik K, Hudak O, Forgo P, Pal M, Turkosi E, et al. Differing metabolic responses to salt stress in wheat-barley addition lines containing different 7H chromosomal fragments. PLoS One. 2017;**12**:1-20. DOI: 10.1371/journal.pone.0174170

[118] Farissi M, Aziz F, Bouizgaren A, Ghoulam C. Legume-rhizobia symbiosis under saline conditions: Agro-physiological and biochemical aspects of tolerance. International Journal of Innovative Science and Research. 2014;**11**:96-104

[119] Navarro-Yepes J, Burns M, Anandhan A, Khalimonchuk O, Del Razo LM, Quintanilla-Vega B, et al. Oxidative stress, redox signaling, and autophagy: Cell death versus survival. Antioxidants and Redox Signaling. 2014;**21**:66-85. DOI: 10.1089/ars.2014.5837

[120] Huang S, Van Aken O, Schwarzländer M, Belt K, Millar AH. The roles of mitochondrial reactive oxygen species in cellular signaling and stress responses in plants. Plant Physiology. 2016;**171**:1551-1559

[121] Kerchev P, Waszczak C, Lewandowska A, Willems P, Shapiguzov A, Li Z. Lack of *GLYCOLATE OXIDASE1*, but not *GLYCOLATE OXIDASE2*, attenuates the photorespiratory phenotype of CATALASE2- deficient Arabidopsis. Plant Physiology. 2016;**171**:1704-1719

[122] Singh A, Kumar A, Yadav S, Singh IK. Reactive oxygen species-mediated signaling during abiotic stress. Plant Gene. 2019;**18**:100-173

[123] Hasanuzzaman M, Bhuyan M, Zulfiqar F, Raza A, Mohsin SM, Mahmud JA, et al. Reactive oxygen species and antioxidant defense in plants under abiotic stress: Revisiting the crucial role of a universal defense regulator. Antioxidants. 2020;**9**:681

[124] Hossain MA, Hoque TS, Zaid A, Wani SH, Mostofa MG, Henry R. Targeting the ascorbate-glutathione pathway and the glyoxalase pathway for genetic engineering of abiotic stress-tolerance in rice. In: Hossain MA, Hassan KM, Iftekharuddaula A, Kumar A, Henry R, editors. Molecular Breeding for Rice Abiotic Stress Tolerance and Nutritional Quality. Hoboken, NJ: Wiley-Blackwell; 2021. pp. 398-427. DOI: 10.1002/9781119633174.ch21

[125] Sachdev S, Ansari SA, Ansari MI, Fujita M, Hasanuzzaman M. Abiotic stress and reactive oxygen species: Generation, signaling, and defense mechanisms. Antioxidants. 2021;**10**:277

[126] Mandhania S, Madan S, Sawhney V. Antioxidant defense mechanism under salt stress in wheat seedlings. Biologia Plantarum. 2006;**50**:227-231

[127] Hollington PA. Technological breakthroughs in screening/breeding wheat varieties for salt tolerance. In: Proceedings of the National Conference 'Salinity Management in Agriculture. Karnal, India: CSSRI; 2000. pp. 273-289

[128] Sharma SK. Success Stories on Development, Spread and Impact of Salt-Tolerant Varieties of Rice, Wheat and Mustard in India. Karnal: Central Soil Salinity Research Institute; 2010

[129] Hollington PA, Royo A, Miller TE, Quarrie SA, Mahmood A, et al. The use of doubled haploid breeding techniques to develop wheat varieties for saline areas. In: Proceedings of the 3rd Congress of the European Society of Agronomy. Netherlands; 1994. pp. 156-157

[130] Dubcovsky J, María GS, Epstein E, Luo MC, Dvořák J. Mapping of the K^+/ Na^+ discrimination locus Kna1 in wheat. Theoretical and Applied Genetics. 1996;**92**:448-454

[131] Munns R, James RA. Screening methods for salinity tolerance: A case study with tetraploid wheat. Plant and Soil. 2003;**253**:201-218

[132] James RA, Munns R, Von Caemmerer S, Trejo C, Miller C, Condon T. Photosynthetic capacity is related to the cellular and subcellular partitioning of Na^+, K^+ and Cl^- in salt-affected barley and durum wheat. Plant, Cell and Environment. 2006;**29**:2185-2197. DOI: 10.1111/j.1365-3040.2006.01592.x

[133] James RA, Blake C, Byrt CS, Munns R. Major genes for Na^+ exclusion, *Nax1* and *Nax2* (wheat *HKT1; 4* and *HKT1; 5*) decrease Na^+ accumulation in bread wheat leaves under saline and waterlogged conditions. Journal of Experimental Botany. 2011;**62**:2939-2947. DOI: 10.1093/jxb/err003

[134] Ilyas N, Amjid MW, Saleem MA, Khan W, Wattoo FM, Rana RM, et al. Quantitative trait loci (QTL) mapping for physiological and biochemical attributes in a Pasban90/Frontana recombinant inbred lines (RILs) population of wheat (*Triticum aestivum*) under salt stress condition. Saudi Journal of Biological Sciences. 2020;**27**:341-351

[135] Mbarki S, Sytar O, Zivcak M, et al. Anthocyanins of coloured wheat genotypes in specific response to salt stress. Molecules. 2018;**23**:1518

[136] Amin AYMAN Y, Diab AA. QTL mapping of wheat (Triticumaestivum L.) in response to salt stress. International Journal of Bio Technology and Research. 2013;**3**:47-60

[137] Asif MA, Garcia M, Tilbrook J, Brien C, Dowling K, Berger B, et al. Identification of salt tolerance QTL in a wheat RIL mapping population using destructive and non-destructive phenotyping. Functional Plant Biology. 2021;**48**:131-140

[138] Devi R, Ram S, Rana V, Malik VK, Pande V, Singh GP. QTL mapping for salt tolerance associated traits in wheat (*Triticum aestivum* L.). Euphytica. 2019;**215**:1-23

[139] Lindsay MP, Lagudah ES, Hare RA, Munns R. A locus for sodium exclusion (Nax1), a trait for salt tolerance, mapped in durum wheat. Functional Plant Biology. 2004;**31**:1105-1114. DOI: 10.1071/FP04111

[140] Masoudi B, Mardi M, Hervan EM, et al. QTL mapping of salt tolerance traits with different effects at the seedling stage of bread wheat. Plant Molecular Biology Reporter. 2015;**33**:1790-1803. DOI: 10.1007/s11105-015-0874-x

[141] Xu YF, An DG, Liu DC, et al. Mapping QTLs with epistatic effects

and QTL 9 treatment interactions for salt tolerance at seedling stage of wheat. Euphytica. 2012;**186**:233-245. DOI: 10.1007/s10681-012-0647-7

[142] Batool N, Ilyas N, Shahzad A, Hauser BA, Arshad M. Quantitative trait loci (QTLs) mapping for salt stress tolerance in wheat at germination stage. Pakistan Journal of Agricultural Sciences. 2018;**55**:47-55

[143] Ghaedrahmati M, Mardi M, Naghavi MR, et al. Mapping QTLs associated with salt tolerance related traits in seedling stage of wheat (*Triticum aestivum* L.). Journal of Agricultural Science and Technology. 2014;**16**:1413-1428

[144] Ma L, Fengzhou E, Huo N. Genetic analysis of salt tolerance in a recombination inbred population of wheat. Euphytica. 2007;**153**:109-117

[145] Mao X, Jia D, Li A, Zhang H, Tian S, Zhang X, et al. Transgenic expression of TaMYB2A confers enhanced tolerance to multiple abiotic stresses in Arabidopsis. Functional and Integrative Genomics. 2011;**11**:445-465

[146] Rong W, Qi L, Wang A, Ye X, Du L, Liang H, et al. The ERF transcription factor Ta ERF 3 promotes tolerance to salt and drought stresses in wheat. Plant Biotechnology Journal. 2014;**12**:468-479

[147] Makhloufi E, Yousfi FE, Pirrello J, Bernadac A, Ghorbel A, Bouzayen M. TdERF1, an ethylene response factor associated with dehydration responses in durum wheat (*Triticum turgidum* L. subsp. durum). Plant Signaling and Behaviour. 2015;**10**(10):e1065366

[148] Brini F, Hanin M, Mezghani I, Berkowitz GA, Masmoudi K. OverexpressionofwheatNa^+/H^+antiporter TNHX1 and H+-pyrophosphatase TVP1 improve salt-and drought-stress tolerance in Arabidopsis thaliana plants. Journal of Experimental Botany. 2007;**58**:301-308

[149] Khoudi H, Nouri-Khemakhem A, Gouiaa S, Masmoudi K. Optimization of regeneration and transformation parameters in tomato and improvement of its salinity and drought tolerance. African Journal of Biotechnology. 2009;**8**(22)

[150] Zhou W, Li Y, Zhao BC, Ge RC, Shen YZ, Wang G, et al. Overexpression of TaSTRG gene improves salt and drought tolerance in rice. Journal of Plant Physiology. 2009;**166**:1660-1671

[151] Huang X, Zhang Y, Jiao B, Chen G, Huang S, Guo F, et al. Overexpression of the wheat salt tolerance-related gene TaSC enhances salt tolerance in Arabidopsis. Journal of Experimental Botany. 2012;**63**:5463-5473

[152] Han Y, Xiu Li A, Li F, Rong ZM, Wang W. Characterization of a wheat (*Triticum aestivum* L.) expansin gene, TaEXPB23, involved in the abiotic stress response and phytohormone regulation. Plant Physiology and Biochemistry. 2012;**54**:49-58

[153] Xu Z, Wang C, Xue F, Zhang H, Ji W. Wheat NAC transcription factor TaNAC29 is involved in response to salt stress. Plant Physiology and Biochemistry. 2015;**96**:356-363

[154] Gouiaa S, Khoudi H. Co-expression of vacuolar Na^+/H^+ antiporter and H^+-pyrophosphatase with an IRES-mediated dicistronic vector improves salinity tolerance and enhances potassium biofortification of tomato. Phytochemistry. 2015;**117**:537-546

[155] Zaidi I, Ebel C, Belgaroui N, Ghorbel M, Amara I, Hanin M. The

wheat MAP kinase phosphatase 1 alleviates salt stress and increases antioxidant activities in Arabidopsis. Journal of Plant Physiology. 2016;**193**:12-21

[156] Zhang M, Zhang GQ, Kang HH, Zhou SM, Wang W. TaPUB1, a putative E3 ligase gene from wheat, enhances salt stress tolerance in transgenic *Nicotiana benthamiana*. Plant & Cell Physiology. 2017;**58**:1673-1688

[157] Chen Y, Han Y, Kong X, Kang H, Ren Y, Wang W. Ectopic expression of wheat expansin gene TaEXPA2 improved the salt tolerance of transgenic tobacco by regulating Na^+/K^+ and antioxidant competence. Physiologia Plantarum. 2017;**159**(2):161-177

[158] Johnson RR, Wagner RL, Verhey SD, Walker-Simmons MK. The abscisic acid-responsive kinase PKABA1 interacts with a seed-specific abscisic acid response element-binding factor, TaABF, and phosphorylates TaABF peptide sequences. Plant Physiology. 2002;**130**:837-846

Chapter 11

Development of Better Wheat Plants for Climate Change Conditions

Saba Akram, Maria Ghaffar, Ayesha Wadood and Mian Abdur Rehman Arif

Abstract

Wheat is a staple food of about 40% of the world population, and continuous improvement is vital to meet the increasing demands of the world population. Climate change, a serious concern of the present time, could strongly affect the wheat crop. To mitigate the climate change effects on wheat, scientists are developing wheat germplasm tolerant to the number of stresses and for this purpose different strategies have been adopted. In this chapter, the effect of climate change on wheat and strategies to develop a better wheat plant for climate change using advance breeding and molecular techniques have been discussed. Conventional breeding including hybridization, mutation breeding and shuttle breeding are some classical approaches which have led to the development of some high yielding wheat varieties but it's a time taking task, the advancement in science has opened the new window for making a better crop for changing climate. Recent achievements in genetic engineering are expected to augment conventional breeding to further increase production. Advances in genome sequencing and molecular breeding have increased the rate of gene discovery. The use of advance genomic technique is a key to overcome the food security issue related to climate change.

Keywords: wheat, climate change, conventional breeding, genetic engineering, CRISPR-Cas9, genome-wide association studies (GWAS)

1. Introduction

Wheat (*Triticum aestivum*), an important commodity since always, is the central pillar of food security like Plato said that 'A true statesman is never ignorant of wheat'. This crop is staple food of about 40% of the world's population and a source of daily protein for about 2.5 billion people in less-developed countries [1]. It is ranked top in terms of area and 2nd in terms of production globally [2, 3]. Wheat is a rich source of carbohydrates: the whole wheat grain and flour contain 60–70% and 65–75% starch respectively [4, 5]. Additionally, it also contains an appreciable amount of protein (20%), dietary fibers and vitamins [5, 6]. It is a multifaceted crop normally used as food, in the guise of bread, macaroni and because of the elasticity of gluten, it is very popular in Asian countries for chapatti making.

Wheat is one of the most widely cultivated cereal crops with the production of 760.93 million tons and an area of 219.01 million hectares of farmland worldwide [7] which is a 15.4% arable area globally. Its production has increased since green revolution in 1961 from 222.35 million tons to an estimated 775.3 million tons in 2021 [8]. In 2017, the global production of wheat was 751.99 million metric tons however it was increased by 8 million tons in 2018 with the total estimated production of 758.02 million metric tons. A similar trend was observed in 2019 where wheat production was 765.76 million metric tons however the global consumption of wheat assessed by World Agricultural Supply and Demand Estimates (WASDE), was 791.1 million tons for the year 2021 [9]. It has been projected that the wheat demand in developing countries will be increased up to 60% by 2050 which is a stern concern related to food security [10]. Wheat is the main rabi crop in Pakistan covering 38% of the cultivated area and accounts for 13.1% of value agricultural products and because of its staple food status, it occupies a central position in agricultural policies. Pakistan ranks 8th in terms of wheat area and production and 58th in terms of average yield (2805.9 kg ha^{-1}) [11]. Wheat productivity is globally increasing only at 1.1% per annum (p.a.) which is not enough to reach the predicted increase in wheat demand at 1.7% (p.a.) rate until 2050, and even in some regions, the productivity is stagnant [12].

Global wheat demand is skyrocketing in recent years because of many factors; change in eating habits, population trends, socio-economic conditions, especially in Asia and Africa. Among these, population explosion and climate change are the most pressing challenges to food availability in the present and future eras. Fast-rising population levels are putting pressure on land due to urbanization and fuelling global food demand [13]. Economic growth and access to food are important factors in alleviating poverty and hunger (hidden and chronic), although mere access to food is not enough to accelerate the reduction in malnutrition and hidden hunger [14]. Another most important factor is the changing climate and extreme weather conditions which are reshaping the whole picture of food security.

To overcome the drastic effects of climate change there is a need to develop such a plant type that can fight the battle against climate change. In this review, we will through light on some classical and advanced techniques which can be helpful to develop a better wheat plant that can win the war against climate change.

2. Threat of climate change on suitable crop production

According to the Intergovernmental Panel on Climate Change (IPCC) climate change refers to "any change in climate over time, whether due to natural variability or as a result of human activity". However, according to the United Nations Framework Convention on Climate Change, it is referred to as a change of climate that is attributed directly or indirectly to human activity that alters the composition of the global atmosphere [15]. Climate change cause very harsh direct, indirect and socio-economic effect on environment, more importantly on crop grown under this type of environment. Different stresses including high temperature, drought, increased salinity level and flooding arise as the result of climate change. These stresses are the most influencing factors which affect the natural system, human health and agricultural production, especially in developing countries [16]. The expeditious increase in world population indirectly affects the demand and supply chain of food which is a great concern for global environment stability [17].

Climate change is a global phenomenon; however, the noticeable changes in rainfall and temperature in recent years have had an impact on wheat productivity. The elevated temperature will change the plant life cycle by inducing early flowering and fruit sets which will shorten the growth period and the developed seeds would be deficient in nutrients due to increased respiration rate. For each °C rise in temperature 6–13% reduction in the potential yield of wheat will happen. Although the exact consequences of climate change are impossible to predict, the general view is that global crop production will be negatively affected [18, 19]. To overcome this pressure and to meet the future demand for wheat international initiatives were taken [20]. Different international policy-making organizations "The Agricultural Ministers of G20" and "the Consultative Group for International Agricultural Research (CGIAR) research centers" keep climate change and food security a key priority area and motivate the need to further see the sights how one key staple food may be influenced by efforts to make the food system more resilient [21, 22].

Adverse effects may happen through increasing levels of CO_2, temperature, pests and diseases [23], and deteriorating quality and yield attributes [24]. The frequency of extreme weather events such as droughts and floods also increase in the response to changing climate [25]. One of the main reasons for changing climate and continuous elevation of CO_2 is deforestation; the level of CO_2 elevated from 280 μmol^{-1} to 400 μmol^{-1} and the prediction tells that might grow into two folds (800 μmol^{-1}) up to the end of this century [26].

Food insecurity is an emerging issue of today's era that is a result of climate change. Almost 815 million people around the globe are facing hunger and malnutrition, hampering viable development programs to accomplish the worldwide goal of stamping out hunger by 2030 [27]. The adverse climatic conditions, mainly elevated temperature, is causing a threat to food security and agricultural yield [28]. The inhabitants are likely to grow up to 9 billion by 2050 and food supplies are expected to accelerate by about 85% [29]. Environmental supremacy is going from bad to worse comprising low variation and high application of inputs, and unbalanced output due to climatic variations in crops [30]. The escalated spells of drought and heavy precipitation, elevation in temperature, salinity, and disease attacks are expected to decrease crop production and leads to higher threats of famine [31]. The best possible way to tackle this problem is the development of climate-smart cultivars.

3. Stresses as an outcome of climate change

In recent years, the environment has been significantly affected due to climate change, the most expected area is agriculture, or the agriculture crops grown in these environments. As a result of climate change, the elevation in CO_2 and temperature was observed by different scientists [32]. These are major limitations that develop a gap between the supply and demand of food and lead most researchers into looking for good adaptation strategies for plants under these conditions [33], by developing climate-smart crops that are resilient against climate change [34]. Sensitivity to this kind of stress causes a serious effect on the plants; like disruption in the plant metabolism processes, thereby resulting in the reduction in felicity and quality of agricultural crop production [35]. There are two types of stresses: biotic and abiotic. Biotic stress in plants occurs by the infestations of living entities like viruses, bacteria, nematodes, fungi, insects and weeds, however, abiotic stresses include drought, eminent CO_2, temperature (low and high) [36], waterlogging, high precipitation, increased sunshine intensity and chemical factors (heavy metals and pH).

3.1 Biotic stresses

When we talk about the wheat crop, different biotic factors (including diseases and pests) come under consideration which limits wheat production. These insect pests and diseases are distributed worldwide and some of these exist in major wheat-growing areas which are destructive for wheat production. Karnal bunt and Russian wheat aphid are more dangerous and cause a heavy loss in yield in their hotspot [37]. Main wheat diseases include yellow rust/stripe rust, tan spot and leaf rust/brown rust caused by *Puccinia striiformis* west, *Pyrenophora tritici-repentis* (Died.) and *Puccinia triticina* respectively. Leaf rust causes considerable yield losses in wheat by disrupting the photosynthesis process of leaves which ultimately result in stunted growth, decreased number of grain per spike, shrined seeds and eventually a huge loss in production [38]. When the onset of the disease is early in the growth cycle of the plant the loss increases up to 50% [39]. Stripe rust is also a major disease prevalent in temperate regions and results in 10–70% yield losses [40]. Different chemical treatments and agronomic practices are available to overcome these diseases, but the development of resistant cultivars is the most economical and effective strategy.

3.2 Abiotic stresses

Abiotic stresses like drought [41], heavy metal stress [42], and salinity significantly affects the average yield of crops including wheat [43]. Approximately 9% area of the globe is under cultivation and 91% of that is affected by different stresses. Statistical models predict that a 10% reduction in wheat yield is due to extreme weather, global warming, resulting in increased evaporation rate and reduction in precipitation [44], and more specifically because of drought [45]. Abiotic stresses contribute 50% loss in yield of different crops which includes temperature (20%), salinity (10%), drought (9%), cold (7%), and other stresses (4%) [46]. Water scarcity is a global issue that takes place in any wheat-growing region which causes osmotic stress. In United states, the losses due to drought reached up to $6–8 billion per year which is a threat to global food security [44, 47]. The most frequent spells of drought are the result of global warming, which is a serious concern for wheat yield [44]. Temperature plays a key role in balancing normal crop growth and development which ultimately regulates the crop yield [48]. Wheat survives in a broad range of temperatures, the upper and lower limits of temperature for wheat survival are -17 ± 1.2°C to 47.5 ± 0.5°C, respectively [49]. The daily high surge in temperature is 25–35°C crossways the world wheat-growing areas [50]. The most affected growth stage by temperature stress is flowering followed by germination which is delayed under heat stress due to the alternation in metabolic activities of nearby soil temperature [51]. The result of a delay in germination is low crop density. The most adverse effect of heat stress occurs during the anthesis and seed setting stage and leads towards significant yield loss [52]. An increase in salt accumulation in soil cause physiological drought which decreases the ability of plants to take up water from the soil, [53]. Similarly, heavy metals also affect wheat plants from germination to growth as well as biochemical mechanism and ultimately the reduction of yield. All above-mentioned stresses in wheat-growing areas bring decisive yield loss of wheat.

4. Impact of climate on wheat productivity

Climate change is a major challenge for wheat productivity, which includes declining water availability, increased temperature and different insect pests which cause a serious reduction in the crop. The first step to mitigate climate change is to assess the possible damages and adaptation strategies to accomplish the size and nature of these effects on crop productivity. As wheat is the staple food of many countries its importance is amplified concerning food security, so, it is a need of time to measure the response of wheat to changing climate. The response of wheat plants to changing climate is different with different stages of growth and development including germination, growth and maturity. High temperature is an imperative variable to study which affects the wheat crop throughout the growth cycle. Similarly, rainfall also has an important positive effect if it occurs at proper time with a proper amount at critical stages of growth. Therefore, the estimation of the effect of climate change on wheat productivity can provide important visions for adaptation [54].

4.1 Germination

Germination is the most sensitive stage in the whole wheat life cycle which affects crop density and uniform maturation, which eventually expedite an important role in yield. Extreme alternation in the immediate environment of a germinating seed can inhibit germination processes, eventually leading to possible yield loss due to a drop in cropping density [52]. Different studies suggested that under salt-affected soil, the germination rate of wheat decreased and the seed took more time than normal to germinate. The scientific reason behind that phenomenon is the reduction in osmotic potential of germination media, which restrict seed imbibition. Salinity also destroys the food reserve of the seed by imbalancing the hormonal status of the seed [55]. Other factors impelling germination include seeds' dormancy, age, seed coat hardness, vigor, polymorphism, temperature, light and gases [56]. The delay in germination may leave crops vulnerable to heat stress at the end of growing seasons or promote uneven maturation of crops [57]. Wheat seed is also sensitive to different chemical and physical conditions such as the presence of heavy metals in the rhizosphere which cause a reduction in germination and affects seedling vigor [58]. Recent studies have documented that heavy metals inhibit the storage of food mobilization, stunt the radical formulation, disrupt the cellular osmoregulation and degrade the proteolytic activities, eventually causing inhibition of seed germination and seedling development [59].

4.2 Physiology

Physiology plays an important role in the growth of any plant. Photosynthesis is an important physiological process for plant development and survival that is greatly affected by environmental conditions. Higher accumulation of salts resulting from climate change primarily lessens the water potential and store Na^+ and Cl^- ions in the chloroplast, which inhibit the photosynthesis process [55]. According to Arfan et al. [60] transpiration rate, stomatal conductance, net CO_2 assimilation and sub stomatal CO_2 concentration were decreased by salinity stress at 150 mM NaCl. Similar to the salt stress, drought/osmotic stress disrupt the photosynthesis process of wheat plants, can damage sugar synthesis essential to drive yield in wheat crops but also leads to stomata closure by turgor loss through reduction of internal water contents. It leads to death of plant by disturbing metabolism [61, 62]. Plant physiological processes are also sensitive to a higher temperature, heat stress cause the deactivation of Rubisco enzyme, reduced photosynthetic capacity, assimilated translocation reduces, brings premature leaf senescence, decrease chlorophyll content and ultimately decrease in yield [63]. High temperature also affects the starch and protein content in grain and induce the production of reactive oxygen species (ROS) which cause a change in membrane stability along with lipid peroxidation, protein oxidation and damage to nucleic acids [64]. Thus, all the stresses are significant variables that emphasis the scientific community to develop climate-smart wheat varieties to tackle food security issues.

4.3 Yield

All the biotic and biotic stresses such as high temperature, water scarcity and frost abate the wheat yield by reducing grain number, grain size and single grain weight. However, how and which yield component will get affected by certain stress depends upon its duration, intensity and timing [65]. For example; if the stress occurs before and during anthesis it reduces the number of grains per ear due to an increased seed abortion. However, when stress occurs after anthesis it does not influence the grain number but effect grain size by shrinking the grain and single grain weight by inhibiting grain filling [66]. Wheat grain yield and number of tillers decreased 53.57% and 15.38% respectively under heat stress [67]. The influence of heat stress is highly significant during the reproductive phase. The increase of 1°C average temperature during the reproductive stage may lead to a higher loss in grain yield [64]. It is also important to note the importance of the flag leaf when looking at yield and grain filling [68]. The flag leaf contributes approximately 30–50% of seed carbohydrates; therefore, any damage to the flag leaf would negatively impact yield [69]. When we talk about the wheat yield loss due to biotic stresses, then the leaf rust (LR) is the main widely spread biotic stress. In the United States, economic losses of $350 million were attributed to LR between 2000 and 2004. In China, annual yield losses due to LR are estimated at 3 million tons [39]. According to a recent estimation, annual yield reductions of 5.47 million tons of wheat are attributable to yellow rust disease, which is equivalent to annual losses of $979 million [70]. A detailed analysis of wheat grain yield and its yield component is crucial to identify genomic regions responsible for grain yield and stress tolerance [71].

5. Approaches to scrimmage against climate change

The fluctuation in climatic conditions directly affects morphology, phenology and physiology of plants and indirectly affects the productivity by alteration in soil biota, fertility, and water and nutrients availability. Keeping in view the current status of wheat production, it is predicted that the wheat productivity will be 1 t/ha short to meet the global demand by 2050. Variation in climate, change in pest and a pathogens life cycle and new variants will further aggravate the situation by threatening global food security. Thus, in future, food security will face a four-fold challenge: upward pressure on demand with downward pressure on supply and the need for sustainable production [72]. All these factors are interlinked and their collective reinforcement will amplify the burden on food demand and require a revolutionized food system [73]. Climate change in short affects plants, their environment and society at large.

Breeding for disease-resistant, climate stress-tolerant and potentially high yielding wheat will improve productivity to meet future demands.

5.1 Conventional approaches

Conventional breeding achieves incremental yield gains by recombining alleles mainly from within elite materials and selecting among thousands of progeny per cross for expression of appropriate agronomic traits, resistance to a spectrum of prevalent diseases and yield based on multi-location trials [1]. Crossbred through conventional breeding is only possible between the same or closely related species. The absence of gene of interest (GOI) in the natural gene pool puts limitations on the introgression for the creation of varieties with desirable traits. Therefore, hunting for an alternate source of GOIs in distantly related plant species and even in microorganisms is necessary [74]. Plant breeding programme's success strongly depends on the climate, market demand and trends. Genomic selection helps in multiple quantitative traits prediction in genotypes from breeding pipelines [75] and by attaining historical phenotypes and adding high-density genotypic information.

5.1.1 Hybridization

In wheat, hybrid cultivar and commercial seed production are still limited to a specific sector as compared to other cereal crops like rice or maize [76]. Conventional breeding by backcrossing is a method to improve an elite line by adding a new trait. An F_1 hybrid is obtained by crossing a donor line carrying GOI with an elite line and then F_1 hybrid is recurrently back-crossed with the elite line until 5-8th generation. The final genotype will be a product of characteristics of the elite line and will carry the introgressed GOI [77]. Wheat is a self-pollinated crop with an out-crossing rate of <1%, so, execution of effective cross-pollination techniques between the wheat elite lines that can overcome the autogamous mode is needed. This can be achieved by crossing between a male-sterile female plant with good pollen recipient properties and a male plant with good pollen shedding properties. Efforts have been made to develop maternal plants with cytoplasmic male-sterility (CMS) for wheat breeding e.g. CMS systems were identified in wheat (i.e. *Triticum timopheevii*) (Angus, 2001) e.g. four new alien CMS (Ae. kotschyi, Ae. uniaristata, Ae. mutica and Hordeum chilense) were discovered [78]. Due to the Bottleneck effect in bread wheat, a potent source of genetic diversity is required. The gene pools of Triticeae, which includes the primary, secondary, and tertiary gene pools are a rich source of genes that can be used to improve traits such as; abiotic and biotic stress tolerance including disease, herbicides and extreme climatic conditions. The plasticity of the wheat genome is depicted by the fact that novel alleles from 52 species have been introgressed into wheat [79]. Landraces, another crucial gene pool, are also reported to contribute genes for yield improvement in irrigated environments or, in drought and heat-stressed environments [80]. *Rh*t dwarfing gene is one of the best examples which originated from a Japanese landrace "Shiro Daruma" and was introgressed into the first dwarf wheat variety "Norin10" [81]. These same genes were utilized by the famous Dr. Norman E. Borlaug to develop the semi-dwarf and high-yielding wheat varieties that triggered the Green Revolution. Breeding efforts to cope with the upcoming foreseeable future and the speeding up of the genetic gain from the current rate of <1% per year will depend on the following four strategies: 1) use of germplasm from exotic sources to broaden the gene pool and overcome the bottleneck effect due to conventional breeding [82]; 2) strategic hybridization to combine Radiation use efficiency (RUE) associated; 3) empirical methods and skill to identify individual plants with desirable traits and extrapolate the observation to increase the efficiency of conventional breeding; 4) advance techniques like molecular-assisted breeding, genome-wide assisted selection (GWAS) and high-throughput phenotyping to permit the efficient utilization of trait-linked markers as they are identified through gene discovery and GWAS modeling [83].

5.1.2 Mutation

Ever since the epoch-making discoveries made by Muller and Stadler [84], the application of mutation techniques by using different chemical and physical agents have played a significant role in modern plant breeding and genetic studies by generating a vast amount of genetic variability [85]. The narrow genetic diversity of the cultivars imposes the prime challenge in the development of varieties with high yield, stress tolerance, and improved traits like early maturity, seed size and nutrition value [86]. Hugo De Vries coined the term mutation, to indicate a sudden change in the genotype that is heritable [87] and these genetic variations provide the raw material for evolution. The rate of spontaneous mutation is relatively very low i.e. 10^{-6} or one out of a million for an individual gene [88], therefore, artificial mutations are necessary to increase the percentage of genetic diversity. The process of inducing desirable mutations and exploiting them for crop improvement is called mutation breeding, which comprises three main steps; using mutagens, screening of mutant candidates for desirable traits, and official release of the new variety [89]. The widespread use of induced mutants in plant breeding programmes across the globe has led to the official release of 265 wheat plant mutant varieties in more than 24 countries throughout the world (**Figure 1**).

According to [82], mutation breeding has a major advantage over other methods is that in this process no genetic material is lost, rather only mutation is induced in the preexisting genome. It offers the possibility of inducing such unique desired traits that were either lost during evolution or do not exist naturally.

5.1.3 Shuttle breeding

The shuttle breeding concept was originally developed by the International Maize and Wheat Improvement Center's (CIMMYT) wheat breeding program and was popularized by Nobel laureate Dr. Norman Borlaug. This system allowed an extra generation to be advanced each year by using different field locations. CIMMYT used two contrasting locations with diverse environmental conditions in Mexico for wheat shuttle breeding: Ciudad Obregón, an irrigated dessert located in Northern Sonora Valley and Central Mexican highlands (2249 m altitude). Since the beginning of this programme in Mexico, segregating populations have been "shuttled" about 130 times representing 200,000 crosses. Off-season breeding activities through shuttle breeding has the advantages of screening segregating material in contrasting locations for developing high yielding, disease-resistant, widely adapted and photoperiod insensitive genotypes of wheat within a limited period [91]. Additionally, Borlaug and his team noticed two more advantages of shuttle breeding: first, breeding in locations with different environmental conditions, soil types and stresses allow selection of breeding materials for broad range disease resistance; secondly, photoperiod-sensitive material is screened and eliminated. In this way, the resulting photoperiod-insensitive

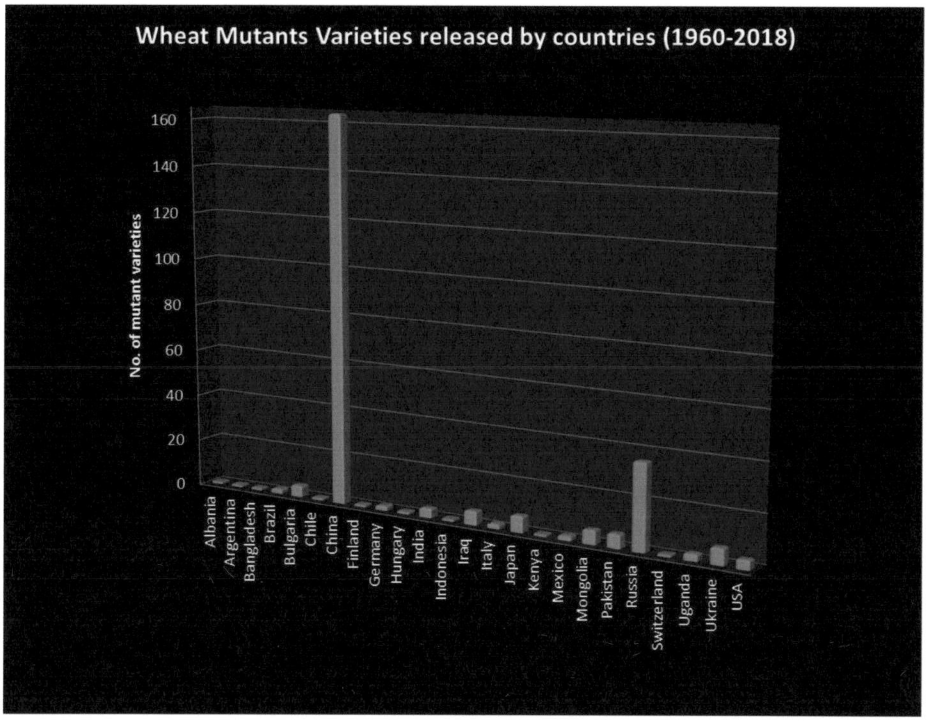

Figure 1.
Wheat mutant varieties released during 1960–2018 [90].

germplasm permitted CIMMYT's semi-dwarf high yielding and disease resistant lines to adapt in multi-range environmental conditions worldwide. Shuttle breeding was the foundation of the success of what we today call "the Green Revolution" [92].

5.2 Chromosome doubling

In the second half of the twentieth century, the emergence of doubled haploid (DH) technology revolutionized the generation of genetically pure and homozygous lines and led to the direct production of completely homozygous lines from heterozygous plants in a single generation. Double haploids production by chromosome doubling, spontaneous or by using colchicine, of haploid cells like pollen grains, which greatly shortens the line fixation stage, at least three to four generations of self-pollination, is a means of accelerating the wheat breeding for development of true breeding lines with desirable traits [93]. This technique includes two main steps: haploid induction and chromosome doubling. Haploid induction attempts to regenerate haploids or spontaneous DH plants, which can be achieved by gynogenesi, androgenesis or parthenogenesis, depending on the species. Antimitotic compounds are used for the chromosomal doubling step, which is mandatory if spontaneous doubling does not occur in haploid plants [94].

This process is performed in tissue-culture laboratories and applies to species that are responsive to tissue culturing this technique could complement the conventional breeding programs to accelerate the release of new varieties. In wheat various methods are employed to develop DHs including; isolated microspore culture (IMC), haploid gene inducer, meiotic restitution genes, doubling chemicals, ovule culture, chromosome removal using hybridization, wide hybridization and anther culture [95]. AC and wide hybridization methods are frequently used in applied research and breeding programs [96] while IMC is still under development [97].

5.3 The genetic and genomic course of action

5.3.1 Omics approaches for amelioration of wheat

Omics approaches are useful in deciphering the whole mechanism and thus providing insight into modification at the molecular level which results from changes in environmental conditions. Omics is a diverse branch that includes genomics, transcriptomics, proteomics, metabolomics, and their interactions with each other. The period of omics has been commenced with the advent of automated sequencing approaches which lead to the first whole-genome sequencing of model plant i.e., *Arabidopsis thaliana*. Advancement in sequencing techniques led to high throughput next-generation sequencing (NGS) followed by a new era of genome-scale molecular analysis with modeling of various molecular and physiological parameters and their correlation provides an accomplished move to deal with different stresses due to climate change. Bread wheat (Allohexaploid, 2n = 6x = 42, AABBDD genomes) has one of the most intricated genomes. The homologous chromosomes containing similar genes mess up the whole biological network. A total of 124,000 gene loci in the wheat genome covering all the three sub-genomes (A, B and D) are involved in a diverse network of biological approaches. Furthermore, transcriptomics (RNA level) and proteomics (protein level) helped in understanding the functions of RNA and proteins respectively. All genes are not transcribed at the same time; therefore, phenotype cannot be fully understood by genomic studies. Thus, the successful combination of genomics (genes), transcriptomics (RNA), proteomics (proteins), and metabolomics (metabolite) will assist in the decoding of diverse metabolism in plants and facilitate breeders to select potential and best traits to improve crop productivity under different stresses due to climate change [42].

5.3.2 Genomics Progress in wheat

Genomics aims at exploring the genome physical structure, studying the whole constitution of the genome including genes and regulatory network. A major milestone in the wheat genome has been achieved in 2012 with the complete de-novo sequencing of bread wheat. Sequencing revealed that A, B and D wheat genomes consist of 28,000, 38,000, and 36,000 genes respectively [98].

5.3.3 Marker assisted breeding perspective in wheat

Upon the advancement in genomics and the advent of the molecular markers era, myriads of shortcomings of conventional breeding approaches are resolved as they are not impacted by environmental factors and can expose variations at the DNA level. Classical breeding is based on the phenotypic selection of genotypes. Genotype X Environment (GE) interaction is the main constraint including the time-consuming and costly procedure of phenotypic selection. By employing molecular markers, desirable genotypic selection can be done at the early generation of the breeding program without the influence of environmental factors. Breeders use molecular markers to enhance the precision of the selection of genetic resources for the best trial combination.

The first study based on molecular markers was initiated in the 1990s when restriction fragment length polymorphism (RFLP) markers were used for identifying genetic diversity, homologous chromosome identification and wheat-rye identification [99]. The use of RFLP is to be sure very successful in the development of linkage groups in wheat. However, it was not so much intriguing due to time consuming, laborious, low frequency and high cost. With more improvement in the marker system, researchers, later on, focused on PCR-based markers including Randomly Amplified Polymorphic DNA (RAPD) and Simple Sequence Repeats (SSR) due to their mapping friendly and cost-effective features. Among PCR-based markers, RAPD was not used extensively due to the availability of scanty information about the location in the genome and lack of reproducibility [100]. Compared to RFLP, SSR markers are reproducible and have a specific location in the genome thereby, more applicable in genome-specific studies. In wheat, the first SSR markers system was reported in 1998 which opens up a new direction for identifying new genetic loci and better yield traits [101]. With time, researchers focused on single nucleotide polymorphism (SNP) and developed trait linked SNP markers. It has higher accuracy than SSR markers. A variety of trait linked DNA markers for wheat were identified for disease resistance and quality of grain. For example, Cre resistance genes (Cre3, Cre1) are used in marker assisted selection (MAS) program of wheat to identify cereal cyst resistant genotypes [102].

5.3.4 Genome-wide association studies (GWAS)

Identification of gene function is a long-standing goal of biology which provides important information for crop improvements. So far, forwards genetics has been the prime approach in which first we mutate the plant, followed by phenotypic screeing to identify the gene function. The identification of genes with major effect is easy as comparision to the gene with minor effect. To overcome this barrier, association mapping (AM) and bi-parental quantitative trait loci (QTL) mapping was introduced with ability to identify genes with subtle effects [103]. Subsequent aim of genetic is to identify a link between a phenotypic function and genotypic data, and AM is one of the approaches to link the phenotype with genotypes. Revolutionary AM orianted approaches were carried out in last decade [169]. Genome-wide association study (GWAS) varies from bi-parental QTL mapping because it is performed on a natural population with a wide genetic base and this wide track of natural variation provides finer resolution of QTL location [103]. The basic apprehension of AM is to identify superior associations (false positive) that can result from population stratification and enigmatic relatedness [104, 105]. To control this issue different statistical methods have been adopted, a mixed linear model (MLM) with population structure and kinship matrix incorporation efficiently eliminate false positive in association mapping [106, 107]. Sequenced-based GWAS has successfully been applied for mapping the agronomic traits and identified the candidate genes inside the significant agronomic regions of wheat [108]. GWAS is a powerful tool to identify the genomic region linked with different traits (linked with biotics and abiotic stress tolerance) in different crops including wheat. It generally highlights linkage among SNPs and traits and is based on GWAS design, genotyping tools, statistical models for examination, and results from interpretation [109]. Using GWAS, Sukumaran et al. [110] detected multiple significant QTL associated with yield and its linked traits of durum wheat grown under drought and heat stress. Similarly, some other studies identified QTL associated with heat and drought tolerance related traits at the seedling stage in wheat [109, 111]. However, limited studies on drought tolerance of wheat have been conducted at the seedling stage.

5.3.5 Genomic selection (GS) in wheat improvement

One of the important technologies utilized in the improvement of the plant is genomic selection along with doubled haploid production, sequencing, QTL mapping, association mapping, genome editing and formation of transgenic plants is genomic selection. In genomic selection, genome-wide markers are used to identify the genotype of a plant and subsequently phenotyped for a particular trait by selection. Contrary to the marker-assisted selection, which utilizes a small number of markers associated with major QTL, GS involves genome-wide markers along with phenotyping data to evaluate genomic estimated breeding values (GEBVs) in one population that will previse the performance of lines in another population only using markers. This technique avoids multiple testing and there is no need to identify marker-trait associations based on arbitrarily chosen significance threshold [112].

Due to the complex genetic makeup of wheat, it requires 10–15 years to transfer new genes into elite germplasm. Genomic selection makes it possible to select parents purely before enter in field trials and nurseries based on genomic estimated breeding values. Annual genetic gain through GS is predicted to be double or triple that of conventional selection due to alleviation in the selection cycle. However, there is still little information regarding GS application in wheat. Improved predictive ability to target traits is cardinal to successful implementation of GS [113]. It is considered that item-based collaborative filtering (IBCF) could be used alternative to conventional predictive model for important target traits in a wheat breeding program [114].

5.3.6 Transcriptome profiling of wheat

RNA sequencing technologies give abundant transcriptomic data which requires expertise in bioinformatics. The wheat hexaploid genome has one of the largest genomes in different crop species constituting 17 Gb in size. Until now, approximately 76% of the wheat genome has been sequenced (International Wheat Genome Sequencing Consortium [IWGSC], [115]. Functional annotation of the wheat genome by homology is becoming very useful but is far from complete as compared to model plants. Transcriptomics in wheat has been facing many challenges due to its complicated genome. Furthermore, RNA sequencing and proteomics study will help in the production of markers associated with particular traits to improve the breeding program. Okada et al., [116] reported that the transcriptional profile of wheat was very useful in the development of molecular markers and was used for the study of wild relative of wheat (Ae. Umbellulata) for population genetics studies. Moreover, many biotic and abiotic stresses can also be studied using expression profiles like drought tolerance mechanisms of two cultivars (Alpowa and Idaho) were studied by Alotaibi [117] using RNA sequencing profiling tool. They identified that differentially expressed genes were 2.32 and 3.9 times more up-regulated and down-regulated respectively in Alpowa as compared to Idaho.

5.3.7 Proteomics in wheat

Proteins play a cardinal role in stress responses, therefore, proteome alterations at different stressed conditions need to be deciphered for the comprehensive understanding of related mechanisms. Stress sensing is the initial pathway to respond to stress conditions followed by the signaling process. For a better understanding of stress coping mechanisms of plants, isolation and characterization of stress-responsive proteins is required. Further, comprehension of post-translational modifications is also needed in plants under stressed conditions. Proteomics plays a very important role in the fine-tuning of pathways that are involved in stress alleviation [42]. For the comprehensive understanding of functional proteomics, there is a dire need to focus on the subcellular proteomics of wheat. To this end, the isolation of proteins from a target organelle is challenging. The conventional approach for the fractionation of subcellular organelles is differential and density gradient centrifugation. Free-flow electrophoresis is also used for the subcellular fractionation based on the isoelectric point of proteins. Despite the diverse application of various proteomics techniques, various subcellular proteins including both stress-induced proteins and housekeeping proteins, remain unclassified. Thus, wheat proteomics data will address the physiological role of the plant under stressed conditions [118].

5.3.8 Metabolomics in wheat

Improvement in genetics is required for the development of new wheat varieties that can work efficiently under stressed conditions. Improvement in the genetics of wheat cultivars would lead to changes in physiological and biochemical responses. Likewise, their change in the metabolic profile that is related to a particular phenotype would result in the development of metabolic markers. Wheat is a crop of higher latitude, therefore, heat stress changes the metabolites in the wheat plant during early summer and terminal heat [119]. Physiological and morphological traits are also important, but they cannot provide the overall picture of the underlying mechanism with changes in metabolic profile under stress conditions. With the advancement in omics techniques, mass spectrometry provides the metabolic profiles of crop genotypes [120]. The metabolic profile of wheat revealed that highly branched amino acids are intolerant in water-deficit stress conditions [121]. It is also reported that different groups of peroxidase genes (TaPrx112-D, TaPrx113-F and TaPrx111-A) were induced by cereal cyst nematodes in some of the resistant wheat lines [122]. Taken together, an amalgamation of wheat "Omics" data including genomics, transcriptomics, proteomics and metabolomics with advanced bioinformatics tools is required to construct a mathematical model that will provide a deep insight into the underlying mechanism of plant undergoing stress condition.

5.3.9 Genetic modification

5.3.9.1 Transformation

Gene transformation is a technique through which the foreign DNA/gene is transferred into target species using molecular methods. Transformation efficiency depends on regeneration frequency of donor tissue (e.g. shoot), the procedure utilized and embryogenesis from somatic or pollen tissue [123]. In monocots, the main challenges for gene transformation are regeneration of explant and difficulties in DNA delivery using monotonous methods of gene transformation [124]. Improvement in DNA delivery methods and advancement in protocols for developing transgenes have led to the expansion of wheat genome sequence information, high-density molecular markers mapping and cloning of several wheat genes [125]. The gene transformation methods can be classified into direct and indirect gene transformation methods (**Figure 2**) [126].

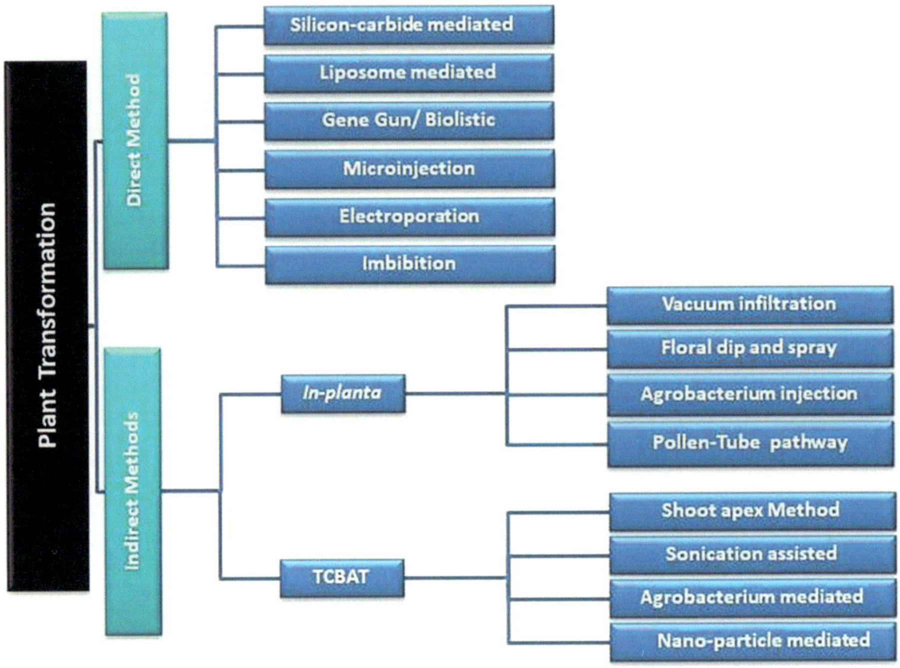

Figure 2.
Gene transformation methods (direct and indirect).

5.3.9.2 Biolistic transformation

The first successful wheat transformation was reported using particle bombardment of embryogenic callus. Particle bombardment, also known as 'Biolistic transformation", is a physical means of forcing DNA molecules into the plant cells and is a most ideal method, only next to *Agrobacterium*-mediated transformation. Klein et al. [127] established the first particle bombardment system for plants, which was later used in various transformation models and for the transformation of crop species such as; maize, rice, onion and wheat [128]. Wheat crop is one of the most challenging crops to transform, with only limited options viable for gene transfer in wheat. Microinjection and PEG (polyethylene glycol)-mediated transformation are not feasible options because regeneration from wheat protoplast is not possible. As a consequence, the discovery of *Agrobacterium tumefaciens* ability to infect monocot species made wheat transformation possible by exploiting this mechanism, biolistic gene transfer was the primary transformation method for wheat [129].

Wheat offers only a few suitable explant tissues for regeneration through tissue culture. The most common explant of choice is the "scutellum" surface of immature embryos, which is responsive to DNA uptake through both AMT and biolistics and can readily form embryogenic callus through regeneration. An integrated method of gene transformation called Agrolistics, have also been reported, that combines biolistics and *Agrobacterium*-mediated transformation [130]. Biolistics gene transfer has become a robust platform for wheat transformation and per 300 immature embryos bombarded; 5–20 independent transgenic plants are produced. Unlike AMT, biolistic-mediated plant transformation does not depend on the receptivity or genotype of the host. Moreover, biolistics transformation is generally more efficient, often results in scrambled and multiple integrations [131] and is less challenging concerning vector requirements because the GOI is co-bombarded with a selectable and separate marker plasmid.

5.3.9.3 Agrobacterium-mediated transformation (AMT)

A. tumefaciens, originally *Bacterium tumefaciens*, is a gram-negative soil born bacteria that has the unique ability to induce tumors in plants [132]. This potential of *Agrobacterium* to genetically transform plants and totipotency of plant kingdom has been exploited and combined to develop a new method for genetic transformation in plants. First *Agrobacterium*-mediated transformed spring wheat was developed by Chen *et al* [129] using embryo-derived immature and regenerable callus. Wounding of the target tissue is an integral step in this method that allows entry of bacterium and stimulates the production of transfer DNA (T-DNA). For this purpose plant tissues are subjected to sonication in presence of *Agrobacterium* carrying the GOI or this can also be done with naked DNA. Once the target tissue is infected, bacterium initiates a unidirectional DNA transfer from their plasmid leading to stable integration of donor DNA into the host nuclear genome [133].

A successful AMT depends on the nature of the explant, *Agrobacterium* strain and plasmids. As mentioned earlier, monocots, i.e. wheat offer a limited choice of explants; shoot apical meristems [134], mature seed callus [135], immature embryos, embryogenic pollen cultures [136] and isolated ovules [137] proved useful for the production of transgenic plants in Triticeae cereals. Moreover, a careful selection of vectors is necessary for cereals' transformation, because most of the plasmids developed for dicot plant species prove to be unsuitable for grasses, especially the marker genes and promoters. Selected plasmids should be highly stable throughout the co-cultivation period, e.g. pVS1-based vector backbones proved particularly valuable in this regard. In addition, hyper-virulent *Agrobacterium* strains such as AGL1 increase the efficiency of cereal transformation because they carry additional copies of virulent genes (Vir) [138].

5.3.9.4 In-planta transformation

In-planta transformation method was developed to avoid the problems associated with regeneration and tissue-culture based transformation. This method allows direct introduction of exogenous DNA into intact plant tissue and has been applied in various plant species such as; tomato (*Solanum lycopersicum*), barrel medic (*Medicago truncatula*) and some cereals [139]. In this method, whole plant, plant tissue or flower can be used as explant. The production of a large number of uniform plants in a short time, fewer labour efforts and minimal reagents requirements are some of the main advantages of in-planta transformation system [140]. The main techniques of in-planta gene transformation are as follows: *Agrobacterium* injection, pollen tube-mediated gene transfer (PTT), vacuum infiltration, floral dip and floral spray methods.

5.3.9.5 Agrobacterium injection

Razzaq et al. [141] developed a rapid and improved in-planta based transformation protocol for wheat variety GA-2002. *A. tumefaciens* strain LBA 4404 harboring pBI121 plasmid carrying GOI was used for direct in-planta transformation. *Agraobacterium* suspension was injected into florets and apical meristem followed by co-cultivation on filter paper. GUS assay and kanamycin was used to screen the transgenic plants which showed that 26 and 27% transgenics gave a positive response to GUS and PCR, respectively.

5.3.9.6 Pollen tube-mediated gene transfer (PTT)

Pollen tube-mediated gene transfer (PTT) was first reported by Zhou et al. [142] in cotton (*Gossypium hirsutum* L.). PTT method is simpler than tissue-culture based transformation techniques and can be performed in three major steps; 1) foreign gene injection into pollen tube, 2) gene integration into the host plant genome, 3) and marker-based selection of transgenic plants. Introduction of a foreign gene into target plant can be done by; direct microinjection, direct application of exogenous DNA on stigma or by co-culturing of foreign gene and pollens and pollination utilizing these pollens [143].

5.3.9.7 Vacuum infiltration-assisted agrobacterium-mediated genetic transformation (VIAAT)

In VIAAT, plant tissues are submerged in a liquid suspension of *A. tumefaciens* and subjected to decreased pressure followed by rapid re-pressurization [144]. Vacuum treatment exposes plant cells, more susceptible to transformation, to GOI carrying *Agrobacterium* and this phenomenon occurs when vacuum is broken and rapid increase in pressure produces a suction effect which leads to force entry of cell suspension into explant to replace the discharged genes with GOI [145]. Transgenic plants selection is done on screening media containing markers such as antibiotics and herbicides [146]. Stable transgenic plants with lower transformation frequency were produced through this method.

Zale et al. [147] devised an efficient in-planta method specifically for wheat to address the regeneration problems. Uninucleated young, mid and late-stage microspores from spikes were immersed in a suspension containing *Agrobacterium* via infiltration method and paromomycin spray was used to select the resulting plantlets. Transformed plantlets stayed green while the non-transgenic plants died in response to the screening marker.

5.3.9.8 Floral dip and floral spray

In this method, the inflorescence of plants is submerged at the early stages of flowering in an *Agrobacterium* suspension with strong optical density to produce transgenic plants. This method is commonly referred to as the 'floral-dip method'. This method is reliable, quick and free from microbial attacks. A slight modification of floral-dip is floral-spray method, where *Agrobacterium* suspension is sprayed on inflorescence shoots instead of immersion [148]. Transformation through floral-dip method can result in more than 100 seeds per reproduction cycle in plants and its efficiency ranges between 0.1 and 5 percent [149]. Supertana et al. [150] developed transgenic wheat variety "Shiranekomugi" by using this method and 33% maximum transformation efficiency was achieved. The one major disadvantages of floral dip method is the random integration of foreign genes into the host genome and their low transformation efficiency [151]. While transformation protocols have improved dramatically, lack of public acceptance and patents have prevented the use of transgenic wheat varieties, but, the hope to get better yielding crops with wide range of adaptability is still there.

5.3.10 Genome editing in wheat

Genome editing is one of the most advanced technologies for crop improvement. The basic mechanism is almost the same in all types of these editing technologies. These technologies involve the generation of double-strand breaks (DSBs) at a target site in a genome using programmable sequence-specific nucleases (SSN) followed by the exploitation of endogenous DSB repair mechanisms to generate a mutation at a particular site. There are two endogenous mechanisms to repair DSBs i.e., non-homologous end joining (NHEJ) and homologous recombination (HR) [152]. In NHEJ, the two broken strands are re-ligated with the generation of insertion and/or deletion. It is error-prone and does not require a homologous template. HR requires a homologous template and is more reliable [153]. However, SSNs use NHEJ frequently as a repair mechanism [153]. Three types of SSNs introduce DSB at a specific site [154]. These include Zinc-Finger Nucleases (ZFN), Transcription Activator-Like Effector Nucleases (TALENs) and CRISPR/Cas9.

5.3.10.1 Zinc-finger nucleases (ZFNs)

ZFNs are artificial endonucleases and consist of designed (according to the target site) zinc finger DNA binding protein (ZFP) fused to the cleavage domain of FokI restriction enzyme. ZFP is generally composed of 3–4 zinc finger arrays. Each array can recognize 3 bp long sequence. The two ZFN monomers are designed in such a way that can recognize 6 bp sequence of a target site and allow the FokI monomer to form an active dimer that can generate DSB at a specific site. Using this genome editing technique, mutation at desirable sites can be created which would lead to the improvement of the plan. However, the presence of very few target sites, difficulty in the engineering of specific zinc finger domains and frequent off-target effects are the main constraint in the application of ZFNs [155].

5.3.10.2 Transcription activator -like effector nucleases (TALENs)

Another DNA binding protein exclusive to plant pathogens is Transcription Activator -Like Effector Nucleases (TALENs). It consists of 33–35 long tandem repeats of amino acids followed by 20 amino acids known as "half repeat" and FokI cleavage domain. IN the TALEN monomer, 12th and 13th position impart specificity to nucleotide recognition. Due to the specificity of these two residues (at 12th and 13th position), these are termed as repeat variable di-residues (RVDs). TALENs works similarly as ZFNs do. They can generate DSBs and introduce mutation at a specific site. Engineering of TALENs is much easier than ZFNs. However, their large size of repetitive sequences, high cost and labor for the construction of novel TALENs are the major drawbacks of this technology.

5.3.10.3 Clustered regularly interspaced short palindromic repeats (CRISPR/Cas9)

CRISPR/Cas9 is simple, cheap and more efficient in contrast to ZFNs and TALENs that require specifically tailored DNA binding protein (**Figure 3**). There are two important components of CRISPR system: guide RNA (gRNA) and CRISPR associated (Cas9) protein. gRNA consists of two components: CRISPR RNA (crRNA) and Trans-activating CRISPR RNA (tracrRNA). crRNA is 18–20 bp in length confers specificity to target DNA. However, trcrRNA is a stretch of loop and acts as a binding

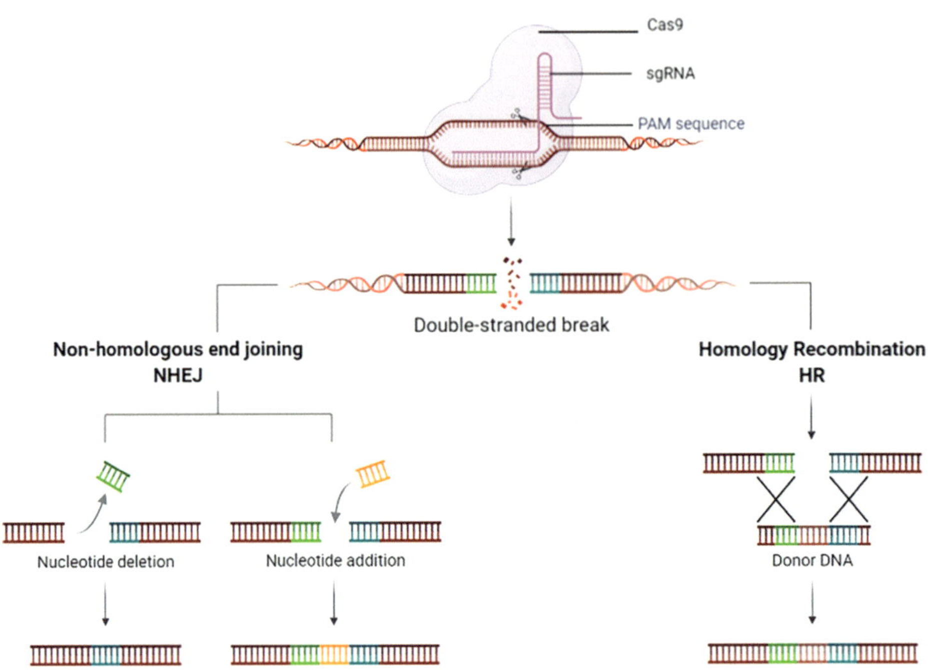

Figure 3.
CRISPR consists of sgRNA and Cas9 protein. Cas 9 protein guided by sgRNA produces double strand break. It would lead to DNA repair either by non-homologous end joining method (NHEJ) or by homology recombination (HR) which require template DNA strand.

scaffold for Cas9. Cas9 protein is an endonuclease consisting of two subparts: 1) Recognition part 2) Nuclease part. The recognition part of Cas protein has two domains i.e., REC1 and REC2 which are responsible for binding with gRNA. Whereas the Nuclease part consists of RuvC, HNH and Protospacer Adjacent Motif (PAM) interacting domain. Former two domains (RuvC and HNH) play role in the cutting of single-stranded DNA. Later domain (PAM interacting domain) confers PAM specificity and initiate the process of binding to target DNA. PAM sequence is 2-5 bp sequence [156]. The mechanism of CRISPR/Cas9 is divided into three steps: recognition, digestion, and repair. gRNA recognizes the specific site on template DNA followed by the generation of DSB at a site 3 bp upstream of the PAM by Cas9. Cas9 can recognize the PAM sequence at 5′-NGG-3′ (where N can be any nucleotide). Finally, DSB is repaired by either NHEJ or HR [157].

5.3.10.4 Base editing and prime editing- a new era of CRISPR/Cas9

Base editing and prime editing are the modified versions of CRISPR/Cas9. In base editing approach, point mutation is created without DSB, foreign donor template or involvement of any repair mechanism. This technique comprises gRNA and catalytically inactive Cas9 (Cas9 nickase) fused with single-stranded DNA deaminase. gRNA directs modified Cas9 deaminase to bind to the locus which produces ssDNA R-loop that exposes the DNA to deaminase. Deaminases are of different types. Based on the type of deaminase, base editing is categorized into two types: Cytidine Base Editor (CBE) and adenine base editor (ABE) [158].

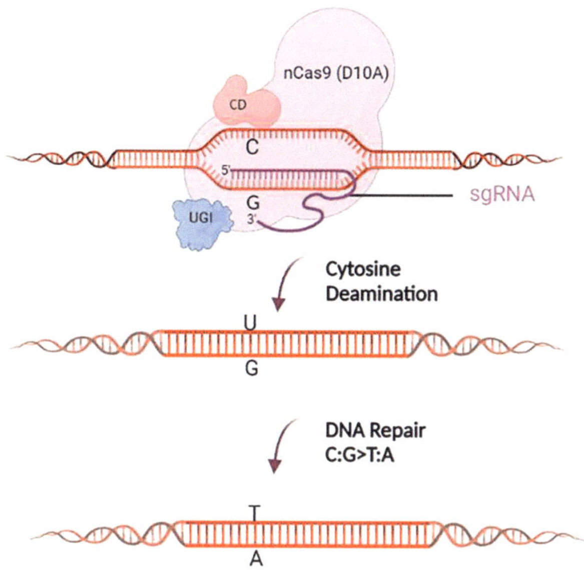

Figure 4.
Cytidine Base editor (CBE) consists of sgRNA, nCas9, cytosine deaminase (CD) and uracil glycosylase inhibitor (UGI). CD causes the deamination of cytosine (C) to uracil (U) which is followed by DNA repair with a result of changing from C:G > T:A.

CBE edit cytidine into uridine. This system is comprised of gRNA, Cas9 nickase (D10A) that is fused with two more proteins viz. cytosine deaminase (CD) and uracil DNA glycosylase inhibitor (UGI) (**Figure 4**). Guided by gRNA, CD converts C into U which is then repaired by the base excision repair pathway and generates C to T substitution. ABE edit adenine into inosine which is treated as guanosine by the polymerase (**Figure 5**). This system is comprised of gRNA, Cas9 nickase fused with adenosine deaminase and also works in the same way as CBE. However, it converts A (Adenosine) into I (Inosine) which is treated as G (Guanine) by DNA polymerase thus generating A to G substitution. Both CBE and ABE can change the base from one purine to another purine or one pyrimidine to another pyrimidine. This is the main shortcoming of this system that purine cannot be replaced by pyrimidine and vice versa [158].

To address this issue, prime editing method was introduced (**Figure 6**). This method consists of Cas9 nickase (H840A) which is fused with reverse transcriptase and prime edited guide RNA (pegRNA). Guided by pegRNA, reverse transcriptase prime new DNA containing the desired editing at the target site. After attaining flat equilibrium, excision, ligation and repairing, DNA is stably edited with desirable incorporation [158]. The main application of CRISPR/Cas9 in wheat was demonstrated in suspension cultures and protoplast. Variety of genes were targeted in wheat protoplast and suspension culture after the publication of the original principle of CRISPR/Cas9 [159]. Generally, *Agrobacterium* or particle bombardment are used as a delivery system of plasmids carrying cassettes for the co-expression of gRNA and Cas9. In the case of wheat genome editing, Cas9 has expressed from a codon-optimized gene under the control of RNA polymerase II promoter (ZmUbi or CaMV35S), whereas gRNA is expressed under the control of polymerase III promoter (mostly U6 and U3) [160].

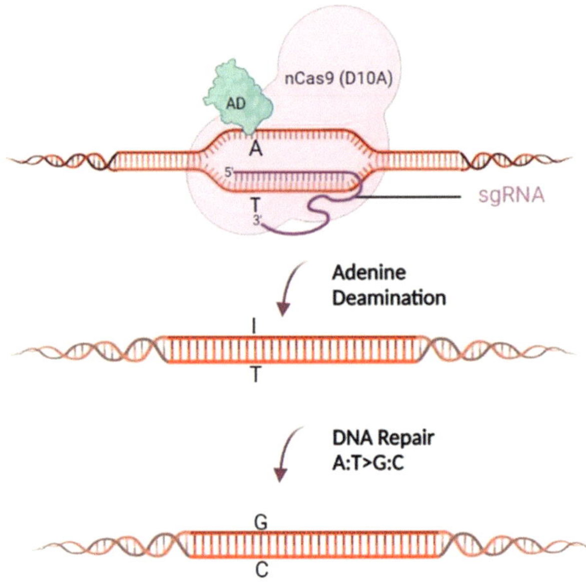

Figure 5.
Adenine base editor (ABE) consists of sgRNA, nCas9 and adenosine deaminase (AD). AD causes the deamination of adenosine (a) to inosine (I) which is treated as guanosine (G) by DNA polymerase. Deamination is followed by DNA repair with a result of changing from a:T > G:C.

5.3.10.5 RNA interference (RNAi) technique in wheat improvement

In eukaryotes, the regulatory mechanism of gene expression is commonly depends upon RNAi. To study functional gene analysis or the development of novel phenotypes, RNAi is a robust tool. This technique involves the expression of antisense or hairpin RNAi constructs to direct gene silencing in a sequence-specific manner [160]. The first wheat gene that was targeted by RNAi was the vernalization gene (TaVRN2). The suppression of this gene provided insight for comprehending the molecular mechanism of flowering time and requirement of vernalization in wheat which is ultimately helpful in varying environments in which wheat can be grown [161].

6. Conclusion

Climate change is a complex of many factors and alarming the world by its destructive effects on crops. Climate change has devastating effects on wheat plant growth and yield. Plants mainly suffer from abiotic stresses. To cope with changing environmental conditions, an integrated management programme is required in addition to crop improvement through conventional and non-conventional methods. To develop better plants under changing climate conditions some bottleneck molecular and physiological encounters present in plants need to be resolved. The rise in temperature and fluctuations in rain fall patterns are very important indicators of climate change. To tackle, this problem different advanced approaches need to be adopted to secure the agriculture future. Climate-resilient crops should be developed using basic breeding approaches. Marker-assisted breeding, omics and proteomics approaches, Genome-wide association studies (GWAS), genomic selection (GS)

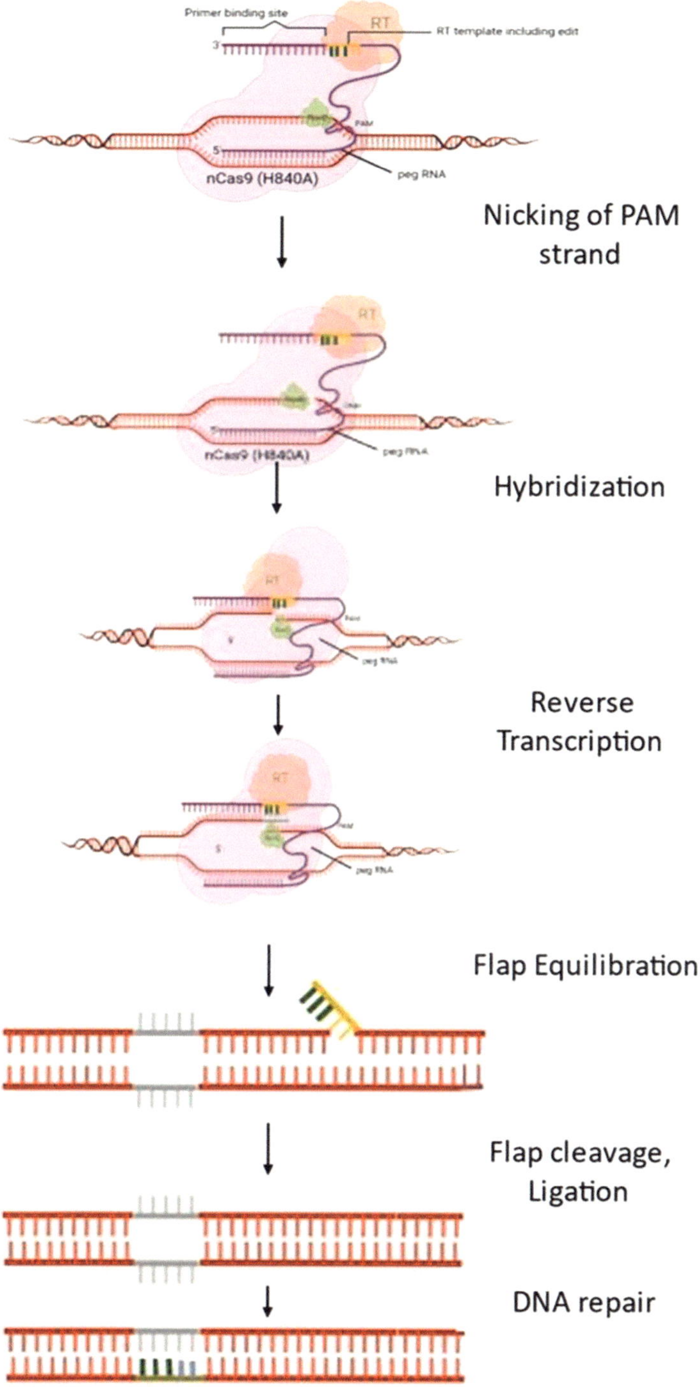

Figure 6.
Prime editing is comprised of Cas9 nickase (H840A) which is fused with reverse transcriptase and prime edited guide RNA (pegRNA). Guided by pegRNA, reverse transcriptase primes new DNA containing the desired editing at the targeted site. After flap equilibration, cleavage, ligation, and DNA repair, the desired editing is incorporated.

genetic modification genome editing, CRISPR/Cas9 and RNA interference techniques all are noteworthy in identifying the different genes linked to tolerance against different stresses. Genetic engineering is a good tool to develop a transgenic plant with improved resistance against stress. CRISPR/Cas9 is the best suitable approach to develop eco-friendly genome-edited wheat plants In future to fight a battle against climate change.

Author details

Saba Akram*, Maria Ghaffar, Ayesha Wadood and Mian Abdur Rehman Arif
Plant Breeding and Genetics Division, Nuclear Institute for Agriculture and Biology, Faisalabad, Pakistan

*Address all correspondence to: sabaakram3463@yahoo.com

References

[1] Braun H-J, Atlin G, Payne T. Multi-location testing as a tool to identify plant response to global climate change. Climate Change and Crop Production. 2010;**1**:115-138

[2] FAO. Crop Prospects and Food Situation. Rome: Food and Agriculture Organization of the United Nations; 2018 Report No.

[3] U.S. Department of Agriculture, Foreign Agricultural Service / Global Market Analysis International Production Assessment Division (IPAD). Washington, DC: World Agricultural Production; 2022

[4] Shewry PR. Wheat. Journal of experimental botany. 2009;**60**(6):1537-1553

[5] Asseng S, Foster I, Turner NC. The impact of temperature variability on wheat yields. Global Change Biology. 2011;**17**(2):997-1012

[6] Giraldo P, Benavente E, Manzano-Agugliaro F, Gimenez E. Worldwide research trends on wheat and barley: A bibliometric comparative analysis. Agronomy. 2019;**9**(7):352

[7] FAOSTAT. Crops/World Total/Wheat/Area Harvested and World Production/2020 (pick list). United Nations, Food and Agriculture Organization, Statistics Division (FAOSTAT). Retrieved on 19 July 2022

[8] USDA. Wheat Outlook/2021. Retrieved on 19 July 2022

[9] Javed MM, Zahoor S, Shafaat S, Mehmooda I, Gul A, Rasheed H, et al. Wheat bran as a brown gold: Nutritious value and its biotechnological applications. African Journal of Microbiology Research. 2012;**6**(4):724-733

[10] Alexandratos N, Bruinsma J. World Agriculture Towards 2030/2050. The 2012 Revision. ESA Working Paper No. 12-03. Rome: FAO of the United Nations; 2012

[11] AMIS (Agriculture Marketing Information Service). Wheat /2019-20. Directorate of Agriculture (Economics and Marketing) Punjab, Lahore, Pakistan. Retrieved on 19 July 2022

[12] Brisson N, Gate P, Gouache D, Charmet G, Oury F-X, Huard F. Why are wheat yields stagnating in Europe? A comprehensive data analysis for France. Field Crops Research. 2010;**119**(1):201-212

[13] United Nations, Department of Economic and Social Affairs (2003). World Urbanization Prospects – The 2003 Revision. New York: United Nations Publications

[14] Gödecke T, Stein AJ, Qaim M. The global burden of chronic and hidden hunger: Trends and determinants. Global Food Security. 2018;**17**:21-29

[15] Change IPOC. Climate change 2007: The physical science basis. Agenda. 2007;**6**(07):333

[16] Andy P. Abiotic stress tolerance in plants. Plant Science. 2016;7:1-9

[17] Noya I, González-García S, Bacenetti J, Fiala M, Moreira MT. Environmental impacts of the cultivation-phase associated with agricultural crops for feed production. Journal of Cleaner Production. 2018;**172**:3721-3733

[18] Zhao C, Liu B, Piao S, Wang X, Lobell DB, Huang Y, et al. Temperature increase reduces global yields of major

crops in four independent estimates. Proceedings of the National Academy of Sciences. 2017;**114**(35):9326-9331

[19] Van Oort PA, Zwart SJ. Impacts of climate change on rice production in Africa and causes of simulated yield changes. Global Change Biology. 2018;**24**(3):1029-1045

[20] Vigani M, Rodríguez-Cerezo E, Gómez-Barbero M. The determinants of wheat yields: The role of sustainable innovation, policies and risks in France and Hungary. JRC Scientific and Policy Reports. 2015. Available from: https://publications.jrc.ec.europa.eu/repository/handle/JRC98031

[21] Enghiad A, Ufer D, Countryman AM, Thilmany DD. An overview of global wheat market fundamentals in an era of climate concerns. International Journal of Agronomy. 2017;**2017**:15

[22] Ministers GA. G20 agriculture Ministers' declaration 2017: Towards food and water security: Fostering sustainability, advancing innovation. German Federal Ministry of Food and Agriculture. 2017. pp. 1-6

[23] Newton AC, Johnson SN, Gregory PJ. Implications of climate change for diseases, crop yields and food security. Euphytica. 2011;**179**(1):3-18

[24] Sreenivasulu N, Butardo VM Jr, Misra G, Cuevas RP, Anacleto R, Kavi Kishor PB. Designing climate-resilient rice with ideal grain quality suited for high-temperature stress. Journal of Experimental Botany. 2015;**66**(7):1737-1748

[25] Hay JE, Easterling D, Ebi KL, Kitoh A, Parry M. Introduction to the special issue: Observed and projected changes in weather and climate extremes. Weather and Climate Extremes. 2016;**11**:1-3

[26] Vaughan MM, Block A, Christensen SA, Allen LH, Schmelz EA. The effects of climate change associated abiotic stresses on maize phytochemical defenses. Phytochemistry Reviews. 2018;**17**(1):37-49

[27] Richardson KJ, Lewis KH, Krishnamurthy PK, Kent C, Wiltshire AJ, Hanlon HM. Food security outcomes under a changing climate: Impacts of mitigation and adaptation on vulnerability to food insecurity. Climatic Change. 2018;**147**(1):327-341

[28] Tito R, Vasconcelos HL, Feeley KJ. Global climate change increases risk of crop yield losses and food insecurity in the tropical Andes. Global Change Biology. 2018;**24**(2):e592-e602

[29] Faostat F. Available online: http://www.fao.org/faostat/en/#data.QC [accessed on: January 2018]. 2017.

[30] Reckling M, Döring TF, Bergkvist G, Chmielewski F-M, Stoddard FL, Watson CA, et al., editors. Grain legume yield instability has increased over 60 years in long-term field experiments as measured by a scale-adjusted coefficient of variation. In: Advances in Legume Science and Practice. Association of Applied Biologists. UK

[31] Dhankher OP, Foyer CH. Climate resilient crops for improving global food security and safety. Plant, Cell & Environment. 2018:**41**(5):877-884

[32] Hirayama T, Shinozaki K. Research on plant abiotic stress responses in the post-genome era: Past, present and future. The Plant Journal. 2010;**61**(6):1041-1052

[33] Rosenzweig C, Elliott J, Deryng D, Ruane AC, Müller C, Arneth A, et al. Assessing agricultural risks of climate

change in the 21st century in a global gridded crop model intercomparison. Proceedings of the National Academy of Sciences. 2014;**111**(9):3268-3273

[34] Wheeler T, Von Braun J. Climate change impacts on global food security. Science. 2013;**341**(6145):508-513

[35] Rejeb IB, Pastor V, Mauch-Mani B. Plant responses to simultaneous biotic and abiotic stress: Molecular mechanisms. Plants. 2014;**3**(4):458-475

[36] Compant S, Van Der Heijden MG, Sessitsch A. Climate change effects on beneficial plant–microorganism interactions. FEMS Microbiology Ecology. 2010;**73**(2):197-214

[37] Braun H-J, Altay F, Kronstad W, Beniwal SO, McNab A. Wheat: Prospects for Global Improvement: Proceedings of the 5th International Wheat Conference, 10-14 June, 1996. Ankara, Turkey: Springer Science & Business Media; 2012

[38] Marasas CN, Smale M, Singh RP. The economic impact in developing countries of leaf rust resistance breeding in CIMMYT-related spring bread wheat. Mexico: CIMMYT; 2004. pp. 1-33

[39] Huerta-Espino J, Singh R, German S, McCallum B, Park R, Chen WQ, et al. Global status of wheat leaf rust caused by Puccinia triticina. Euphytica. 2011;**179**(1):143-160

[40] Chen X. Epidemiology and control of stripe rust [Puccinia striiformis f. sp. tritici] on wheat. Canadian Journal of Plant Pathology. 2005;**27**(3):314-337

[41] Anwaar HA, Perveen R, Mansha MZ, Abid M, Sarwar ZM, Aatif HM, et al. Assessment of grain yield indices in response to drought stress in wheat (*Triticum aestivum* L.). Saudi journal of Biological Sciences. 2020;**27**(7):1818-1823

[42] Shah T, Xu J, Zou X, Cheng Y, Nasir M, Zhang X. Omics approaches for engineering wheat production under abiotic stresses. International Journal of Molecular Sciences. 2018;**19**(8):2390

[43] Barlow K, Christy B, O'leary G, Riffkin P, Nuttall J. Simulating the impact of extreme heat and frost events on wheat crop production: A review. Field Crops Research. 2015;**171**:109-119

[44] Dai A. Drought under global warming: A review. Wiley Interdisciplinary Reviews: Climate Change. 2011;**2**(1):45-65

[45] Lesk C, Rowhani P, Ramankutty N. Influence of extreme weather disasters on global crop production. Nature. 2016;**529**(7584):84-87

[46] Kajla M, Yadav VK, Khokhar J, Singh S, Chhokar R, Meena RP, et al. Increase in wheat production through management of abiotic stresses: A review. Journal of Applied and Natural Science. 2015;7(2):1070-1080

[47] Fontaine MM, Steinemann AC, Hayes MJ. State drought programs and plans: Survey of the Western United States. Natural Hazards Review. 2014;**15**(1):95-99

[48] Gray SB, Brady SM. Plant developmental responses to climate change. Developmental Biology. 2016;**419**(1):64-77

[49] Abhinandan K, Skori L, Stanic M, Hickerson N, Jamshed M, Samuel MA. Abiotic stress signaling in wheat–an inclusive overview of hormonal interactions during abiotic stress responses in wheat. Frontiers in Plant Science. 2018;**9**:734

[50] Porch T, Jahn M. Effects of high-temperature stress on microsporogenesis in heat-sensitive and heat-tolerant genotypes of Phaseolus vulgaris. Plant, Cell & Environment. 2001;**24**(7):723-731

[51] Jame Y, Cutforth H. Simulating the effects of temperature and seeding depth on germination and emergence of spring wheat. Agricultural and Forest Meteorology. 2004;**124**(3-4):207-218

[52] Almansouri M, Kinet J-M, Lutts S. Effect of salt and osmotic stresses on germination in durum wheat (*Triticum durum* Desf.). Plant and Soil. 2001;**231**(2):243-254

[53] Hasanuzzaman M, Nahar K, Rahman A, Anee TI, Alam MU, Bhuiyan TF, et al. Approaches to enhance salt stress tolerance in wheat. Wheat improvement, management and utilization. IntechOpen. 2017:151-187

[54] Ashfaq M, Zulfiqar F, Sarwar I, Quddos A, Baig I. Impact of climate Change on wheat productivity in mixed cropping system of Punjab. Soil and Environment. 2011;**30**:110-114

[55] Hasanuzzaman M, Nahar K, Fujita M. Plant response to salt stress and role of exogenous protectants to mitigate salt-induced damages. In: Ecophysiology and Responses of Plants under Salt Stress. New York: Springer; 2013. pp. 25-87

[56] Hasanuzzaman M, Nahar K, Rahman A, Anee TI, Alam MU, Bhuiyan TF, et al. Approaches to enhance salt stress tolerance in wheat. Wheat improvement, management and utilization. 2017:151-187

[57] Muhammad K. Performance of wheat genotypes under osmotic stress at germination and early seedling growth stage. Caderno de Pesquisa Serie Biologia. 2010;**22**(1):5-12

[58] Adrees M, Ali S, Rizwan M, Ibrahim M, Abbas F, Farid M, et al. The effect of excess copper on growth and physiology of important food crops: A review. Environmental Science and Pollution Research. 2015;**22**(11):8148-8162

[59] Karmous I, Chaoui A, Jaouani K, Sheehan D, El Ferjani E, Scoccianti V, et al. Role of the ubiquitin-proteasome pathway and some peptidases during seed germination and copper stress in bean cotyledons. Plant Physiology and Biochemistry. 2014;**76**:77-85

[60] Arfan M, Athar HR, Ashraf M. Does exogenous application of salicylic acid through the rooting medium modulate growth and photosynthetic capacity in two differently adapted spring wheat cultivars under salt stress? Journal of Plant Physiology. 2007;**164**(6):685-694

[61] Subrahmanyam D, Subash N, Haris A, Sikka A. Influence of water stress on leaf photosynthetic characteristics in wheat cultivars differing in their susceptibility to drought. Photosynthetica. 2006;**44**(1):125-129

[62] Jaleel CA, Sankar B, Murali P, Gomathinayagam M, Lakshmanan G, Panneerselvam R. Water deficit stress effects on reactive oxygen metabolism in Catharanthus roseus; impacts on ajmalicine accumulation. Colloids and Surfaces B: Biointerfaces. 2008;**62**(1):105-111

[63] Hossain A, Sarker M, Saifuzzaman M, Teixeira da Silva J, Lozovskaya M, Akhter M. Evaluation of growth, yield, relative performance and heat susceptibility of eight wheat (*Triticum aestivum* L.) genotypes grown under heat stress. International Journal of Plant Production. 2013;**7**(3):615-636

[64] Poudel PB, Poudel MR. Heat stress effects and tolerance in wheat: A review. Journal of Biology and Today's World. 2020;**9**(3):1-6

[65] Weldearegay D, Yan F, Jiang D, Liu F. Independent and combined effects

of soil warming and drought stress during anthesis on seed set and grain yield in two spring wheat varieties. Journal of Agronomy and Crop Science. 2012;**198**(4):245-253

[66] Blum A. Improving wheat grain filling under stress by stem reserve mobilisation. Euphytica. 1998;**100**(1):77-83

[67] Riaz-ud-Din M, Ahmad N, Hussain M, Rehman AU. Effect of temperature on development and grain formation in spring wheat. Pakistan Journal of Botany. 2010;**42**(2):899-906

[68] Borrill P, Fahy B, Smith AM, Uauy C. Wheat grain filling is limited by grain filling capacity rather than the duration of flag leaf photosynthesis: A case study using NAM RNAi plants. PLoS One. 2015;**10**(8):e0134947

[69] Farooq M, Hussain M, Siddique KH. Drought stress in wheat during flowering and grain-filling periods. Critical Reviews in Plant Sciences. 2014;**33**(4):331-349

[70] Beddow JM, Pardey PG, Chai Y, Hurley TM, Kriticos DJ, Braun H-J, et al. Research investment implications of shifts in the global geography of wheat stripe rust. Nature Plants. 2015;**1**(10):1-5

[71] Schmidt J, Claussen J, Wörlein N, Eggert A, Fleury D, Garnett T, et al. Drought and heat stress tolerance screening in wheat using computed tomography. Plant Methods. 2020;**16**(1):1-12

[72] Smith P. Delivering food security without increasing pressure on land. Global Food Security. 2013;**2**(1):18-23

[73] Reardon T, Timmer CP. Five inter-linked transformations in the Asian agrifood economy: Food security implications. Global Food Security. 2014;**3**(2):108-117

[74] Tohidfar M, Khosravi S. Transgenic Crops with an Improved Resistance to Biotic Stresses. A Review. Wallonia, Belgium: BASE, University of Liege; 2015

[75] Meuwissen TH, Hayes BJ, Goddard ME. Prediction of total genetic value using genome-wide dense marker maps. Genetics. 2001;**157**(4):1819-1829

[76] Whitford R, Fleury D, Reif JC, Garcia M, Okada T, Korzun V, et al. Hybrid breeding in wheat: Technologies to improve hybrid wheat seed production. Journal of Experimental Botany. 2013;**64**(18):5411-5428

[77] Forster BP, Heberle-Bors E, Kasha KJ, Touraev A. The resurgence of haploids in higher plants. Trends in Plant Science. 2007;**12**(8):368-375

[78] Lukaszewski AJ. Chromosomes 1BS and 1RS for control of male fertility in wheats and triticales with cytoplasms of Aegilops kotschyi, ae. Mutica and ae. Uniaristata. Theoretical and Applied Genetics. 2017;**130**(12):2521-2526

[79] Cucinotta M, Colombo L, Roig-Villanova I. Ovule development, a new model for lateral organ formation. Frontiers in Plant SCIENCE. 2014;**5**:117

[80] Lopes MS, El-Basyoni I, Baenziger PS, Singh S, Royo C, Ozbek K, et al. Exploiting genetic diversity from landraces in wheat breeding for adaptation to climate change. Journal of Experimental Botany. 2015;**66**(12):3477-3486

[81] Reitz L, Salmon S. Origin, history, and use of Norin 10 wheat. Crop Science. 1968;**8**(6):686-689

[82] Parry MA, Madgwick PJ, Bayon C, Tearall K, Hernandez-Lopez A, Baudo M, et al. Mutation discovery for crop improvement. Journal of Experimental Botany. 2009;**60**(10):2817-2825

[83] Tester M, Langridge P. Breeding technologies to increase crop production in a changing world. Science. 2010;**327**(5967):818-822

[84] Muller HJ. The production of mutations by X-rays. Proceedings of the National Academy of Sciences of the United States of America. 1928;**14**(9):714

[85] Laskar RA, Khan S, Deb CR, Tomlekova N, Wani MR, Raina A, et al. Lentil (Lens culinaris Medik.) diversity, cytogenetics and breeding. In: Advances in Plant Breeding Strategies: Legumes. New York: Springer; 2019. pp. 319-369

[86] Chaudhary J, Alisha A, Bhatt V, Chandanshive S, Kumar N, Mir Z, et al. Mutation breeding in tomato: Advances, applicability and challenges. Plants. 2019;**8**(5):128

[87] van Harten AM. Mutation Breeding: Theory and Practical Applications. England: Cambridge University Press; 1998

[88] Saravanan K, Sabesan T. Physical and chemical mutagenesis methods for development of insect-resistant crop varieties. In: Experimental Techniques in Host-Plant Resistance. Singapore: Springer; 2019. pp. 295-301

[89] Jankowicz-Cieslak J, Tai TH, Kumlehn J, Till BJ. Biotechnologies for Plant Mutation Breeding: Protocols. Switzerland: Springer Nature; 2017

[90] [IAEA] International Atomic Energy Agency. Mutant Variety Database/ Wheat. 2022. Retrieved on 19 July 2022

[91] Ortiz R, Trethowan R, Ferrara GO, Iwanaga M, Dodds JH, Crouch JH, et al. High yield potential, shuttle breeding, genetic diversity, and a new international wheat improvement strategy. Euphytica. 2007;**157**(3):365-384

[92] Trethowan RM, Reynolds MP, Ortiz-Monasterio JI, Ortiz R. The genetic basis of the green revolution in wheat production. Plant Breeding Reviews. 2007;**28**:39

[93] Mishra R, Rao GJN. In-vitro androgenesis in rice: Advantages, constraints and future prospects. Rice Science. 2016;**23**(2):57-68

[94] Hooghvorst I, Nogués S. Chromosome doubling methods in doubled haploid and haploid inducer-mediated genome-editing systems in major crops. Plant Cell Reports. 2021;**40**(2):255-270

[95] Mahato A, Chaudhary HK. Relative efficiency of maize and Imperata cylindrica for haploid induction in Triticum durum following chromosome elimination-mediated approach of doubled haploid breeding. Plant Breeding. 2015;**134**(4):379-383

[96] Puntel LA, Sawyer JE, Barker DW, Dietzel R, Poffenbarger H, Castellano MJ, et al. Modeling long-term corn yield response to nitrogen rate and crop rotation. Frontiers in plant Science. 2016;**7**:1630

[97] Asif M, Eudes F, Randhawa H, Amundsen E, Spaner D. Induction medium osmolality improves microspore embryogenesis in wheat and triticale. In Vitro Cellular & Developmental Biology-Plant. 2014;**50**(1):121-126

[98] Kim SK, Kim J-H, Jang W-C. Past, present and future molecular approaches to improve yield in wheat. Wheat Improvement, Management and Utilization. IntechOpen. 2017. pp. 17-37

[99] Gupta P, Varshney R, Sharma P, Ramesh B. Molecular markers and their applications in wheat breeding. Plant Breeding. 1999;**118**(5):369-390

[100] Devos K, Gale MD. The use of random amplified polymorphic DNA markers in wheat. Theoretical and Applied Genetics. 1992;**84**(5-6):567-572

[101] Röder MS, Korzun V, Wendehake K, Plaschke J, Tixier M-H, Leroy P, et al. A microsatellite map of wheat. Genetics. 1998;**149**(4):2007-2023

[102] Moustafa KA, Al-Doss AA, Motawei MI, Al-Otayk S, Dawabah AA, Abdel-Mawgood AL, et al. Selection of spring bread wheat genotypes for resistance to cereal cyst nematode ('Heterodera avenae'Woll.) based on field performance and molecular markers. Plant Omics. 2015;**8**(5):392-397

[103] Burghardt LT, Young ND, Tiffin P. A guide to genome-wide association mapping in plants. Current Protocols in plant Biology. 2017;**2**(1):22-38

[104] Astle W, Balding DJ. Population structure and cryptic relatedness in genetic association studies. Statistical Science. 2009;**24**(4):451-471

[105] Balding DJ. A tutorial on statistical methods for population association studies. Nature Reviews Genetics. 2006;7(10):781-791

[106] Chen H, Wang C, Conomos MP, Stilp AM, Li Z, Sofer T, et al. Control for population structure and relatedness for binary traits in genetic association studies via logistic mixed models. The American Journal of Human Genetics. 2016;**98**(4):653-666

[107] Yu J, Pressoir G, Briggs WH, Bi IV, Yamasaki M, Doebley JF, et al. A unified mixed-model method for association mapping that accounts for multiple levels of relatedness. Nature genetics. 2006;**38**(2):203-208

[108] Akram S, Arif MAR, Hameed A. A GBS-based GWAS analysis of adaptability and yield traits in bread wheat (*Triticum aestivum* L.). Journal of Applied Genetics. 2021;**62**(1):27-41

[109] Maulana F, Ayalew H, Anderson JD, Kumssa TT, Huang W, Ma X-F. Genome-wide association mapping of seedling heat tolerance in winter wheat. Frontiers in Plant Science. 2018;**9**:1272

[110] Sukumaran S, Reynolds MP, Sansaloni C. Genome-wide association analyses identify QTL hotspots for yield and component traits in durum wheat grown under yield potential, drought, and heat stress environments. Frontiers in plant Science. 2018;**9**:81

[111] Gahlaut V, Jaiswal V, Singh S, Balyan H, Gupta P. Multi-locus genome wide association mapping for yield and its contributing traits in hexaploid wheat under different water regimes. Scientific Reports. 2019;**9**(1):1-15

[112] Haile JK, N'Diaye A, Sari E, Walkowiak S, Rutkoski JE, Kutcher HR, et al. Potential of genomic selection and integrating "omics" data for disease evaluation in wheat. Crop Breeding, Genetics and Genomics. 2020;**4**(3):01-29

[113] Bassi FM, Bentley AR, Charmet G, Ortiz R, Crossa J. Breeding schemes for the implementation of genomic selection in wheat (Triticum spp.). Plant Science. 2016;**242**:23-36

[114] Lozada DN, Carter AH. Genomic selection in winter wheat breeding using a recommender approach. Genes. 2020;**11**(7):779

[115] Alotaibi F, Alharbi S, Alotaibi M, Al Mosallam M, Motawei M, Alrajhi A. Wheat omics: Classical breeding to new breeding technologies. Saudi Journal of Biological Sciences. 2021;**28**(2):1433

[116] Okada M, Yoshida K, Nishijima R, Michikawa A, Motoi Y, Sato K, et al. RNA-seq analysis reveals considerable genetic diversity and provides genetic markers saturating all chromosomes in the diploid wild wheat relative Aegilops umbellulata. BMC Plant Biology. 2018;**18**(1):1-13

[117] Alotaibi FS. changes in gene expression under water stress in the flag leaf and seed head during the grain-filling stage of two spring wheat cultivars. Stillwater, United States: Oklahoma State University; 2018

[118] Komatsu S, Kamal AH, Hossain Z. Wheat proteomics: Proteome modulation and abiotic stress acclimation. Frontiers in Plant Science. 2014;**5**:684

[119] Farooq M, Bramley H, Palta JA, Siddique KH. Heat stress in wheat during reproductive and grain-filling phases. Critical Reviews in Plant Sciences. 2011;**30**(6):491-507

[120] Khakimov B, Jespersen BM, Engelsen SB. Comprehensive and comparative metabolomic profiling of wheat, barley, oat and rye using gas chromatography-mass spectrometry and advanced chemometrics. Food. 2014;**3**(4):569-585

[121] Krugman T, Peleg Z, Quansah L, Chagué V, Korol AB, Nevo E, et al. Alteration in expression of hormone-related genes in wild emmer wheat roots associated with drought adaptation mechanisms. Functional & Integrative Genomics. 2011;**11**(4):565-583

[122] Al-Doss A, Al-Hazmi AS, Dawabah AA, Abdel-Mawgood AA, Al-Rehiayani SM, Al-Otayk S, et al. Impact of'Cre'and peroxidase genes of selected new wheat lines on cereal cyst nematode ('Heterodera Avenae'Woll) resistance. Australian Journal of Crop Science. 2010;**4**(9):737-743

[123] Sobhanian N, Habashy A, Far F, Tohid F. Optimizing regeneration and reporter gene (gus) transformation of alfalfa (Medicago sativa). Annals of Biological Research. 2012;**3**(5):2419-2427

[124] Mrízová K, Holasková E, Öz MT, Jiskrová E, Frébort I, Galuszka P. Transgenic barley: A prospective tool for biotechnology and agriculture. Biotechnology Advances. 2014;**32**(1):137-157

[125] Somers DJ, Isaac P, Edwards K. A high-density microsatellite consensus map for bread wheat (Triticum aestivum L.). Theoretical and Applied Genetics. 2004;**109**(6):1105-1114

[126] Rao AQ, Bakhsh A, Kiani S, Shahzad K, Shahid AA, Husnain T, et al. RETRACTED: The myth of plant transformation. Biotechnology Advances. 2009;**27**(6):753-763

[127] Klein R. High-velocity microprojectiles for delivering nucleic acids into living cells. Biotechnology. 1992;**24**:384-386

[128] Klein TM, Kornstein L, Sanford JC, Fromm ME. Genetic transformation of maize cells by particle bombardment. Plant Physiology. 1989;**91**(1):440-444

[129] Cheng M, Fry JE, Pang S, Zhou H, Hironaka CM, Duncan DR, et al. Genetic transformation of wheat mediated by agrobacterium tumefaciens. Plant Physiology. 1997;**115**(3):971-980

[130] Mohanty D, Chandra A, Tandon R. Germline transformation for crop improvement. In: Molecular Breeding for Sustainable Crop Improvement. Cham, Denmark: Springer; 2016. pp. 343-395

[131] Altpeter F, Baisakh N, Beachy R, Bock R, Capell T, Christou P, et al. Particle bombardment and the genetic enhancement of crops: Myths and realities. Molecular Breeding. 2005;**15**(3):305-327

[132] Smith EF, Townsend CO. A plant-tumor of bacterial origin. Science. 1907;**25**(643):671-673

[133] Gelvin SB. Agrobacterium-mediated plant transformation: The biology behind the "gene-jockeying" tool. Microbiology and Molecular Biology Reviews. 2003;**67**(1):16-37

[134] Hamada H, Linghu Q, Nagira Y, Miki R, Taoka N, Imai R. An in planta biolistic method for stable wheat transformation. Scientific Reports. 2017;7(1):1-8

[135] Medvecká E, Harwood WA. Wheat (*Triticum aestivum* L.) transformation using mature embryos. In: Agrobacterium protocols. New York: Springer; 2015. pp. 199-209

[136] Kumlehn J, Serazetdinova L, Hensel G, Becker D, Loerz H. Genetic transformation of barley (*Hordeum vulgare* L.) via infection of androgenetic pollen cultures with agrobacterium tumefaciens. Plant Biotechnology Journal. 2006;**4**(2):251-261

[137] Holme IB, Brinch-Pedersen H, Lange M, Holm PB. Transformation of barley (*Hordeum vulgare* L.) by agrobacterium tumefaciens infection of in vitro cultured ovules. Plant Cell Reports. 2006;**25**(12):1325-1335

[138] Goedeke S, Hensel G, Kapusi E, Gahrtz M, Kumlehn J. Transgenic barley in fundamental research and biotechnology. Transgenic Plant Journal. 2007;**1**(1):104-117

[139] Rod-in W, Sujipuli K, Ratanasut K. The floral-dip method for rice (Oryza sativa) transformation. Journal of Agriculture Technology. 2014;**10**:467-474

[140] Bent AF. Arabidopsis in planta transformation. Uses, mechanisms, and prospects for transformation of other species. Plant Physiology. 2000;**124**(4):1540-1547

[141] Razzaq A, Hafiz IA, Mahmood I, Hussain A. Development of in planta transformation protocol for wheat. African Journal of Biotechnology. 2011;**10**(5):740-750

[142] Zhou G-Y, Weng J, Zeng Y, Huang J, Qian S, Liu G. Introduction of exogenous DNA into cotton embryos. Methods in Enzymology. 1983;**101**:433-481

[143] Wang M, Zhang B, Wang Q. Cotton Transformation Via Pollen Tube Pathway. Transgenic Cotton: Springer; 2013. pp. 71-77

[144] Bechtold N. In planta agrobacterium-mediated gene transfer by infiltration of adult Arabidopsis plants. CR Academic Science Series III ScienceView. 1993;**316**:1194-1199

[145] Mariashibu TS, Subramanyam K, Arun M, Mayavan S, Rajesh M, Theboral J, et al. Vacuum infiltration enhances the agrobacterium-mediated genetic transformation in Indian soybean cultivars. Acta Physiologiae Plantarum. 2013;**35**(1):41-54

[146] Clough SJ, Bent AF. Floral dip: A simplified method for agrobacterium-mediated transformation of Arabidopsis thaliana. The Plant Journal. 1998;**16**(6):735-743

[147] Zale JM, Agarwal S, Loar S, Steber C. Evidence for stable transformation of wheat by floral dip in

agrobacterium tumefaciens. Plant Cell Reports. 2009;**28**(6):903-913

[148] Chung M-H, Chen M-K, Pan S-M. Floral spray transformation can efficiently generate Arabidopsis. Transgenic Research. 2000;**9**(6):471-486

[149] Tague BW, Mantis J. In planta agrobacterium-mediated transformation by vacuum infiltration. In: Arabidopsis protocols. New Jersey: Springer; 2006. pp. 215-223

[150] Supartana P, Shimizu T, Shioiri H, Nogawa M, Nozue M, Kojima M. Development of simple and efficient in planta transformation method for rice (Oryza sativa L.) using agrobacterium tumefaciens. Journal of Bioscience and Bioengineering. 2005;**100**(4):391-397

[151] Li J, Tan X, Zhu F, Guo J. A rapid and simple method for Brassica napus floral-dip transformation and selection of transgenic plantlets. International Journal of Biology. 2010;**2**(1):127

[152] Symington LS, Gautier J. Double-strand break end resection and repair pathway choice. Annual review of Genetics. 2011;**45**:247-271

[153] Gao C. Genome engineering for crop improvement and future agriculture. Cell. 2021;**184**(6):1621-1635

[154] Zhan X, Lu Y, Zhu JK, Botella JR. Genome editing for plant research and crop improvement. Journal of Integrative Plant Biology. 2021;**63**(1):3-33

[155] Aglawe SB, Barbadikar KM, Mangrauthia SK, Madhav MS. New breeding technique "genome editing" for crop improvement: Applications, potentials and challenges. 3 Biotech. 2018;**8**(8):1-20

[156] Asmamaw M, Zawdie B. Mechanism and applications of CRISPR/Cas-9-mediated genome editing. Biologics: Targets & Therapy. 2021;**15**:353

[157] Demirci Y, Zhang B, Unver T. CRISPR/Cas9: An RNA-guided highly precise synthetic tool for plant genome editing. Journal of Cellular Physiology. 2018;**233**(3):1844-1859

[158] Li J, Li Y, Ma L. Recent advances in CRISPR/Cas9 and applications for wheat functional genomics and breeding. aBIOTECH. 2021:1-11

[159] Shan Q, Wang Y, Li J, Gao C. Genome editing in rice and wheat using the CRISPR/Cas system. Nature protocols. 2014;**9**(10):2395-2410

[160] Borisjuk N, Kishchenko O, Eliby S, Schramm C, Anderson P, Jatayev S, et al. Genetic modification for wheat improvement: From transgenesis to genome editing. BioMed Research International. 2019;**2019**:18

[161] Yan L, Loukoianov A, Blechl A, Tranquilli G, Ramakrishna W, SanMiguel P, et al. The wheat VRN2 gene is a flowering repressor down-regulated by vernalization. Science. 2004;**303**(5664):1640-1644

Chapter 12

Microwave Soil Treatment Alleviates Arsenic Phytotoxicity and Reduces Wheat Grain Arsenic Concentration

Mohammad Humayun Kabir, Graham Brodie, Dorin Gupta and Alexis Pang

Abstract

Arsenic (As) contamination in soil and accumulation in food crops has raised much concern worldwide due to its phytotoxicity and possible human health risk. This study was conducted to determine whether microwave (MW) soil treatment could alleviate As phytotoxicity and reduce wheat grain As concentration or not. Experimental soils were spiked to five levels of As concentration (As-0, As-20, As-40, As-60, and As-80 mg kg^{-1}) prior to applying three levels of MW treatment (MW-0, MW-3, and MW-6 minute). Significantly higher plant growth and grain yield and lower grain As concentration was recorded in MW treatments compared with the control treatment. For instance, significantly higher grain yield (28.95 g pot^{-1}) and lower grain As concentration (572.03 μg kg^{-1}) were recorded in MW-6 treatment compared with MW-0 (22.03 g pot^{-1} and 710.45 μg kg^{-1}, respectively) at the same soil As concentration. Hence, MW soil treatment has the potential to alleviate As phytotoxicity and to reduce the grain As concentration. Ultimately, MW soil treatment will reduce As bioaccumulation in the human body even if wheat is grown in As contaminated soil. Nevertheless, further validation experiments are needed to explore the effectiveness of MW treatment in field conditions.

Keywords: microwave, soil heating, arsenic mitigation, wheat, grain arsenic

1. Introduction

Arsenic (As) is the most devastatingly toxic heavy metal and is raising global concerns for sustainable agriculture and human health, due to its ultimately toxic effect, persistence in nature, and ability to bio-accumulate in the ecosystem [1]. Inorganic As is considered to be a Group I human carcinogen and responsible for different types of cancer [2]. It is estimated that 220 million people, worldwide, are exposed to elevated concentrations of As in drinking water, which are above the World Health Organization (WHO) standard limit (10 mg l^{-1}) [3]. In addition to geogenic sources of As, which mainly contaminate drinking water, it can build up in soil because of long-term excessive use of As contaminated irrigation water and ultimately results in As uptake by crops [4]. Thus, the presence of As in the food chain, through the water-soil-crop pathway is triggering concerns about human health [1]. Excessively high As pollution in water, soil, and crops has already been identified in many countries [5–8]. Even though rice is a good accumulator of As, concerns are mounting about the amount of As being found in other crops, like vegetables, tubers, fruits, and even wheat [9].

Although there is no worldwide standard safe limit of As in food grains, the European Commission, recently (January 2016), set the maximum limits for As in milled rice (polished or white rice) as 200 $\mu g\ kg^{-1}$ [10]. A wheat field experiment with a 12.00 $mg\ kg^{-1}$ soil As concentration, reported 2.00–17.00 $\mu g\ kg^{-1}$ of As in grain samples [11]. While 5.00–285.00 $\mu g\ kg^{-1}$ [12], 4.00–362.00 $\mu g\ kg^{-1}$ [13], and 1.00–500.00 $\mu g\ kg^{-1}$ of As were also reported in wheat grain collected from an As contaminated site, where soil As concentration ranged from 3.00 to 201.00 $mg\ kg^{-1}$ [14]. Thus, besides rice, wheat could be a major source of dietary As. Wheat is the second most-produced (771.72 million tonnes) cereal crop throughout the world, with the highest harvested area being 218.54 million ha [15]. Therefore, to feed the rising global population, wheat will stay as a vital component of human nutrition. Hence, increasing its quality of production, free from toxic heavy metals, is an important requirement for sustainable agriculture and food security. Therefore, As remediation techniques, not only for drinking water but also for soil, are crucial to avoid food As contamination through crop uptake.

Different physical, chemical, and biological techniques are being used for remediation of As contamination in soil. These remediation methods include vitrification, electrokinetic treatment, soil flushing and solidification, phytoremediation by hyper accumulative plants, etc. [16]. Hitherto, these methods have been frequently revealed to be ineffective, costly, or too lengthy, with usage being restricted to smaller-scale operations with lower efficiency, selectivity, and disposal of materials after remediation [17]. Thus, alternative options or combinations of technologies, for reducing soil As pollution, are required.

Recent research has revealed that pre-sowing microwave (MW) soil heating application in agricultural systems is a promising technique, which not only has potential to control weeds by deactivating the soil weed seed bank [18, 19], but it can also significantly increase crop growth and yield of rice and wheat by increasing some soil nutrients (N, P, K, and S) availability as a result of soil humification processes and nutrient recovery from dead microorganisms after exposure to MW heating [20–23].

Microwave energy is a form of electromagnetic radiation, with wavelengths ranging from 1 m to 1 mm and frequencies between 300 MHz to 300 GHz, which can induce the rotation of the dipoles of polar molecules (e.g. water), due to the oscillating electromagnetic field, which results in the generation of heat by intermolecular friction [24]. This produces a fast heating rate, since soil moisture is considered to be an efficient absorber of MW radiation [25]. Thus, MW heating has major advantages over other heating processes. These advantages include short start-up, selective heating, precise control, no direct contact with heated materials, and volumetric heating [26, 27]. Because of these advantages, MW has been used in diversified fields including removal of organic contaminants [28] and immobilization of some toxic metals (Cu, Mn, Th, Zn, Ni, Cd, Cr, and Pb) in soil [29] and solid sediments [30, 31]. However, no study has been found to address soil As immobilization using MW heating to alleviate As phytotoxicity in wheat. Therefore, this study was designed to explore whether MW soil treatment could alleviate As phytotoxicity and reduce wheat grain As concentration or not. Thus, the hypotheses of this study were (i) microwave soil treatment increases the soil organic matter, increases soil organic carbon, nitrogen mineralization, and nutrients availability that might be favorable for better plant growth and development, (ii) better plant growth and higher grain yield as the result of MW soil treatment will reduce the grain As concentration by means of dilution effect and (iii) microwave soil treatment synthesized the macromolecular organic substance that possesses a higher number of functional groups and organometallic and coordination compounds that are able to retain As, decrease mobility, and reduce bioavailability by adsorbing As.

2. Materials and methods

2.1 Experimental site, soil collection, and preparation

Experimental soils were collected from a wheat production paddock of the Dookie agricultural farm (36°37′ S; 145°70′ E) at a depth of 0–15 cm. The soil was a brownish-gray loam and classified as a Major Clay Loam [32] or a red mesotrophic-haplic dermosol [33]. Some important soil properties are given in **Table 1**. The collected soils were dried and sieved through a 4 mm mesh to minimize the undesired effects of stones, sticks, and clods. This operation did not reflect the true field situation, where the distribution of coarse material is highly irregular; however, it was essential to ensure a uniform experimental condition for MW soil heating. After sieving, 8.5 kg of soil was thoroughly mixed and shifted into pots (diameter 27 cm and height 30 cm). Unperforated pots were used to prevent the loss of water-soluble As from the pots [34].

2.2 Physicochemical properties of soil

Soil samples were sent to the Nutrient Advantage Laboratory, a NATA (National Association of Testing Authorities, Australia) accredited laboratory (Lab number: 11958, ISO/IEC 17025), for analysis of soil properties. The physicochemical properties of the soil are presented in **Table 1**.

2.3 Arsenic application

Five different levels of As concentration (0, 20, 40, 60, and 80 mg kg^{-1} soil) as sodium arsenate heptahydrate ($Na_2HAsO_4{\cdot}7H_2O$) [35] were mixed with the initial soil. Respective amounts of sodium arsenate were mixed with deionized water to prepare the As solutions. Then, the As solution was mixed with the soil by spraying and homogenizing thoroughly by hand mixing. The background As concentration in soil varies depending on the extent of As pollution in that area. For example, in highly naturally contaminated agricultural soil, the concentration of As in Bangladesh is 20–83 mg kg^{-1} [36], in China it is 40–70 mg kg^{-1} [37], and in India, it is 9–105 mg kg^{-1} [38]. Therefore, the As treatments in this study represented the different extent of contamination from several countries. To establish an equilibrium condition between soil and applied As, soil moisture was maintained at field capacity for 2 weeks prior to applying the MW treatment.

Soil properties	Analytical method	Units	Microwave treatments		
			MW-0	MW-3	MW-6
Organic carbon (OC)	Walkley & Black	%	1.41	1.35	1.34
Organic matter (OM)	Walkley & Black	%	2.43	2.32	2.30
Electrical conductivity (EC)	Saturated extract	dS/m	1.00	1.20	1.70
pH	1:5 $CaCl_2$	N/A	5.60	5.60	5.60
Cation exchange capacity (CEC)	$BaCl_2$ exchange	cmol(+)/kg	10.10	9.31	9.29
Nitrate nitrogen (NO_3^--N)	Kjeldahl	mg kg^{-1}	49.00	45.00	43.00
Ammonium nitrogen (NH_4^+-N)	Kjeldahl	mg kg^{-1}	150.00	230.00	310.00
Available Potassium (K)	Atomic emission	mg kg^{-1}	610.00	610.00	600.00
Sulfur (S)	0.25 M KCl at 40°C	mg kg^{-1}	12.00	18.00	39.00
Phosphorus (P)	Colwell	mg kg^{-1}	120.00	170.00	190.00
Calcium (Ca)	Ammonium acetate	cmol(+)/kg	6.70	6.00	5.80
Magnesium (Mg)	Ammonium acetate	cmol(+)/kg	1.80	1.80	1.80
Potassium (K)	Ammonium acetate	cmol(+)/kg	1.60	1.60	1.50
Sodium (Na)	Ammonium acetate	cmol(+)/kg	<0.02	0.03	0.06
Aluminum (Al)	Ammonium acetate	cmol(+)/kg	<0.10	<0.10	0.11
Copper (Cu)	DTPA	mg kg^{-1}	5.10	4.70	4.80
Zinc (Zn)	DTPA	mg kg^{-1}	4.80	4.60	5.00
Manganese (Mn)	DTPA	mg kg^{-1}	58.00	62.00	72.00
Iron (Fe)	DTPA	mg kg^{-1}	130.00	120.00	120.00
Boron (B)	DTPA	mg kg^{-1}	0.78	0.77	0.85
Silicon (Si)	$CaCl_2$ soluble	mg kg^{-1}	80.00	82.00	110.00
Arsenic	HG-AFS	µg kg^{-1}	<0.01	<0.01	<0.01

Table 1.
Physicochemical properties of pre and post microwave (MW) treated soils before sowing.

2.4 Microwave application

Three levels of MW energy were applied for 0, 3, and 6 min to attain soil temperatures of around room temperature, 60 and 90°C, respectively. The duration of MW irradiation to heat the soil at the desired temperature was determined by following the method of previous research work [23, 39]. Soil heating at around 90°C has been found to be effective for controlling weed infestations and destroying weed seed banks in the topsoil without significantly changing the soil properties [39]. Therefore, the same soil heating temperature was also used in this experiment to explore the potentiality in As phytotoxicity alleviation. Soil heating at around 60°C was included as an intermediate treatment between 90°C and control.

An MW chamber, consisting of six magnetrons (1 kW each), operating at a frequency of 2.45 GHz, was used for soil treatment (**Figure 1**). Energy dissipated in the soil sample after MW treatment (**Figure 1b**), and chamber electric field (**Figure 1c**)

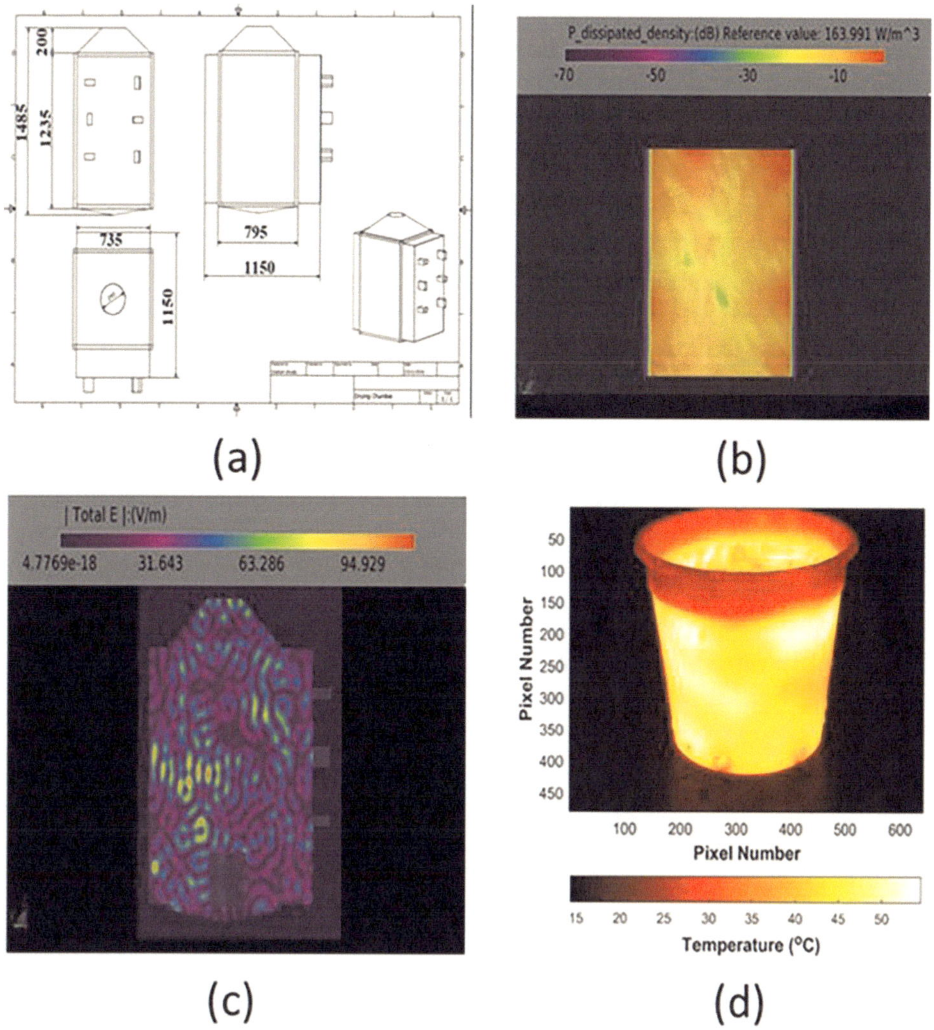

Figure 1.
(a) Schematic diagram of 6-kW microwave (MW) chamber (internal capacity of approximately 1.0 m^3) [23, 40], (b) energy dissipated in the soil sample after MW treatment, (c) chamber electric field, and (d) thermal images captured with an infrared camera (FLIR C3) after 3 min of MW irradiation of soil showing the temperature of 60 ± 5°C.

was modeled. The modeling was done using XFdtd software version 7.9 produced by Remcom (USA). **Figure 1(b)** illustrates the microwave's electric field distribution in the soil sample only, which **Figure 1(c)** illustrates the microwave's electric field distribution throughout the whole chamber. Both illustrations are on a vertical plane that passes through the center of the soil sample and the soil sample was in the center of the chamber, resting on the floor of the chamber. The soil temperature was measured for each MW treatment at a depth of 10–15 cm, immediately after MW energy exposure, by using liquid-in-glass thermometers [41]. An infrared camera was also used for taking thermal images to show the energy dissipated and temperature distribution across the MW treated soil. Due to the very high dependence of the dielectric properties on moisture content [25], the moisture content in the soil will greatly affect the heating effect of MW energy on the soil. In this experiment, the moisture content was maintained at around 15% (w/w) at the time of MW soil treatment.

2.5 Experiment setup

The experiment was conducted in a glasshouse, at Dookie campus, The University of Melbourne, Australia, by following a completely randomized design (CRD) with four replications. To describe the treatment combination more conveniently, abbreviated forms have been used for As treatments (As-0, As-20, As-40, As-60, and As-80) and MW treatments (MW-0, MW-3, and MW-6). Before the seed sowing, mono ammonium phosphate (MAP) fertilizer was applied (equivalent to 150 kg ha^{-1}) to each pot as a basal dose, as per standard practices for Australian wheat cultivation. The rest of the N requirement was calculated (based on a total 150 kg N ha^{-1}) and applied, as urea, in two split doses viz. at early stem elongation (GS30-32) stage and booting (GS45-49) stage. Twelve seeds of the EGA Gregory wheat variety (*Triticum aestivum* L.) were sown per pot on the 6th of June 2017. Tap water was used for crop irrigation purposes. This water source contained As below the detection limit (<0.01 μg l^{-1}); thus, there were no possibilities of As addition from the tap water to the pot soil. After 180 days of the growing period, at the physiological maturity stage, the crop was harvested on the 5th of December 2017.

2.6 Recording of crop agronomic data

The plant height was measured as a distance from the soil surface to the top of a plant using a measuring scale. Plant vigor data was recorded by an ordinal scale ranging from 1 (low vigor) to 9 (high vigor). Leaf chlorophyll content was measured as SPAD (Soil-Plant Analysis Development) value using the Chlorophyll Meter-SPAD-502Plus [42] at the tillering stage. To get the plant height, plant vigor, and leaf chlorophyll content data, five plants within a pot were selected randomly and data recorded as the mean value of these five plants. At the tillering stage, plant samples (3 hills per pot) were collected to determine the shoot biomass and measure the leaf area, width, and length of the last fully expanded leaf by using a leaf area meter (LASER Leaf Area Meter, CI-202, CID Bio-Science, USA). At the physiological maturity stage, the crop was harvested, and shoot biomass, total number of spikes, root biomass, and grain yield were recorded. Both the shoot and root samples were dried at 60°C in an oven for 48 h to determine the dry biomass.

2.7 Grain total arsenic analysis

Grain total As analysis was performed as per the method described in the user manual of atomic fluorescence spectrometry (AFS; PSA 10.055 Millennium Excalibur, 2009) [43]. Since the method is generalized for solid materials, some modifications were made for the wheat grain As analysis. The modifications were, (i) a 0.5 g sample used for analysis instead of 0.25 g because generally wheat grain As concentration is lower than in soil; (ii) heating time was extended up to 90–100 min until a clear solution appeared (as an indication of good digestion), whereas 40 min was suggested in the original method;

and (iii) digested liquid was filtered with Whatman 42 (ashless, 2.7 μm) filter paper as it is better than the 541 (ashless, 20–25 μm) and usually used in heavy metal analysis.

2.7.1 Sample preparation

The whole grain sample was oven-dried at 105°C for 24 h prior to grinding and homogenizing with the ultra-centrifugal mill (RETSCH, ZM 200) [44]. The powdered sample was stored in a polypropylene pot for further analysis. Sample digestion was performed by hydrochloric-nitric (HCl:HNO_3 = 3:1) di-acid (aqua regia) with a block digester (VELP Scientifica, DK-42). For pre-digestion, a 0.5 g powdered sample was taken into a 100 ml digestion tube (Ø 26 mm). After that, 12 ml of concentrated HCl (37%, 12 M) and 4 ml of concentrated HNO_3 (70%, 15.8 M) were added and left overnight to allow the vigorous initial reaction to subside. Excessive foaming was reduced by adding 2–3 drops of n-dodecane into the mixture. The mixture was heated at 140°C for 90–100 min until the appearance of a clear solution. After cooling the tube, the liquid was filtered with Whatman 42 filter paper.

2.7.2 Atomic fluorescence spectrometry

Total As analysis was performed using atomic fluorescence spectrometry (AFS; PSA 10.055 Millennium Excalibur) [43]. Prior to total As determination, all the samples were pre-reduced with potassium iodide (1% m/v) and ascorbic acid (0.2% m/v) to reduce As(V) to As(III). For analysis standard preparation, 1000 $\pm$ 2 mg l^{-1} CRM (certified reference material) As standard supplied in 2% HNO_3 (Sigma-Aldrich) was used as a standard stock solution. A working standard solution of 10 mg l^{-1} was prepared weekly from the stock solution and used to prepare calibration standards (0–10 μg l^{-1}). The standards and samples were prepared by following the same analytical matrix of 25% (v/v) HCl, 1% (m/v) potassium iodide, and 0.2% (m/v) ascorbic acid.

Properties	Unit	Value
Carrier gas (Ar) flow rate	l min^{-1}	0.25
Carrier gas pressure	Psi	35–45
Dryer gas (H_2) flow rate	l min^{-1}	2.50
Dryer gas pressure	Psi	35–45
$NaBH_4$ concentration in 0.1 $moll^{-1}$ NaOH	% (m/V)	0.70
HCl concentration for hydride generation	mol l^{-1}	3.00
$NaBH_4$ flow rate (reductant)	ml min^{-1}	4.50
HCl flow rate (reagent blank)	ml min^{-1}	9.00
Sample flow rate	ml min^{-1}	9.00
Lamp current	Ma	27.50 (primary), 35.00 (boost)
Lamp wavelength	Nm	197.30
Analysis period	Sec	15 (delay), 30 (analysis), and 30 (memory)
Lower limit of detection (LOD)	ng l^{-1}	10.00

Table 2.
Operating environment of atomic fluorescence spectrometer (AFS) for total arsenic analysis [43].

Sodium tetra hydroborate (0.7% m/v in 0.1 mol l^{-1} NaOH) was continuously added to the sample during the analysis to produce gaseous arsine (AsH_3), which was atomized using a hydrogen diffusion flame. The overall reactions are represented in the following Eqs. (1)–(4) [45]. Atomic fluorescence was measured after excitation using an As boosted discharge hollow cathode lamp (Photron) [46]. The operating states of AFS for As determination are given below in **Table 2**.

$$\text{Sample} + \text{KI} + \text{Ascorbic acid} + \text{HCl} \rightarrow \text{KCl} + 2\text{HI} \rightarrow 2\text{H}^{+} + 2\text{I}^{-} + 2\text{e}^{-} \quad (1)$$

$$\text{AsO}_4{}^{3-} + 2\text{H}^{+} + 2\text{I}^{-} + 2\text{e}^{-} \rightarrow \text{AsO}_3{}^{3-} + \text{I}_2 + \text{H}_2\text{O} \quad (2)$$

$$\text{NaBH}_4 + \text{HCl} \rightarrow \text{NaCl} + \text{H}_3\text{BO}_3 + 8\text{H} \quad (3)$$

$$\text{AsO}_4{}^{3-} + \text{H} \rightarrow \text{AsH}_3 + \text{H}_2 \quad (4)$$

2.7.3 Quality assurance of arsenic analysis

For quality control, the appropriate procedures and safety measures were taken to ensure the consistency of the results by following the techniques described by Thompson and Walsh [47]. Samples were handled carefully to avoid cross-contamination. All glassware was cleaned with the laboratory dishwashing machine followed by a 10% HNO_3 solution and rinsing with deionized water. High purity analytical grade chemicals and gases (99.99% pure) were used for the analysis to ensure the minimal blank concentration value. Deionized water was used for all dilutions and preparation of chemicals during the analysis. To ensure good recovery of sample As, a 1568b rice flour standard reference material (SRM), from NIST (National Institute of Standard and Technology), was used at the time of digestion. Therefore, the block digestion set consisted of one blank, one SRM, one duplicate and with the remaining 39 tubes being the main samples with three replications each. Data was deemed to be acceptable if recovery of SRM As was $\pm 10\%$ and the calculated relative standard deviation (RSD) of duplicate samples was no greater than 5%. To provide measurement clarification regarding the response of the Millennium Excalibur, the background equivalent concentration (BEC) was calculated using Eq. (5) to determine the performance of the instrument.

$$\text{BEC} = \frac{\text{Background value}}{\text{Peak height}} \times \text{standard concentration} \quad (5)$$

The lower the BEC value the more sensitive the instrument. If the BEC value was below 0.5, the instrument was considered to be operating correctly.

2.8 Statistical analysis

Statistical analysis of recorded data was performed using GenStat (16th Edition, VSN International) software. Normality and homogeneity of variance of the experimental data were tested. The analysis of variance (ANOVA) test was performed to determine the significance of tested treatments on variables. The Least Significant Difference (LSD) test was performed to compare the treatments' means at a 5% level of significance. The Pearson correlation test was performed to determine the correlation coefficient among the variables. For grain As concentration data, Grubb's test was performed to identify outlier values, which were replaced by the other replicates' average values, if found. After MW soil heating, thermal images were captured with an infrared camera (FLIR C3) and post-processed in MATLAB (MathWorks, Inc., USA) software.

3. Results

3.1 Plant growth and grain yield

The results revealed that the addition of As to the soil had a significant negative impact on plant growth and grain yield, and MW soil treatments provided a beneficial effect compared with non-MW treated soils, irrespective of soil As concentration. To describe the plant growth some growth parameter results are given below.

3.1.1 Plant height

Plant height decreased significantly ($p < 0.001$) with increasing soil As concentration. This trend was observed up to 60 days after sowing (DAS) of plant growth. After that, the effect of As on plant height was not statistically significant. Plant height increased significantly ($p < 0.001$) in MW treatments irrespective of soil As concentration throughout the growing period. Greater plant height was recorded in MW-6 compared with MW-3 and MW-0 treatment (**Table 3**).

3.1.2 Plant vigor

With the increase of soil As concentration, plant vigor decreased significantly ($p < 0.001$), while significantly ($p < 0.001$) higher plant vigor was found in the MW treatments. For instance, at As-80 the plant vigor was lowest (4.00) in MW-0, whereas significantly higher plant vigor was observed in the MW-6 (7.00) treatment (**Figure 2a**).

3.1.3 Leaf chlorophyll content

Leaf chlorophyll content increased significantly ($p < 0.001$) in the MW treatments compare with the control. For example, at As-80 significantly higher leaf chlorophyll content (45.88) was recorded in the MW-6 treatment compared with the MW-0 treatment (39.10). In the MW-6 treatment, no significant changes were observed in

Soil As (mg kg $^{-1}$)	30 DASS			60 DASS			90 DASS		
	MW soil treatments (min)								
	0	3	6	0	3	6	0	3	6
0	35.46^{ab}	35.87^{a}	37.65^{a}	43.25^{bcd}	45.85^{ab}	46.40^{ab}	68.60^{ab}	66.65^{abc}	71.20^{a}
20	34.74^{ab}	36.00^{a}	38.22^{a}	43.45^{bcd}	45.10^{abc}	45.45^{ab}	61.50^{bcd}	68.40^{ab}	71.25^{a}
40	34.47^{bc}	34.96^{ab}	37.69^{a}	44.30^{bcd}	45.25^{ab}	47.70^{a}	61.95^{bcd}	$65.85^{a\text{–}d}$	72.55^{a}
60	32.40^{cd}	33.15^{cd}	36.57^{ab}	41.95^{d}	44.35^{bcd}	46.20^{ab}	60.35^{cd}	67.35^{abc}	69.40^{ab}
80	28.40^{e}	30.49^{d}	35.57^{ab}	38.10^{e}	42.10^{cd}	45.50^{ab}	58.00^{d}	$65.20^{a\text{–}d}$	73.25^{a}
$LSD_{0.05}$		2.09			1.36			6.96	

Mean data with superscript same letter are not significantly different. Least Significant Difference (LSD) test performed at a 5% level of significance to determine the difference between the arsenic (As) and microwave (MW) treatments. DAS (Days after sowing) indicates the sampling time.

Table 3.
Mean plant height (cm) in response to microwave (MW) soil heating and soil arsenic (As) treatments.

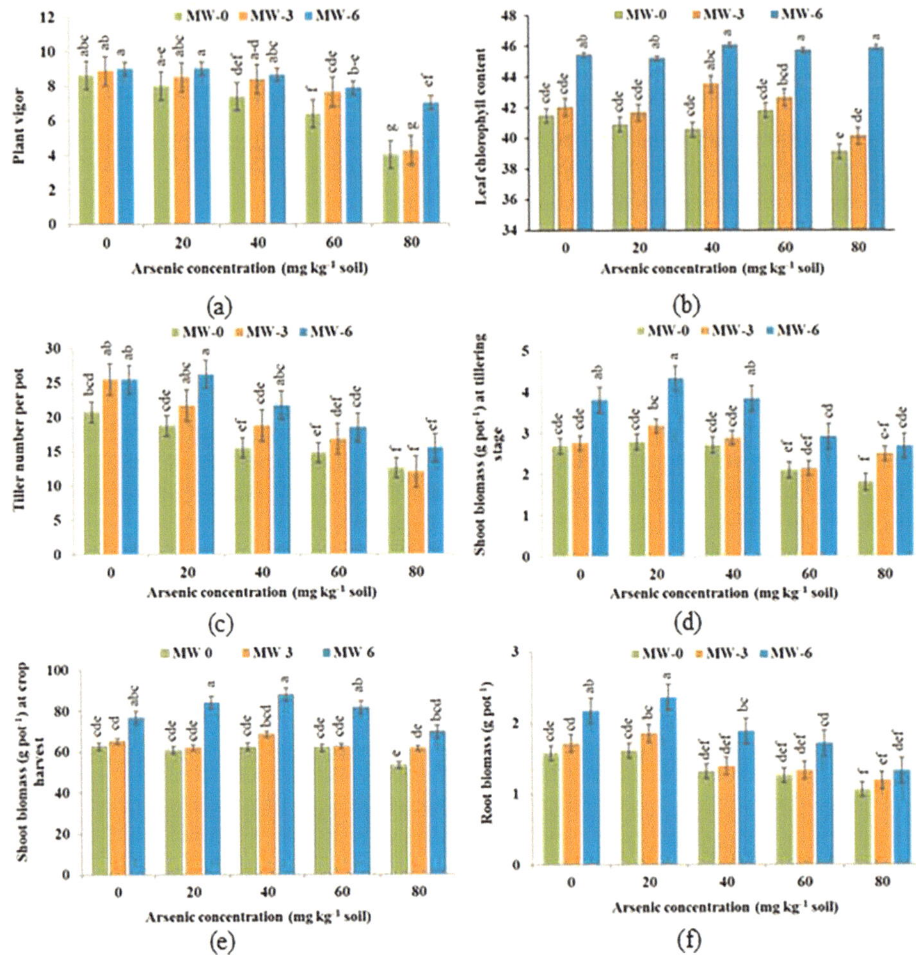

Figure 2.
Effect of microwave (MW) soil treatment on wheat plant growth in arsenic (As) contaminated soils. (a) Plant vigor, (b) leaf chlorophyll content, (c) tiller number, (d) shoot biomass at tillering stage, (e) shoot biomass at crop harvest, and (f) root biomass. Bar represents the mean value with standard error and different letters indicate the significant difference (LSD at p = 0.05) among the treatments.

chlorophyll content across all soil As concentrations. Although the effect of soil As concentration on the leaf chlorophyll content was not significant ($p = 0.187$), a decreasing trend was observed in the MW-0 treatment (**Figure 2b**).

3.1.4 Tiller number

Tiller number reduced significantly ($p < 0.001$) across the treatments with increasing soil As concentration. However, a significantly ($p = 0.005$) higher tiller number was obtained in the MW treatments. This was especially so in the MW-6 treatment where the tiller number was higher than the MW-0 and the MW-3 treatment. The highest tiller number (26.25) was recorded in the MW-6 treatment at As-20 soil As concentration, while it was 21.75 and 18.75 in MW-3 and MW-0 treatment, respectively (**Figure 2c**).

3.1.5 Plant biomass

At the tillering stage, shoot biomass was reduced significantly ($p < 0.001$) with increasing soil As concentration, while in the MW treated pots, significantly ($p < 0.001$) higher biomass was recorded. In view of the MW treatments, higher biomass was harvested from the MW-6 treatment compared with the MW-3 and MW-0 treatments (**Figure 2d**). At crop harvest stage, shoot biomass was reduced significantly ($p < 0.008$) in response to increased soil As concentration, whereas significantly ($p < 0.001$) higher biomass was recorded in the MW treatments. In the MW-6 treatment, higher biomass was harvested compared with the MW-0 and MW-3 treatment (**Figure 2e**). Like shoot biomass, similar results were observed for root biomass (**Figure 2f**).

3.1.6 Leaf area, width, and length

With increasing soil As concentration, leaf area, width, and length were reduced, although the effect was not statistically significant. On the other hand, leaf area ($p < 0.001$), width ($p < 0.001$), and length ($p = 0.048$) increased significantly in MW soil treatments. The lowest value for all leaf parameters was found at the highest As concentration with no MW treatment, while the highest value was found in the MW-6 treatment, irrespective of soil As concentration (**Table 4**).

3.1.7 Total number of spikes

There was no significant ($p = 0.064$) effect of As on total spike number, but a significantly ($p < 0.001$) higher number of spikes was found in the MW treatments. The highest spike number (17.00) was found in the MW-6 treatment, while it was 12.00 and 9.00 in MW-3 and MW-0 treatment respectively at As-20 treatment (**Figure 3a**).

3.1.8 Grain yield

The wheat grain yield increased significantly ($p < 0.001$) in the MW treated pots, while the effect of As on grain yield was not significant ($p = 0.210$). Higher grain yield

Soil As (mg kg^{-1})	Leaf area (cm^2).			Leaf width (cm).			Leaf length (cm).		
	MW soil treatment (min)								
	0	3	6	0	3	6	0	3	6
0	21.12^b	20.60^b	24.52^a	1.03^b	1.09^b	1.13^b	25.94^c	25.66^b	27.88^b
20	21.12^b	25.04^a	21.99^b	1.10^b	1.12^b	1.11^b	25.12^b	28.91^b	27.26^b
40	20.25b^c	23.71^a	27.92^a	1.01^b	1.11^b	1.25^a	26.46^c	27.73^b	29.21^a
60	18.85c^d	22.50^b	27.91^a	0.97^b	1.09^b	1.30^a	25.52^d	26.26^b	29.67^a
80	16.51^e	23.62^a	22.24^b	0.82^c	1.14^a	1.03^b	24.98^e	27.16^b	27.10^c
LSD $_{(0.05)}$		4.70			0.17			4.65	

Mean data with superscript same letter are not significantly different. Least Significant Difference (LSD) test performed at a 5% level of significance to determine the difference between the arsenic (As) and microwave (MW) treatments.
#Data recorded at tillering stage.

Table 4.
Effect of microwave (MW) treatment on leaf area, length, and width at different soil arsenic (As) concentration.#

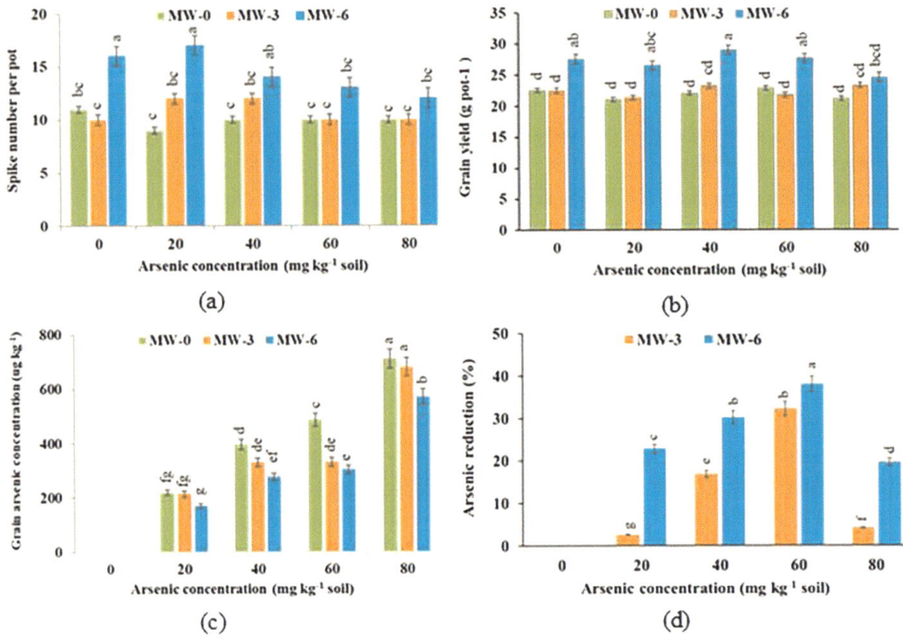

Figure 3.
Effect of microwave (MW) soil treatment on (a) spike number, (b) grain yield, (c) grain arsenic (As) concentration in response to different soil As concentration, and (d) grain As concentration reduction in MW treatments. Bar represents the mean value with standard error and different letters indicate the significant difference (LSD at p = 0.05) among the treatments.

was found in the MW-6 treatment compared with the MW-0 and MW-3 treatments. For instance, a significantly higher grain yield (28.95 g pot^{-1}) was recorded in the MW-6 treatment compared with the MW-3 (23.21 g pot^{-1}) and MW-0 (22.03 g pot^{-1}) treatments at As-40 treatment (**Figure 3b**).

3.1.9 Grain total arsenic concentration

Grain As concentration increased significantly ($p < 0.001$) with increasing soil As concentration while it was significantly ($p < 0.001$) lower in the MW treatments compare with the control (**Figure 3c**). At As-80 the highest grain As concentration (710.45 µg kg^{-1}) was recorded in MW-0 while, it was significantly lower (572.03 µg kg^{-1}) in the MW-6 treatment. The highest grain As concentration reduction (37.98%) was observed in the MW-6 treatment at As-60 followed by the MW-3 treatment (32.20%) compared with the MW-0 treatment (**Figure 3d**).

3.2 Grain mineral content

Grain P ($p < 0.001$) content decreased significantly, and K ($p = 0.008$) and Na ($p < 0.001$) content increased significantly with the increase of soil As treatment, while there was no significant effect of MW treatment on grain P, K and Na content. On the other hand, Mn ($p = 0.012$) and Zn ($p < 0.001$) content decreased significantly with the increasing soil As concentration, while Mn ($p < 0.001$) and Zn ($p < 0.001$)

Soil As (mg kg^{-1})	P (mg kg^{-1})			K (mg kg^{-1})			S (mg kg^{-1})			Ca (mg kg^{-1})			Mg (mg kg^{-1})		
	MW soil treatment (min)														
	0	3	6	0	3	6	0	3	6	0	3	6	0	3	6
0	3568^{a}	3529^{a}	3571^{a}	4481^{bc}	4443^{c}	4543^{bc}	1118^{a}	1086^{ab}	1090^{ab}	341^{a}	318^{a-d}	309^{b-e}	999^{abc}	983^{bc}	1065^{a}
20	3380^{ab}	3355^{ab}	3387^{ab}	4497^{bc}	4549^{bc}	4490^{bc}	1023^{ab}	1098^{ab}	1030^{ab}	319^{a-d}	313^{b-e}	294^{def}	991^{abc}	1012^{abc}	1058^{ab}
40	3131^{bc}	3056^{c}	3016^{c}	4611^{abc}	4546^{bc}	4583^{bc}	1065^{ab}	1048^{ab}	992^{b}	331^{ab}	316^{a-e}	283^{f}	1029^{abc}	1022^{bc}	1019^{abc}
60	2881^{c}	2849^{c}	2912^{c}	4745^{ab}	4684^{abc}	4484^{abc}	1037^{ab}	1028^{ab}	1008^{ab}	321^{abc}	313^{b-e}	296^{c-f}	996^{abc}	981^{c}	1032^{abc}
80	2423^{d}	2431^{d}	2432^{d}	4695^{abc}	4819^{a}	4633^{abc}	1012^{ab}	985^{b}	1051^{ab}	329^{ab}	318^{a-d}	290^{ef}	967^{c}	969^{abc}	1006^{abc}
$LSD_{0.05}$		270			230			103			23			68	

Mean data with superscript same letter are not significantly different. Least significant difference (LSD) test was performed at a 5% level of significance to determine the difference between the treatments.

Table 5.
Grain macronutrient content in response to microwave (MW) soil treatment at different soil arsenic (As) concentration.

Soil As (mg kg^{-1})	Fe (mg kg^{-1})			Cu (mg kg^{-1})			Zn (mg kg^{-1})			Mn (mg kg^{-1})			Na (mg kg^{-1})		
	MW soil treatment (min)														
	0	3	6	0	3	6	0	3	6	0	3	6	0	3	6
0	19.00^{ef}	20.00^{def}	26.00^{a}	4.20^{abc}	3.90^{abc}	4.50^{a}	21.00^{cd}	22.00^{cd}	32.00^{a}	50.00$^{b–e}$	58.00^{bcd}	97.00^{a}	7.60^{b}	6.90^{b}	6.60^{b}
20	21.00$^{b–f}$	19.00^{f}	24.00^{ab}	3.70^{bc}	3.90^{abc}	4.10^{abc}	19.00^{def}	21.00^{cd}	27.00^{b}	59.00^{bcd}	61.00^{bc}	109.00^{a}	6.50^{b}	6.90^{b}	7.00^{b}
40	20.00^{def}	20.00$^{c–f}$	24.00$^{a–d}$	3.90^{abc}	3.90^{abc}	4.20^{abc}	18.00^{efg}	20.00^{cde}	23.00^{c}	43.00^{cde}	64.00^{b}	105.00^{a}	7.20^{b}	7.10^{b}	9.40^{b}
60	20.00^{def}	20.00^{def}	23.00$^{a–e}$	3.70^{bc}	3.70^{bc}	4.30^{ab}	16.00^{fg}	17.00^{efg}	22.00^{c}	42.00^{de}	50.00$^{b–e}$	99.00^{a}	8.60^{b}	8.20^{b}	8.20^{b}
80	23.00$^{a–e}$	23.00$^{a–d}$	24.00^{abc}	3.70^{bc}	3.50^{c}	4.00^{abc}	15.00^{g}	15.00^{g}	22.00^{c}	39.00^{e}	47.00$^{b–e}$	94.00^{a}	9.10^{b}	13.20^{a}	9.00^{b}
$LSD_{0.05}$		3.20			0.58			2.80			16.10			2.95	

Mean data with superscript same letter are not significantly different. Least significant difference (LSD) test was performed at a 5% level of significance to determine the difference between the treatments.

Table 6.
Grain micronutrient and sodium content in response to microwave (MW) soil treatment at different soil arsenic (As) concentration.

Variables		**r–value**							
Above-ground biomass	1	—							
Grain As concentration	2	-0.28^{*}	—						
Grain yield	3	0.84^{***}	-0.15^{ns}	—					
Leaf area	4	0.48^{***}	-0.20^{ns}	0.45^{***}	—				
Plant height	5	0.62^{***}	-0.55^{***}	0.4^{***}	0.55^{***}	—			
Plant vigor	6	0.48^{***}	-0.80^{***}	0.32^{*}	0.25^{ns}	0.74^{***}	—		
Leaf chlorophyll content	7	0.67^{***}	-0.23^{ns}	0.62^{***}	0.52^{***}	0.66^{***}	0.47^{***}	—	
Spike number	8	0.62^{***}	-0.30^{*}	0.71^{***}	0.28^{*}	0.41^{**}	0.45^{***}	0.50^{***}	—
Tiller number	9	0.52^{***}	-0.67^{***}	0.43^{***}	0.21^{ns}	0.57^{***}	0.77^{***}	0.39^{**}	0.58^{***}
		1	2	3	4	5	6	7	8

ns indicate non-significant.
**Significance at $p < 0.05$.*
***Significance at $p < 0.01$.*
****Significance at $p < 0.001$.*

Table 7.
Pearson's correlation matrix of different growth and yield parameters with grain arsenic concentration and accumulation.

Variables		**r–value**									
Grain As concentration	1	—									
P	2	-0.86^{***}	—								
K	3	0.46^{***}	0.11^{ns}	—							
Ca	4	0.01^{ns}	0.16^{ns}	0.16^{ns}	—						
Mg	5	-0.24^{ns}	0.21^{ns}	-0.02^{ns}	0.05^{ns}	—					
S	6	-0.24^{ns}	-0.30^{*}	-0.09^{ns}	0.38^{**}	0.27^{*}	—				
Fe	7	0.12^{ns}	0.01^{ns}	-0.07^{ns}	-0.17^{ns}	0.38^{**}	-0.10^{ns}	—			
Cu	8	-0.30^{*}	-0.07^{ns}	-0.41^{***}	-0.13^{ns}	0.58^{***}	0.37^{**}	0.32^{*}	—		
Mn	9	-0.29^{*}	-0.10^{ns}	-0.27^{*}	-0.63^{***}	0.42^{***}	-0.07^{ns}	0.47^{***}	0.39^{**}	—	
Zn	10	-0.62^{***}	-0.09^{ns}	-0.42^{***}	-0.30^{*}	0.57^{***}	0.34^{**}	0.46^{***}	0.62^{***}	0.78^{***}	—
Na	11	0.48^{***}	-0.11^{ns}	0.41^{**}	-0.04^{ns}	-0.33^{*}	-0.26^{*}	0.21^{ns}	-0.35^{**}	-0.14^{ns}	-0.37^{**}
		1	2	3	4	5	6	7	8	9	10

ns indicate non-significant.
**Significance at $p < 0.05$.*
***Significance at $p < 0.01$.*
****Significance at $p < 0.001$.*

Table 8.
Pearson's correlation matrix of different grain minerals with grain arsenic concentration and accumulation.

content increased significantly in the MW treated soil compared with the control. The effect of As on grain Ca, Mg, Fe, and Cu was statistically non-significant. However, grain Mg ($p = 0.012$), Fe ($p < 0.001$), and Cu ($p = 0.003$) content increased significantly while, Ca ($p < 0.001$) content decreased significantly in the MW treated soil compared with the control treatment (**Tables 5** and **6**).

3.3 Correlation of grain arsenic with plant growth and yield parameters and grain mineral content

Pearson's correlation coefficient (r value) showed that all the growth parameters were positively correlated with the yield parameters, and all the growth and yield parameters were negatively correlated with grain As concentration. Although the grain yield was negatively correlated with grain As concentration ($r = -0.1511$), the correlation coefficient was statistically non-significant (**Table 7**). Also, Pearson's correlation coefficient (r value) showed that, grain K ($r = 0.46^{***}$) and Na ($r = 0.48^{***}$) were positively correlated, while grain P ($r = -0.86^{***}$), S ($r = -0.34^{**}$), Cu ($r = -0.30^{*}$), Mn ($r = -0.29^{*}$), and Zn ($r = -0.62^{***}$) were negatively correlated with the grain As concentration (**Table 8**).

4. Discussion

It is well known that soil As has adverse effects on plant growth and development. Previous research revealed that plant growth traits such as plant height, tiller number, and, the number of grains per spike can decrease significantly with increasing soil As concentration [48]. Pigna et al. [49] reported a 60% plant biomass and 83.6% root biomass reduction of wheat in As contaminated soil. Also, several other experiments reported that the reduction of plant growth was ultimately the result of As phytotoxicity at high soil As concentrations [50–52]. This experiment also found a significant reduction in plant growth represented by plant height, plant vigor, tiller number, shoot, and root biomass with increasing soil As concentration, which agrees with these other studies. Like other growth parameters, similar results were observed in leaf chlorophyll content (measured as SPAD), leaf area, leaf width, and length, which also reflect the lower plant growth at higher soil As concentration. Chlorophyll consists of mainly N, as a core component, and its content in the leaf represents the chlorophyll content. Higher soil As can reduce the N content in various crops [53, 54]. Thus, it was anticipated that higher soil As concentrations may also decrease N content in wheat plants, which may lead to a decrease in chlorophyll content. The results of the present experiment revealed that higher soil As concentrations decreased the chlorophyll content, therefore leading to a lower photosynthesis rate, which might have reduced plant growth and grain yield.

In relation to grain yield, a decreasing trend was observed as As concentration in the soil increased, although it was not statistically significant. From the results, it is clear that, at the early growth stage, As had more effect on the plant growth, whereas, at the mature stage As had less effect (**Figure 2**), which can be correlated with the final grain yield (**Figure 3b**). A similar result was observed in the previous experiment conducted on wheat grown in As contaminated soil (0–40 mg kg^{-1}), by Asaduzzaman et al., where less impact of As was reported on grain yield [48]. Some other previous experiments also revealed a greater effect of As on the shoots but showed less effect on the yield and yield contributing characters [48, 52], which is similar to the current results of this study.

On the other hand, the results showed that the MW soil treatment had a significantly beneficial effect on wheat plant growth and grain yield irrespective of soil As concentration (**Figures 2** and **3b**). A previous study by Khan et al. reported a 33.1% increase in plant dry biomass and a 39.2% increase in grain yield in MW treated soil [55]. Similarly, some other studies demonstrated that increased plant growth and grain yield resulted from MW treatment of soil as well [18, 19]. The above findings agree with the results obtained in this experiment. One of the possible reasons for the increased growth and yield of wheat is the higher availability of nutrients for the plants in MW-treated soil. Increased N and S availability in soil was reported after application of MW [56]. Speir et al. reported increased N levels in MW treated soil compared with the control [57]. Additionally, another previous study showed increased indigenous soil N after MW soil heating [21]. A similar result was observed in this present study where, N, P, and S increased after MW soil heating (**Table 1**). Using SPAD as a method for leaf chlorophyll measurement, a higher value (58–64) was reported in MW treated soil as compared with the control (42–56) pots [58]. In this present study, leaf chlorophyll content was significantly higher in the MW treated pots which could correlate with the higher photosynthesis rate and ultimately contributed to the higher crop growth and grain yield.

Alternatively, MW also has negative impacts on the microbial community in the soil. By generating heat, MW energy would kill certain microbes. As a result, MW irradiation-induced disintegration of the cell walls can release the intracellular and extracellular macromolecules, which may increase the soluble OM in the soil and release some nutrients [27]. Previous research reported three pathways of organic N (org-N) transformation: (1) microorganisms based org-N mineralization to ammonium, (2) release of org-N due to cell lysis, and (3) ammonium excreted from the bacteria grazing on soil fauna [59]. Research has shown that the org-N mineralization following MW irradiation of soil is of microbial origin [57]. Furthermore, nurturing and/or shaping the soil microbial communities was reported as the result of humification and thermal denaturation of soil organic compounds after soil heating [20, 60]. These microbial communities accelerate the availability of different soil nutrients that could be responsible for better plant growth and grain yield [22].

From **Figure 3c**, it was evident that the grain As concentration increased significantly with increased soil As concentration. Several studies described a higher concentration of wheat grain As when cultivated in soil containing higher As [48, 61]. A significantly high concentration of As in wheat grain (220.00–620.00 $\mu g\ kg^{-1}$), grown in soil containing 2.00–70.00 $mg\ kg^{-1}$ As, compared with those (0–50.00 $\mu g\ kg^{-1}$) grown in 1.00–13.40 $mg\ kg^{-1}$ soil As has been reported [62]. A similar trend was observed in this current experiment where, the grain As concentration was lower at low soil As concentrations, whereas it was higher at high soil As concentrations. However, MW soil treatment significantly reduced the As concentration in wheat grain (**Figure 3c**). This reduced grain As concentration could be explained by a couple of changes after MW soil heating. One of the possible reasons could be increased plant biomass and grain yield in MW treated soil, where there was a dilution effect on As concentration in the plants grown in the MW treated soil. From the Pearson's correlation study (**Table 7**) it is evident that the grain As concentration ($r = -0.1511^{ns}$) was negatively correlated with grain yield but non-significantly. This also explains the diminishing effect of As on grain yield. Another reason could be the increased soil P and Si concentration after MW soil heating (**Table 1**). The PO_4^{3-} level in the soil is known to control plant growth and development, and As(V) is a PO_4^{3-} analog [63]. Therefore, increasing PO_4^{3-} in the soil results in enhanced competition between PO_4^{3-} and As(V) for sorption sites on soil particle surfaces and for plant uptake because of the similar uptake mechanism of PO_4^{3-} and As(V) through PO_4^{3-} transporter present in the plant root [63]. Furthermore, Si can compete with As(III) for plant uptake due to a similar uptake mechanism through aquaglyceroporins the more likely nodulin 26-like intrinsic protein (NIP) class of aquaporin channels [64, 65]. Thus, the increased soil P and Si concentration after MW soil heating could complete with As(V) and As(III) for plant uptake and reduce the accumulation in the grain.

However, according to the Steindorf-Rebhun-Sheintuch equation, ligand exchange theory, and a share charge hypothesis, PO_4^{3-} has more probability to replace As(V) from soil adsorption site depending on the concentration of PO_4^{3-} and As(V). Nevertheless, PO_4^{3-} might also be desorbed by As(V) due to a mass action effect of a high ratio of As(V):P concentrations in the soil solution [66, 67]. Therefore, PO_4^{3-} and As (V) interaction needs to be considered for applying PO_4^{3-} amendment as As remediation technique. Some researchers reported that, at low soil As concentration, displacement of soil PO_4^{3-} by As(V) increased the availability of PO_4^{3-} to the plant, which resulted in the increase of plant growth parameters [49, 68]. In this experiment, it was also found that shoot biomass at both tillering stage and final crop harvest, root biomass, and spike number increased at low soil As concentrations (As-20 and As-40). However, all these traits decreased again at higher As concentrations (As-60 and As-80). Previous researches also reported plant growth and yield increases due to small additions of As in tomato, potato, rye, corn, and wheat [69–71] which agrees with the findings of this present study in wheat. Although, As is not an essential element for plants, small amounts of As can stimulate plant growth and increase plant biomass by releasing some P from soil adsorption sites and making it available for plant uptake [72].

Furthermore, As concentration in wheat grain also depends on the genetic differences of different varieties. Previous research reported that different wheat varieties, grown in the same soil As concentration, can tolerate, accumulate and translocate different concentrations of As due to phytoextraction or phyto-morphological potential of the varieties [73, 74]. The wheat variety used in this study could accumulate less As due to genetic constituents. However, further experiments are needed with different wheat varieties to explore the varietal effect on As accumulation. In addition, it has been demonstrated that MW soil heating markedly altered the physical and chemical properties of SOM [75] and enhanced the humification of SOM [76]. It has also increased the soil organic carbon and N mineralization [77], macromolecular organic substances that possess a higher number of functional groups [76], and syntheses of organometallic and coordination compounds [78]. These organic substances can retain, decrease mobility, reduce bioavailability, and adsorb soil heavy metals [79]. Therefore, more As could be adsorbed by the adsorption sites in the soil and become unavailable for plant uptake which ultimately could reduce the grain As concentration.

From **Table 5**, it was evident that grain P content decreased significantly, while Na content increased significantly with increasing soil As concentration. A similar result was found in another study [80] where, P uptake decreased in As treatments. Since, sodium arsenate was used in this study to artificially contaminate the soil, the addition of this Na might contribute to the higher plant uptake in higher As treated soil and ultimately more accumulation of Na in the grain. The result also shows that grain Mn, Zn, and Cu content decreased significantly with increasing soil As concentration, while they increased significantly in the MW treatments compared with the control. Addition of As can reduce Mn content in shoots and roots [80], which results in lower Mn translocation to the grain. By contrast, the opposite findings have also been reported [81] where increased Mn content with increased As was observed. It is known that divalent Mn is absorbed by facilitated diffusion across the plasmalemma [82]. It is possible that As phytotoxicity may hamper the activity of the root plasmalemma and reduce Mn^{2+} absorption. Similarly, decreased Zn and Cu content in shoots and roots [80] can facilitate the lower translocation to the grain. Similar findings were reported in another study [83], where an antagonistic relation between As and Zn was described. Another study also reported lower Zn content in rice grain, where higher As was present in the soil [84]. However, MW soil treatment can increase the grain Zn and Cu by reducing the As phytotoxicity. From the Pearson's correlation (**Table 8**) it was also evident that grain K and Na content were significantly positively correlated with grain As concentration and significantly negatively correlated with grain P, Cu, Mn, and Zn content.

5. Conclusions

Although the elevated concentration of soil As can reduce the plant growth and grain yield of wheat due to the As phytotoxicity, MW soil treatment can mitigate this As phytotoxicity. The MW-6 treatment showed a better influence than the MW-0 and MW-3 treatments. Furthermore, MW soil treatment can reduce grain As concentration. Although the soil remains contaminated after MW treatment, wheat grain As concentration was lower in the MW treated pots, which results in lower As accumulation in humans through wheat consumption. Nevertheless, further experiments are needed to explore the effectiveness of MW treatment with different types of soils in field conditions.

Acknowledgements

We thank Dr. Ravneet Kaur Jhajj, laboratory manager for her assistance with chemical analysis in the laboratory.

Conflict of interest

The authors declare no conflict of interest.

Author details

Mohammad Humayun Kabir[1,2], Graham Brodie[1*], Dorin Gupta[1] and Alexis Pang[1]

1 Faculty of Veterinary and Agricultural Sciences, The University of Melbourne, Dookie, Australia

2 Department of Soil Science, Bangabandhu Sheikh Mujibur Rahman Agricultural University, Gazipur, Bangladesh

*Address all correspondence to: grahamb@unimelb.edu.au

References

[1] Naidu R, Smith E, Owens G, Bhattacharya P. Managing Arsenic in the Environment: From Soil to Human Health. Collingwood, VIC: CSIRO Publishing; 2006

[2] Abdul KSM, Jayasinghe SS, Chandana EP, Jayasumana C, De Silva PMC. Arsenic and human health effects: A review. Environmental Toxicology and Pharmacology. 2015;**40**(3):828-846

[3] Podgorski J, Berg M. Global threat of arsenic in groundwater. Science. 2020; **368**(6493):845-850

[4] Williams PN, Islam M, Adomako E, Raab A, Hossain S, Zhu Y, et al. Increase in rice grain arsenic for regions of Bangladesh irrigating paddies with elevated arsenic in groundwaters. Environmental Science & Technology. 2006;**40**(16):4903-4908

[5] Shankar S, Shanker U, Shikha. Arsenic contamination of groundwater: A review of sources, prevalence, health risks, and strategies for mitigation. The Scientific World Journal. 2014;**2014**: 304524

[6] Singh R, Singh S, Parihar P, Singh VP, Prasad SM. Arsenic contamination, consequences and remediation techniques: A review. Ecotoxicology and Environmental Safety. 2015;**112**: 247-270

[7] Upadhyay MK, Shukla A, Yadav P, Srivastava S. A review of arsenic in crops, vegetables, animals and food products. Food Chemistry. 2019;**276**: 608-618

[8] Kabir MH, Rahman GKMM, Ahmed ZU, Rahman MM. Spatial variability of rice grain arsenic in confined and unconfined basins of Ganges River floodplain soils of Bangladesh. Journal of Environmental Science and Natural Resources. 2016;**8**(2):51-55

[9] Chandiramani M, Kosina P, Jones J. Fact Sheet of the IRRI-CIMMYT—Cereal Knowledge Bank. Mexico: The International Maize and Wheat Improvement Center (CIMMYT); 2007

[10] Union E. Commission Regulation (EU) 2015/1006 Amending Regulation (EC) No. 1881/2006 as Regards Maximum Levels of Inorganic Arsenic in Foodstuffs. 2015. Available from: http://eur-lex.europa.eu/homepage.html [cited: June 26, 2015]

[11] Zhao F-J, Lopez-Bellido FJ, Gray CW, Whalley WR, Clark LJ, McGrath SP. Effects of soil compaction and irrigation on the concentrations of selenium and arsenic in wheat grains. The Science of the Total Environment. 2007;**372**(2–3):433-439

[12] Wiersma D, Van Goor BJ, Van der Veen NG. Cadmium, lead, mercury and arsenic concentrations in crops and corresponding soils in the Netherlands. Journal of Agricultural and Food Chemistry. 1986;**34**(6):1067-1074

[13] Roychowdhury T, Uchino T, Tokunaga H, Ando M. Survey of arsenic in food composites from an arsenic-affected area of West Bengal, India. The Science of the Total Environment. 2002; **40**(11):1611-1621

[14] Williams PN, Villada A, Deacon C, Raab A, Figuerola J, Green AJ, et al. Greatly enhanced arsenic shoot assimilation in rice leads to elevated grain levels compared to wheat and barley. Environmental Science & Technology. 2007;**41**(19): 6854-6859

[15] FAO. Food and Agriculture Organisation. FAOSTAT Database. 2017. Available from: http://www.fao.org/faostat/en/#data/QC/visualize [cited: November 4, 2019]

[16] Yang L, Donahoe RJ, Redwine JC. In situ chemical fixation of arsenic-contaminated soils: An experimental study. Science of the Total Environment. 2007;**387**(1):28-41

[17] Lim KT, Shukor MY, Wasoh H. Physical, chemical, and biological methods for the removal of arsenic compounds. BioMed Research International. 2014;**2014**:503784

[18] Brodie G, Khan MJ, Gupta D, Foletta S, Bootes N. Microwave weed and soil treatment in agricultural systems. AMPERE Newsletter (Association for Microwave Power in Europe for Research and Education). 2017;(93)

[19] Jamal M, Brodie G, Gupta D. The effect of microwave soil treatment on rice production under field conditions. Transactions of the ASABE. 2017;**60**(2): 517-525

[20] O'Brien PL, DeSutter TM, Casey FX, Khan E, Wick AF. Thermal remediation alters soil properties—A review. Journal of environmental management. 2018; **206**:826-835

[21] Khan MJ, Brodie GI, Gupta D, He J. Microwave soil treatment increases soil nitrogen supply for sustained wheat productivity. Transactions of the ASABE. 2019;**62**(2):355-362

[22] Maynaud G, Baudoin E, Bourillon J, Duponnois R, Cleyet-Marel JC, Brunel B. Short-term effect of 915-MHz microwave treatments on soil physicochemical and biological properties. European Journal of Soil Science. 2019;**70**(3):443-453

[23] Kabir MH, Brodie G, Gupta D, Pang A. Microwave soil treatment along with biochar application alleviates arsenic Phytotoxicity and reduces rice grain arsenic concentration. Energies. 2021;**14** (23):8140

[24] Falciglia PP, Bonifacio A, Federico GA, Vagliasindi. An overview on microwave heating application for hydrocarboncontaminated soil and groundwater remediation. Oil and Gas Research. 2016;**2**(1):110

[25] Kabir H, Khan MJ, Brodie G, Gupta D, Pang A, Jacob MV, et al. Measurement and modelling of soil dielectric properties as a function of soil class and moisture content. Journal of Microwave Power and Electromagnetic Energy. 2020;**54**(1):3-18

[26] Menéndez JA, Arenillas A, Fidalgo B, Fernández Y, Zubizarreta L, Calvo EG, et al. Microwave heating processes involving carbon materials. Fuel Processing Technology. 2010;**91**(1):1-8

[27] Zhou BW, Shin SG, Hwang K, Ahn J-H, Hwang S. Effect of microwave irradiation on cellular disintegration of Gram positive and negative cells. Applied Microbiology and Biotechnology. 2010;**87**(2):765-770

[28] Falciglia P, Vagliasindi F. Remediation of hydrocarbon-contaminated soils by ex situ microwave treatment: Technical, energy and economic considerations. Environmental Technology. 2014;**35**(18):2280-2288

[29] Abramovitch RA, ChangQing L, Hicks E, Sinard J. In situ remediation of soils contaminated with toxic metal ions using microwave energy. Chemosphere. 2003;**53**(9):1077-1085

[30] Chen C-L, Lo S-L, Kuan W-H, Hsieh C-H. Stabilization of Cu in

acid-extracted industrial sludge using a microwave process. Journal of Hazardous Materials. 2005;**123**(1–3): 256-261

[31] Hsieh C-H, Lo S-L, Chiueh P-T, Kuan W-H, Chen C-L. Microwave enhanced stabilization of heavy metal sludge. Journal of Hazardous Materials. 2007;**139**(1):160-166

[32] Downes R. Soil, Land-use, and Erosion Survey around Dookie, Victoria. Melbourne, Australia: CSIRO, Aust. Bull.; 1949. p. 243

[33] Isbell R. The Australian Soil Classification. Melbourne, Australia: CSIRO Publishing; 2016

[34] Rahman MA, Hasegawa H, Rahman MM, Rahman MA, Miah M. Accumulation of arsenic in tissues of rice plant (*Oryza sativa* L.) and its distribution in fractions of rice grain. Chemosphere. 2007;**69**(6):942-948

[35] Bhattacharya P, Samal AC, Santra SC. A greenhouse pot experiment to study arsenic accumulation in rice varieties selected from Gangetic Bengal, India. In: Safe and Sustainable Use of Arsenic-Contaminated Aquifers in the Gangetic Plain. Cham, Switzerland: Springer; 2015. pp. 265-274

[36] Ullah S. Arsenic contamination of groundwater and irrigated soils of Bangladesh. In: International Conference on Arsenic Pollution of Ground Water. 1998

[37] Hua X, Xiao-Yong L, Tong-Bin C, Jiao-zhao L. Arsenic in plants of farmland and its healthy risk: A case study in an As-contaminated site in Dengjiatang, Chenzhou City, Hunan Province. 地理研究. 2005; **24**(1):151-159

[38] Patel K, Shrivas K, Brandt R, Jakubowski N, Corns W, Hoffmann P. Arsenic contamination in water, soil, sediment and rice of Central India. Environmental Geochemistry and Health. 2005;**27**(2):131-145

[39] Khan MJ, Brodie G, Gupta D. Effect of microwave (2.45 GHz) treatment of soil on yield components of wheat (*Triticum aestivum* L.). Journal of Microwave Power and Electromagnetic Energy. 2016;**50**(3):191-200

[40] Harris G, Brodie G, Ozarska B, Taube A. Design of a microwave chamber for the purpose of drying of wood components for furniture. Transactions of the ASABE. 2011;**54**(1): 363-368

[41] Cooper A, Brodie G. The effect of microwave radiation and soil depth on soil pH, N, P, K, SO_4 and bacterial colonies. Plant Protection Quarterly. 2009;**24**(2):67

[42] Yuan Z, Cao Q, Zhang K, Ata-Ul-Karim ST, Tian Y, Zhu Y, et al. Optimal leaf positions for SPAD meter measurement in rice. Frontiers in Plant Science. 2016;7:719-719

[43] Analytical P, PSA 10.055 millennium Excalibur users manual and PSA customer technical information file. P S Analytical Ltd.: UK. 2009; 9.4.

[44] Chen B, Corns WT, Stockwell PB, Huang J-H. Accurate fast screening for total and inorganic arsenic in rice grains using hydride generation atomic fluorescence spectrometry (HG-AFS). Analytical Methods. 2014;**6**(18):7554-7558

[45] Iqbal M, Rahman GKMM, Panaullah GM, Kabir H, Biswas JC. Influence of soil arsenic levels on biomass production and relationship the concentration of arsenic

between Rice straw and grain. Asian Journal of Soil Science and Plant Nutrition. 2019;**5**(2):1-10

[46] Warren T. Development of an atomic fluorescence spectrometer for the hydride-forming elements. Journal of Analytical Atomic Spectrometry. 1993;**8** (1):71-77

[47] Thompson M, Walsh JN. Handbook of Inductively Coupled Plasma Spectrometry. London, UK: Blackie Academic & Professional; 2012

[48] Asaduzzman M, Hossain M, Masum SM. Effect of arsenic in three wheat varieties of Bangladesh. IJBSM. 2010;**1** (2):115-120

[49] Pigna M, Cozzolino V, Giandonato Caporale A, Mora M, Di Meo V, Jara A, et al. Effects of phosphorus fertilization on arsenic uptake by wheat grown in polluted soils. Journal of Soil Science and Plant Nutrition. 2010;**10**(4):428-442

[50] Rahman M, Rahman M, Miah MM, Khaled H. Influence of soil arsenic concentrations in rice (*Oryza sativa* L.). Journal of Subtropical Agricultural Research and Development. 2004;**2**(3): 24-31

[51] Liu X, Zhang S, Shan X, Zhu Y-G. Toxicity of arsenate and arsenite on germination, seedling growth and amylolytic activity of wheat. Chemosphere. 2005;**61**(2):293-301

[52] Zhang W, Liu D, Tian J, He F. Toxicity and accumulation of arsenic in wheat (*Triticum aestivum* L.) varieties of China. Phyton (Buenos Aires). 2009;**78**: 147-154

[53] Paivoke A. The long-term effects of lead and arsenate on the growth and development, chlorophyll content and nitrogen fixation of the garden pea. Annales Botanici Fennici. 1983;**20**(3): 297-306

[54] Merry RH, Tiller K, Alston A. The effects of contamination of soil with copper, lead and arsenic on the growth and composition of plants. Plant and Soil. 1986;**91**(1):115-128

[55] Khan MJ, Brodie GI, Gupta D, Foletta S. Microwave soil treatment improves weed management in Australian dryland wheat. Transactions of the ASABE. 2018;**61**(2):671-680

[56] Wainwright M, Killham K, Diprose M. Effects of 2450 MHz microwave radiation on nitrification, respiration and S-oxidation in soil. Soil Biology and Biochemistry. 1980;**12**(5):489-493

[57] Speir T, Cowling J, Sparling G, West A, Corderoy D. Effects of microwave radiation on the microbial biomass, phosphatase activity and levels of extractable N and P in a low fertility soil under pasture. Soil Biology and Biochemistry. 1986;**18**(4):377-382

[58] Ling Q, Huang W, Jarvis P. Use of a SPAD-502 meter to measure leaf chlorophyll concentration in *Arabidopsis thaliana*. Photosynthesis Research. 2011; **107**(2):209-214

[59] Schimel JP, Bennett J. Nitrogen mineralization: Challenges of a changing paradigm. Ecology. 2004;**85** (3):591-602

[60] Voort VDM, Kempenaar M, van Driel M, Raaijmakers JM, Mendes R. Impact of soil heat on reassembly of bacterial communities in the rhizosphere microbiome and plant disease suppression. Ecology Letters. 2016;**19** (4):375-382

[61] Tong J, Guo H, Wei C. Arsenic contamination of the soil–wheat system

irrigated with high arsenic groundwater in the Hetao Basin, Inner Mongolia, China. The Science of the Total Environment. 2014;**496**:479-487

[62] Baig JA, Kazi TG, Shah AQ, Afridi HI, Kandhro GA, Khan S, et al. Evaluation of arsenic levels in grain crops samples, irrigated by tube well and canal water. Food and Chemical Toxicology. 2011;**49**(1):265-270

[63] Anawar HM, Rengel Z, Damon P, Tibbett M. Arsenic-phosphorus interactions in the soil-plant-microbe system: Dynamics of uptake, suppression and toxicity to plants. Environmental Pollution. 2018;**233**:1003-1012

[64] Zhao F-J, McGrath SP, Meharg AA. Arsenic as a food chain contaminant: Mechanisms of plant uptake and metabolism and mitigation strategies. Annual Review of Plant Biology. 2010; **61**:535-559

[65] Mosa KA, Kumar K, Chhikara S, Mcdermott J, Liu Z, Musante C, et al. Members of rice plasma membrane intrinsic proteins subfamily are involved in arsenite permeability and tolerance in plants. Transgenic Research. 2012;**21**(6): 1265-1277

[66] Roy W, Hassett J, Griffin R. Competitive interactions of phosphate and molybdate on arsenate adsorption. Soil Science. 1986;**142**(4):203-210

[67] McBride M. Environmental Chemistry of Soils. New York: Oxford Univ. Press; 1994

[68] Pigna M, Cozzolino V, Violante A, Meharg AA. Influence of phosphate on the arsenic uptake by wheat (*Triticum durum* L.) irrigated with arsenic solutions at three different concentrations. Water, Air, and Soil Pollution. 2009;**197**(1–4):371-380

[69] Carbonell A, Aarabi M, DeLaune R, Gambrell R, Patrick W Jr. Arsenic in wetland vegetation: Availability, phytotoxicity, uptake and effects on plant growth and nutrition. Science of the Total Environment. 1998;**217**(3): 189-199

[70] Gulz P, Gupta S. Arsenaufnahme von Kulturpflanzen. Agrarforschung. 2000;**7**:360-365

[71] Gulz PA, Gupta S-K, Schulin R. Arsenic accumulation of common plants from contaminated soils. Plant and Soil. 2005;**272**(1–2):337-347

[72] Onken B, Hossner L. Plant uptake and determination of arsenic species in soil solution under flooded conditions. Journal of Environmental Quality. 1995; **24**(2):373-381

[73] Kamrozzaman M, Khan M, Ahmed S, Sultana N, Quddus AR. Evaluation of Five wheat varieties on arsenic contaminated soils of Bangladesh. International Journal of Applied Sciences and Biotechnology. 2016;**4**(3): 330-337

[74] Shi GL, Zhu S, Meng JR, Qian M, Yang N, Lou LQ, et al. Variation in arsenic accumulation and translocation among wheat cultivars: The relationship between arsenic accumulation, efflux by wheat roots and arsenate tolerance of wheat seedlings. Journal of Hazardous Materials. 2015;**289**:190-196

[75] Hur J, Park S-W, Kim MC, Kim HS. Enhanced binding of hydrophobic organic contaminants by microwave-assisted humification of soil organic matter. Chemosphere. 2013;**93**(11): 2704-2710

[76] Kim MC, Kim HS. Artificial and enhanced humification of soil organic

matter using microwave irradiation. Environmental Science and Pollution Research. 2013;**20**(4):2362-2371

[77] Zagal E. Effects of microwave radiation on carbon and nitrogen mineralization in soil. Soil Biology and Biochemistry. 1989;**21**(4):603-605

[78] Taylor M, Atri BS, Minhas S, Team I. Developments in Microwave Chemistry. Zug, Switzerland: Evalueserve; 2005

[79] Yin Y, Impellitteri CA, You S-J, Allen HE. The importance of organic matter distribution and extract soil: Solution ratio on the desorption of heavy metals from soils. Science of the Total Environment. 2002;**287**(1):107-119

[80] Shaibur MR, Kitajima N, Huq SI, Kawai S. Arsenic–iron interaction: Effect of additional iron on arsenic-induced chlorosis in barley grown in water culture. Soil Science and Plant Nutrition. 2009;**55**(6):739-746

[81] Yamane T. The mechanisms and countermeasures of arsenic toxicity to rice plant. Bulletin of the Shimane Agricultural Experiment Station (Japan). 1989;**24**:1-95

[82] Fox TC, Guerinot ML. Molecular biology of cation transport in plants. Annual Review of Plant Biology. 1998;**49** (1):669-696

[83] Liu Y-M, Liu D-Y, Zhang W, Chen X-X, Zhao Q-Y, Chen X-P, et al. Health risk assessment of heavy metals (Zn, Cu, Cd, Pb, As and Cr) in wheat grain receiving repeated Zn fertilizers. Environmental Pollution. 2020;**257**:113581

[84] Duan G, Liu W, Chen X, Hu Y, Zhu Y. Association of arsenic with nutrient elements in rice plants. Metallomics. 2013;**5**(7):784-792

Section 4

Modern Techniques in Wheat Breeding

Chapter 13

Genomic Approaches in Wheat Breeding for Sustainable Production under Changing Climate

Zahid Manzoor, Junwei Liu, Muhammad Sheeraz Qadir, Muhammad Ahsan Jamil, Zeshan Hassan, Muhammad Shah Jahan and Amir Shakeel

Abstract

Wheat is the most important cereal crop, a great source of dietary protein. It is grown worldwide for its consumption in the form of different products. Wheat production faces a lot of biotic and abiotic stresses that hinder growth and yield. Changing climate is a worse scenario to be adopted for sustainable production. Food demand is rapidly increasing by a drastic increase in the world population. Conventional breeding techniques are time-consuming and ineffective in attaining high yield goals under changing climates. Next-generation sequencing revolutionized wheat breeding through molecular approaches for effective selection. The use of genomic approaches in wheat breeding is the need of time for sustainable production. Several genomic approaches, such as use of genome-wide markers for gene mapping, genomic selection and recurrent selection through QTL and meta-QTL analysis, markers-assisted selection in haploid breeding, heterosis breeding through genomic tools, and biotechnological tools, are currently used as modern techniques for developing climate-resilient wheat cultivars. This chapter illustrated the challenges of changing climate, molecular techniques in wheat breeding to develop climate-resilient genotypes, sustainable wheat production to cope with food demand, and future breeding strategies.

Keywords: genomic approaches, wheat breeding, sustainable goals, climate change, resilient cultivars, marker-assisted breeding

1. Introduction

The one-third population of the world mainly relies on wheat (*Triticum aestivum* L.) for their daily diet and it is becoming more important with a great increase in the world's population [1–3]. Wheat is grown in an area of about 220 million hectares worldwide [4], with an average annual production of 729 million tons [5]. Wheat is also used as an industrial material and renewable feed resource [6]. Wheat demand is increasing due to increased population thought the world [5]; thus, it is influencing market prices [7], competition, and growing demand [8]. The considerable challenge in sustainable wheat production is to increase the yield with demand in continuously changing environmental conditions. It has been suggested that wheat yield should be increased by 1.7% per annum globally for the next 30 years [9]. Still, its production rate is 0.9% per annum [10], insufficient to meet global hunger and even in main wheat-producing countries, this percentage is gradually decreasing. The goal of sustainable wheat production is only possible to achieve by growing wheat in the best environmental condition since the world is facing the massive challenge of climate change; therefore, this could not be possible [11]. Yield and yield stability are highly affected by climate change [12].

Although agronomic practices and conventional breeding contributed to sustainable wheat production, now it is time to boost wheat production with the latest introduced technologies. So, breeders are looking for highly efficient methods to increase yield in a limited time [13]. Newly developed technologies, such as phenomics [14], marker-assisted selection (MAS) [15], genomic breeding [16], and biotechnological (cis-genic and transgenic) techniques [17], are promising technologies to fight hunger in the future.

Marker-assisted breeding is based on gene linkage and recombination events in meiosis [18]. Different molecular markers are being used to detect the variations in wheat germplasm [19] and resistant genes are being identified in various lines and used in marker-assisted breeding [20, 21]. Genomic selection (GS) is an advanced form of MAS [13]. Initially, a panel of genotypes, training population, is selected in GS, then genotyping is performed with genome-wide markers and lastly, phenotyping is done for the trait of interest. Genome-estimated breeding values (GEBVs) are calculated with the help of a training population for all genotypes included in the panel, newly developed lines, and the validation population [22]. GS is more effective based on computing more variations with the help of GEBVs without phenotyping [23]. Modified methods in GS significantly increase genetic gain and accuracy. Several recent genomic approaches have been illustrated in **Figure 1**.

This chapter focuses on genomic approaches in wheat breeding to combat hunger in changing climate. Genomic breeding in wheat is an efficient way to increase wheat production in changing climate and global warming scenarios. Using molecular markers in QTL-mapping and its combination with genome with

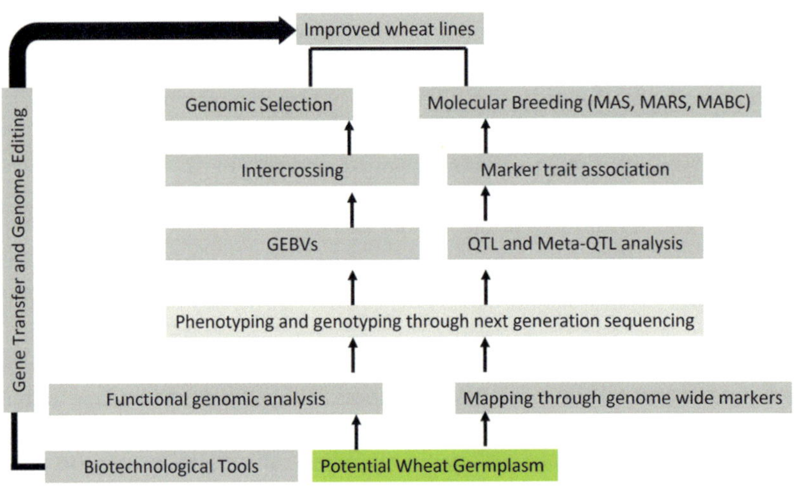

Figure 1.
Schematic diagram for recent genomic approaches in wheat breeding.

association study (GWAS) could significantly improve yield to achieve sustainable wheat production goals.

2. Gene mapping through genome-wide markers

Initially, restriction fragment polymorphism (RFLP) markers were introduced to identify cultivars [24] and gene mapping, but the frequency of markers was not impressive due to the extremely low level of polymorphism for the D genome of bread wheat [25]. International Triticeae Mapping Initiative (ITMI) was very efficient because of generating high-density linkage groups [26]. With the advancement in technology, PCR-based markers were developed. There were two broad categories [27], simple sequence repeat (SSR) [28] and randomly amplified polymorphic DNA (RAPD), considered far better than RFLP markers. These markers proved to be time-saving and cost-effective, especially SSR markers were extensively used in wheat due to reproducibility. RAPD markers were used to make sequence characterized amplified regions (SCAR) or sequence-tagged sites (STS) markers [29] that were more reliable in wheat, for example, Lr24 and Lr2 QTL for Russian wheat aphid [30].

The first discovery of simple sequence repeats (SSR) markers, also known as microsatellites [28], opened a new era of wheat breeding due to their extensive use because they are highly reproducible, genome-specific, highly polymorphic, and relatively abundant [31]. The yield and yield-related traits in wheat were exploited with more resolution on their respective loci in the genome [32]. SSR markers also had limitations due to high cost, random distribution in the genome, finite motifs, and difficulty obtaining exact information [33]. Single nucleotide polymorphism (SNP) markers analyze variations at a single nucleotide level; therefore, they are not very effective in marker-assisted selection of wheat [34–37].

3. Genome based breeding strategies in wheat

The reciprocal recurrent genomic selection in wheat is important to increase wheat yield [38]. Genomic selection of desired traits speeds up the breeding program with the accurate selection of traits of interest with the help of QTL and GWAS [39, 40]. When additive effects of focused QTL are not determined, they can be estimated by genome-wide prediction. The Ridge regression model has been proposed to be used in the genomic selection of desired traits with more accuracy and unbiased decisions [41, 42]. $F\infty$ metric is implemented to determine additive effects and additive-by-additive epistasis [43]. The superior plants are selected for the next generation with the help of the estimation of additive effects in recurrent genomic selection [22]. The QTL study with the ridge regression model revealed that nonadditive effects were related to grain yield [44]; therefore, they are suggested to be included in the genome prediction model to increase wheat yield. Persistency in predicting models is mainly dependent upon the adequate size of parental population and linkage disequilibrium, size of training population, use of the statistical model to estimate markers effects in recurrent genomic selection, and density of markers [45–47].

Estimating haploid breeding and genomic breeding values are two important strategies for assessing long-term genetic gain and maintenance in recurrent genomic breeding [48]. Based on these values, optimal population value selection is performed to make blocks of genotypes exhibiting maximum haploid or genomic breeding values. Genotypes with maximum haploid value would fall in the block with minimal segregation. In contrast, minimal value haploid value will lead to minimal population value selection with a haplotype block with maximum segregation for the desired trait [48]. Genomic recurrent selection and its modified form reciprocal recurrent selection increase the efficacy of wheat breeding programs [49] to develop high-yielding, climate-resilient genotypes. Gene pyramiding is a novel concept in modern breeding to accumulate desired genes in a single genotype to develop an ideotype [50]. Implications of this breeding strategy via genomic breeding can be a way forward to achieve a landmark in wheat breeding programs for sustainable production in changing climate.

4. Marker-assisted selection

Genetic linkage between loci of the same chromosome and their recombination events during meiosis are the main basis of MAS [18]. The transfer of two loci together or separately in the next generation is dependent on how closely these loci are located on the chromosome [51]. There will be more chances to be inherited together if they are located closely on the same chromosome. Molecular markers are used to identify a specific region on the chromosome for a gene of interest [52]. Different alleles are detected in several lines, known as a polymorphism for a trait of interest in different lines. Molecular markers detect the presence of linked alleles on the base of genetic linkage [27]. Marker-assisted selection could prove a very effective technique for developing climate-resilient wheat cultivars in changing climates.

Single sequence repeat (SSR) markers were used to detect the cell membrane stability of wheat cultivated under drought stress. SSR markers were significantly linked with cell membrane stability, but the association was weak. SSR markers were suggested to detect increased frequency in progenies with drought tolerance [53] and used Xwmc273.3 marker was used to detect QTL associated with higher grain yield of wheat under drought conditions. Different wheat cultivars in irrigated and rainfed regions were selected with the help of QTL mapping and phenotypic selection based on higher yield in water stress conditions [54]. Marker-assisted backcross breeding was employed to detect and transfer three drought-tolerant QTLs in high-yielding cultivars. QTLs were detected in drought-tolerant cultivar HI5100 and HD2733 was used as a recurrent cultivar. They identified 29 lines having drought tolerant QTLs; further background selection resulted in five varieties for evaluation in the national breeding program [55]. QTL expression in common wheat for cold tolerance was detected in the region of CBF and Cor/Lea gene families located at 5AL [56]. Soriano et al. [40] performed a QTL meta-analysis to identify QTLs for biotic and abiotic stress tolerance in durum wheat. They identified 315 of 85 MQTL, while 71 corresponded to biotic stress and 127 to abiotic stresses.

5. Heterosis breeding through genome based strategies

Only <1% of the wheat cultivated area is under the cultivation of hybrid wheat due to the highly self-pollinated wheat and other technical problems. Meanwhile, genomic breeding has helped resolve the technical issues of hybrid wheat production. Male sterility is not well-understood in wheat because it is highly self-pollinated, and hybrid seed production is very cost-effective. Male sterility is of great importance in the hybridization of wheat. Male sterility II (*ms2*) has been identified in wheat for 40 years, but its corresponding genes were unknown. It was determined through mapped-based cloning experiments in 2017 that the promoter region of *ms2* has *TRIM* element, activates *Ms2* allele in anther, which induces male sterility in wheat [57, 58].

Furthermore, it was also investigated by map-based cloning studies that male sterility 1 (ms1) also prevailed in wheat [59, 60]. Functional analysis of MS1 has revealed a newly introduced protein in the wheat and Poaceae family. It is localized in mitochondria and plastids, associated with the phospholipid-binding activity to induce male sterility. The split gene system also inserts male sterility in wheat by expressing the phytotoxic gene barnase, controlled by two alleles and its activation induces male sterility. This system maintains male sterile female plants while, after crossing, sterile hybrids are produced because it does not need male storer lines and entirely relies on the genetically modified female plants [61].

The utilization of genomics to predict heterotic patterns is another strategy for hybridizing wheat. These patterns are used mainly to characterize parents and the hybrids population on a large scale. Heterotic patterns were used to assess 1604 wheat hybrids for disease resistance and morphological parameters. It was demonstrated that 69 hybrids performed better than the best commercial line by 7.2–10.7% in production [62]. Zhao et al. [63] described a three-way genome-based strategy to predict heterotic wheat patterns in 1604 hybrids and 135 parents. In another experiment, 135 parents and their 1604 hybrids were assessed through genomic prediction of heterotic patterns and a complete performance of hybrids was evaluated to suggest high yielding heterotic patterns.

Heterotic patterns were used to assess a well-defined population and their effectiveness, limitations, and representation were estimated to measure their success. Identifying and exploiting major genes is essential and helpful in hybridization, such as genes responsible for dwarf wheat and very important for selecting the parental population to make hybrids. Further, these genes greatly influence pollen mass and anther extrusion in wheat [64]. Hybrids D1b and B1b with reduced height showed poor anther extrusion due to the expression of genes related to dwarfness [65, 66]. While developing high-yielding cultivars with lodging resistance must be considered for sustainable goals. Using another gene *Rht24* for male sterility in hybridization could be very effective because it has no effects on male floral parts and anther extrusion [67]. These studies predicted heterotic patterns via a quantitative genetic framework and laid the basis for the hybridization of wheat, having fine-tuned major genes of plant stature and floral traits [68]. The development of hybrid cultivars at a commercial scale could be feasible through genome-based prediction of these genes.

6. Developing climate-resilient cultivars through biotechnology

In the last 15 years, recombinant DNA technology, genetic manipulation technologies, and culturing methodologies have enabled the efficient transformation and development of transgenics in a wide variety of crop plants [69]. Moreover, transgenesis can be a supplementary technique for single-gene or transgenic plants development [70]. Despite traditional breeding, this method introduces only the cloned gene(s) of agronomic significance without the hazard of transferring any additional undesired genes from the donor [71]. In transgenesis, the backcrossing is unnecessary because the recipient plant/crop genotype is least affected [72]. Furthermore, this genetic transformation method opens the door to a vast array of genetic material, sourced from viruses and bacteria to fungus, insects to animals and human beings to unrelated plants, and even from chemical synthesis [73–76]. Plant transgenics have been developed and tested for various crops [77], fruits, and trees with surprising speed and success. However, the breeders focus on the gradual enhancement of commercial cultivars by introducing cloned genes of important agronomic values. The

Breeding techniques	Output	References
Marker-assisted selection (600k SNP marker)	Targeted genotyping and genetic improvement	[15]
Marker-assisted breeding (SNPs)	Evaluation of multiple elite traits	[93]
Marker-assisted back-crossing	Drought tolerance in bread wheat	[55]
Marker-assisted back-crossing	Introgression of drought-tolerant QTLs	[94]
Marker-assisted back-cross selection	Improvement in rust resistance through a selection of *Yr59*	[95]
Marker-assisted back-crossing	Transfer of recessive *skr* cross-ability trait	[96]
Marker-assisted back-crossing	Adaptation of a variety of Unnat PBW 343 in diverse environments	[97]
Marker-assisted back-crossing	Enhanced rust resistance Two genes (*Lr19/Sr25* and *Lr24/Sr24*) for leaf rust resistance One gene (*Yr15*) for stripe rust resistance	[98]
Marker-assisted back-crossing	Development of near-isogenic lines for grain softness	[99]
Marker-assisted back-crossing	Development of advanced lines for grain softness	[100]
Marker-assisted recurrent selection	Improved crown rot resistance	[101]
Marker-assisted recurrent selection	Enhanced genetic gains	[49]
Reciprocal recurrent selection	Hybridization	[38]
S1 recurrent selection, early generation genomic selection, marker-assisted back-crossing, and gene pyramiding	Introgression of Ms3 gene for genetic male sterility in hybrid wheat	[102]
Double haploid breeding	Development of thermos-sensitive genic male sterile lines	[103]
Biotechnology (Horizontal gene transfer)	Fhb7 from fungus expression in wheat for Fusarium head blight resistance	[75]
CRISPR-Cas9-based multiplexed gene editing	heritable mutations in the *TaGW2*, *TaLpx-1*, and *TaMLO* genes	[104]
Development of the GlutEnSeq (Gluten gene Enrichment and Sequencing)	Homozygous deletions for the α-gliadins on 6A and the γ-gliadins on 1B in two γ-irradiated lines of cultivar Homozygous deletions of the γ-gliadins on 1B and heterozygous deletions for the α-gliadins on 6A in four Fielder CRISPR/Cas9 gliadin gene-edited lines	[105]
CRISPR/Cas9 system delivered via *Agrobacterium tumefaciens*	Obtained thirteen mutant lines by targeting seven sites of three genes (*Pinb*, *waxy*, and *DA1*)	[106]

Table 1.
Recent approaches for modern wheat breeding in the era of genomic studies.

gene-transfer methods and strategies have been used to create important agronomic features in numerous crop varieties.

The enormous size and structural complexity of the polyploid wheat genome initially hindered the genomic study. By the passage of time, the development of new genomic technologies has enabled the breeders to map the bread wheat and its ancestors. However, the introduction of modern genomic technologies like next-generation sequencing has resulted in draught genomes for bread wheat and its progenitors [78], paving the door for developing novel crop enhancement strategies. Diverse germplasms are evaluated in pre-breeding for several physiological, agronomical, and biochemical traits [79], then crossing [80] and high throughput phenotyping [81] are required, while marker-based next-generation sequencing [80], genomic prediction [82], and validation of climate-resilient lines [83] can be very helpful in developing high yielding cultivars in changing environment. The *CRISPR-Cas* (Clustered Regularly Interspaced Short Palindromic Repeats) technology makes breeding time conservative and helps to find and transfer the same gene of interest in the host wheat genotype [84]. These RNA-guided nucleases are used for genome editing and isolated from the microbial adaptive immune system [85]. A wheat line has been developed that has reduced gluten in the grain by using Cas9 protein with 20 nucleotides in its sequence [86]. A gene *Mildew locus O* (*MLo*) was inserted in the wheat genome to introduce resistance against powdery mildew, and this was the first successful attempt of *CISPER/Cas9* technology in wheat [87]. Transcription activator-like effector nucleases, an earlier technology, were used with the combination of *CRISPER/Cas9* to achieve this goal.

Since the development of *CRISPR-Cas* technology, wheat genes of agronomical and fundamental scientific interest have been targeted, such as -gliadin genes to reduce gluten grain content [88], *TaGW2* to increase grain weight [89], *TaZIP4-B2* to understand meiotic homologous crossover [90], *TaQsd1* to minimize preharvest sprouting [91], *TaMTL* and *CENH3* for haploid plant induction [92]. The Wheat CRISPR tool, which is freely accessible at https://crispr.bioinfo.nrc.ca/WheatCrispr/, identifies efficient sgRNAs that are anticipated high on-target and low off-target activity scores (enabling researchers to explore all potential sgRNAs inside a target gene or sequence of interest). The Wheat CRISPR tool considers hexaploidy in bread wheat, allowing the researcher to target either a single gene copy or all three homeologs by checking a box.

The recent approaches for modern wheat breeding in the era of genomic studies have been summarized in **Table 1**.

7. Conclusion and future perspective

Wheat is one of the most important food crops and a basic source of calories globally. The rising population necessitates an increase in wheat production for food security. On the other hand, its production faces great challenges under changing climate and global warming. Wheat production is still lower than its demand and conventional methods proved inefficient to cope with this gap. Modern plant breeding techniques need time to be adopted for sustainable wheat production. Genomic breeding and biotechnological tools are more precise and time-conserving techniques with maximum efficiency to increase wheat production. The discovery of molecular markers-initiated, marker-assisted breeding while QTL and meta-QTL analysis improved the technique's efficacy. Genomic breeding is considered an advanced form of MAS and the genome-wide study provides quick backcrossing and recurrent selection in wheat breeding. Genomic heterotic patterns have been used in molecular hybridization, while different statistical models have also been made to make selection and hybridization more and more precise. Combining genomic knowledge with biotechnological tools makes it quick to breed wheat with sustainable goals in a limited time in changing climate. In the future, the adoption of genomic breeding and biotechnological techniques to develop climate-resilient wheat cultivars at commercial scales will only be the way to achieve sustainable wheat production.

Acknowledgements

We acknowledge all colleagues in the respective departments who guided and helped in the collection of material and outlines.

We do not have any funding source for this book chapter

Conflict of interest

The authors declare no conflict of interest.

Acronyms and abbreviations

MAS	(Marker-assisted selection)
GS	(Genomic selection)
QTL	(Quantitative trait loci)
GEBVs	(Genome-estimated breeding values)
GWAS	(genome-wide association study)
RFLP	(Restriction fragment length polymorphism)
ITMI	(International triticeae mapping initiative)
SSR	(Simple sequence repeat)
RAPD	(Random amplified polymorphic DNA)
SCAR	(Sequence characterized amplified regions)
STS	(Sequence tagged sites)
CRISPER	(Clustered regularly interspaced short palindromic repeats)

Author details

Zahid Manzoor[1,2]*, Junwei Liu[1], Muhammad Sheeraz Qadir[3],
Muhammad Ahsan Jamil[4], Zeshan Hassan[5], Muhammad Shah Jahan[5]
and Amir Shakeel[2]

1 Key Lab of Horticultural Plant Biology, Huazhong Agricultural University, Wuhan, China

2 Faculty of Agriculture, Department of Plant Breeding and Genetics, University of Agriculture Faisalabad, Pakistan

3 Ayub Agricultural Research Institute, Vegetable Research Station Karor, Layyah, Pakistan

4 Department of Agronomy, University of Agriculture Faisalabad, Pakistan

5 College of Agriculture, Bahauddin Zakariya University Multan, Pakistan

*Address all correspondence to: zahidmanzoor041@gmail.com

References

[1] Hassan NA. Afghan Wheat and Flour Market

[2] Hossain A, Skalicky M, Brestic M, Maitra S, Ashraful Alam M, Syed MA, et al. Consequences and mitigation strategies of abiotic stresses in wheat (*Triticum aestivum* L.) under the changing climate. Agronomy. 2021;**11**(2):241

[3] Jasrotia P, Kashyap PL, Bhardwaj AK, Kumar S, Singh G. Scope and applications of nanotechnology for wheat production: A review of recent advances. Wheat Barley and Research. 2018;**10**(1):1-14

[4] Altaf A, Gull S, Zhu X, Zhu M, Rasool G, Ibrahim MEH, et al. Study of the effect of peg-6000 imposed drought stress on wheat (*Triticum aestivum* L.) cultivars using relative water content (RWC) and proline content analysis. Pakistan Journal of Agricultural Sciences. 2021;**58**(1):357-367

[5] Khatodia S, Bhatotia K. Prospects of advanced genomics for development of climate resilient wheat genotypes. Ekin Journal of Crop Breeding and Genetics. 2019;**5**(1):54-55. DOI: 10.1007/978-3-319-59379-1

[6] Barbieri V, Gualtieri ML, Siligardi C. Wheat husk: A renewable resource for bio-based building materials. Construction and Building Materials. 2020;**251**:118909. DOI: 10.1016/j.conbuildmat.2020.118909

[7] Mahmood N, Arshad M, Kächele H, Ma H, Ullah A, Müller K. Wheat yield response to input and socioeconomic factors under changing climate: Evidence from rainfed environments of Pakistan. Science of the Total Environment. 2019;**688**:1275-1285. DOI: 10.1016/j.scitotenv.2019.06.266

[8] Dong C, Yin Q, Lane KJ, Yan Z, Shi T, Liu Y, et al. Competition and transmission evolution of global food trade: A case study of wheat. Physica A: Statistical Mechanics and its Applications. 2018;**509**:998-1008. DOI: 10.1016/j.physa.2018.06.052

[9] Tadesse W, Sanchez-Garcia M, Assefa SG, Amri A, Bishaw Z, Ogbonnaya FC, et al. Genetic gains in wheat breeding and its role in feeding the world. Crop Breeding Genetics and Genomics. 2019;**1**:e190005

[10] Ray DK, Mueller ND, West PC, Foley JA. Yield trends are insufficient to double global crop production by 2050. PLOS ONE. 2013;**8**(6):e66428. DOI: 10.1371/journal.pone.0066428

[11] Kahiluoto H, Kaseva J, Balek J, Olesen JE, Ruiz-Ramos M, Gobin A, et al. Decline in climate resilience of European wheat. Proceedings of the National Academy of Sciences. 2019;**116**(1):123-128. DOI: 10.1073/pnas.1804387115

[12] Sehgal D, Rosyara U, Mondal S, Singh R, Poland J, Dreisigacker S. Incorporating genome-wide association mapping results into genomic prediction models for grain yield and yield stability in CIMMYT spring bread wheat. Frontiers in Plant Science. 2020;**11**:197. DOI: 10.3389/fpls.2020.00197

[13] Larkin DL, Lozada DN, Mason RE. Genomic selection—considerations for successful implementation in wheat breeding programs. Agronomy. 2019;**9**(9):479. DOI: 10.3390/agronomy9090479

[14] Robert P, Auzanneau J, Goudemand E, Oury F-X, Rolland B, Heumez E, et al. Phenomic selection in wheat breeding: Identification and optimisation of factors influencing prediction accuracy and comparison to genomic selection. Theoretical and Applied Genetics. 2022;**35**:895-914. DOI: 10.1007/s00122-021-04005-8

[15] Sun C, Dong Z, Zhao L, Ren Y, Zhang N, Chen F. The Wheat 660K SNP array demonstrates great potential for marker-assisted selection in polyploid wheat. Plant Biotechnology Journal. 2020;**18**(6):1354-1360. DOI: 10.1111/pbi.13361

[16] Rasheed A, Xia X. From markers to genome-based breeding in wheat. Theoretical and Applied Genetics. 2019;**132**(3):767-784. DOI: 10.1007/s00122-019-03286-4

[17] Munawar S, Mustafa G, Khan MS, Joyia FA. Role of biotechnology in climate resilient agriculture. In: Environment, Climate, Plant and Vegetation Growth. Cham: Springer; 2020. pp. 339-365

[18] Pandurangan S, Workman C, Nilsen K, Kumar S. Introduction to Marker-Assisted Selection in Wheat Breeding: Accelerated Breeding of Cereal Crops. Humana, New York: Springer; 2022. pp. 77-117. DOI: 10.1007/978-1-0716-1526-3_3

[19] Rasheed A, Jin H, Xiao Y, Zhang Y, Hao Y, Zhang Y, et al. Allelic effects and variations for key bread-making quality genes in bread wheat using high-throughput molecular markers. Journal of Cereal Science. 2019;**85**:305-309. DOI: 10.1016/j.jcs.2018.12.004

[20] Iquebal MA, Sharma P, Jasrotia RS, Jaiswal S, Kaur A, Saroha M, et al. RNAseq analysis reveals drought-responsive molecular pathways with candidate genes and putative molecular markers in root tissue of wheat. Scientific reports. 2019;**9**(1):1-18. DOI: 10.1038/s41598-019-49915-2

[21] Zhang LY, Liu DC, Guo XL, Yang WL, Sun JZ, Wang DW, et al. Genomic distribution of quantitative trait loci for yield and yield-related traits in common wheat. Journal of Integrative Plant Biology. 2010;**52**(11):996-1007. DOI: 10.1111/j.1744-7909.2010.00967.x

[22] Desta ZA, Ortiz R. Genomic selection: Genome-wide prediction in plant improvement. Trends in Plant Science. 2014;**19**(9):592-601. DOI: 10.1016/j.tplants.2014.05.006

[23] Poland JA, Endelman J, Dawson J, Rutkoski J, Wu S, Manes Y, et al. Genomic selection in wheat breeding using genotyping-by-sequencing. Plant Genome. 2012;**5**(3):103-113. DOI: 10.3835/plantgenome2012.06.0006

[24] Lin L-C, Tsai T-H, Kuo C-L. Chemical constituents comparison of Codonopsis tangshen Codonopsis pilosula var. modesta and Codonopsis pilosula. Natural Product Research. 2013;**27**(19):1812-1815

[25] lhan D. Molecular marker technologies in wheat. In: Current Academic Studies in Science and Mathematics Sciences-II. Lyon, France: Livre de Lyon; 2021. pp. 17-31

[26] Yu M, Mao SL, Hou DB, Chen GY, Pu ZE, Li W, et al. Analysis of contributors to grain yield in wheat at the individual quantitative trait locus level. Plant Breeding. 2018;**137**(1):35-49. DOI: 10.1111/pbr.12555

[27] Nadeem MA, Nawaz MA, Shahid MQ, Doğan Y, Comertpay G, Yıldız M, et al. DNA molecular markers in plant breeding: Current status and recent

advancements in genomic selection and genome editing. Biotechnology & Biotechnological Equipment. 2018;**32**(2):261-285. DOI: 10.1080/13102818.2017.1400401

[28] Kikuchi T, Hara M, Terao K. Development of a microsatellite marker set applicable to genome-wide screening of cynomolgus monkeys (*Macaca fascicularis*). Primates. 2007;**48**(2):140-146. DOI: 10.1007/s10329-006-0008-z

[29] Paran I, Michelmore RW. Development of reliable PCR-based markers linked to downy mildew resistance genes in lettuce. Theoretical and Applied Genetics. 1993;**85**(8):985-993. DOI: 10.1007/BF00215038

[30] Feuillet C, Keller B. Molecular markers for disease resistance: The example wheat. Molecular Marker Systems in Plant Breeding and Crop Improvement. 2006;**55**:353-370. DOI: 10.1007/3-540-26538-4_21

[31] Jaiswal S, Sheoran S, Arora V, Angadi UB, Iquebal MA, Raghav N, et al. Putative microsatellite DNA marker-based wheat genomic resource for varietal improvement and management. Frontiers in Plant Science. 2017;**8**:2009. DOI: 10.3389/fpls.2017.02009

[32] Haque MS, Saha NR, Islam MT, Islam MM, Kwon S-J, Roy SK, et al. Screening for drought tolerance in wheat genotypes by morphological and SSR markers. Journal of Crop Science and Biotechnology. 2021;**24**(1):27-39. DOI: 10.1007/s12892-020-00036-7

[33] Jiang Y, Zhao Y, Rodemann B, Plieske J, Kollers S, Korzun V, et al. Potential and limits to unravel the genetic architecture and predict the variation of Fusarium head blight resistance in European winter wheat (*Triticum aestivum* L.). Heredity. 2015;**114**(3):318-326. DOI: 10.1038/hdy.2014.104

[34] Ren J, Sun D, Chen L, You FM, Wang J, Peng Y, et al. Genetic diversity revealed by single nucleotide polymorphism markers in a worldwide germplasm collection of durum wheat. International Journal of Molecular Sciences. 2013;**14**(4):7061-7088. DOI: 10.3390/ijms14047061

[35] Rufo R, Alvaro F, Royo C, Soriano JM. From landraces to improved cultivars: Assessment of genetic diversity and population structure of Mediterranean wheat using SNP markers. PloS one. 2019;**14**(7):e0219867. DOI: 10.1371/journal.pone.0219867

[36] Chao S, Zhang W, Akhunov E, Sherman J, Ma Y, Luo M-C, et al. Analysis of gene-derived SNP marker polymorphism in US wheat (*Triticum aestivum* L.) cultivars. Molecular Breeding. 2009;**23**(1):23-33. DOI: 10.1007/s11032-008-9210-6

[37] Gao L, Jia J, Kong X. A SNP-based molecular barcode for characterization of common wheat. PloS one. 2016;**11**(3):e0150947. DOI: 10.1371/journal.pone.0150947

[38] Rembe M, Zhao Y, Jiang Y, Reif JC. Reciprocal recurrent genomic selection: An attractive tool to leverage hybrid wheat breeding. Theoretical and Applied Genetics. 2019;**132**(3):687-698. DOI: 10.1007/s00122-018-3244-x

[39] Gao L, Turner MK, Chao S, Kolmer J, Anderson JA. Genome wide association study of seedling and adult plant leaf rust resistance in elite spring wheat breeding lines. PLoS One. 2016;**11**(2):e0148671. DOI: 10.1371/journal.pone.0148671

[40] Soriano JM, Colasuonno P, Marcotuli I, Gadaleta A. Meta-QTL

analysis and identification of candidate genes for quality, abiotic and biotic stress in durum wheat. Scientific Reports. 2021;**11**(1):1-15. DOI: 10.1038/s41598-021-91446-2

[41] Endelman JB. Ridge regression and other kernels for genomic selection with R package rrBLUP. The Plant Genome. 2011;**4**(3):250-255. DOI: 10.3835/plantgenome2011.08.0024

[42] Heffner EL, Jannink JL, Sorrells ME. Genomic selection accuracy using multifamily prediction models in a wheat breeding program. The Plant Genome. 2011;**4**(1):65-75. DOI: 10.3835/plantgenome2010.12.0029

[43] Zhao Y, Mette M, Gowda M, Longin C, Reif J. Bridging the gap between marker-assisted and genomic selection of heading time and plant height in hybrid wheat. Heredity. 2014;**112**(6):638-645. DOI: 10.1038/hdy.2014.1

[44] Hill WG, Goddard ME, Visscher PM. Data and theory point to mainly additive genetic variance for complex traits. PLoS Genetics. 2008;**4**(2):e1000008. DOI: 10.1371/journal.pgen.1000008

[45] Schopp P, Müller D, Technow F, Melchinger AE. Accuracy of genomic prediction in synthetic populations depending on the number of parents, relatedness, and ancestral linkage disequilibrium. Genetics. 2017;**205**(1):441-454. DOI: 10.1534/genetics.116.193243

[46] Müller D, Schopp P, Melchinger AE. Persistency of prediction accuracy and genetic gain in synthetic populations under recurrent genomic selection. G3: Genes, Genomes, Genetics. 2017;**7**(3):801-811. DOI: 10.1534/g3.116.036582

[47] Varona L, Legarra A, Toro MA, Vitezica ZG. Non-additive effects in genomic selection. Frontiers in Genetics. 2018;**9**(78). DOI: 10.3389/fgene.2018.00078

[48] Müller D, Schopp P, Melchinger AE. Selection on expected maximum haploid breeding values can increase genetic gain in recurrent genomic selection. G3: Genes, Genomes, Genetics. 2018;**8**(4):1173-1181. DOI: 10.1534/g3.118.200091

[49] Singh M, Nara U, Kumar A, Thapa S, Jaswal C, Singh H. Enhancing genetic gains through marker-assisted recurrent selection: From phenotyping to genotyping. Cereal Research Communications. 2021:1-16. DOI: 10.1007/s42976-021-00207-4

[50] Ali Y, Khan MA, Atiq M, Hussain M. Novel gene pyramiding to combat rusts in global wheat varieties against prevalent virulence: A review. G3: Genes, Genomes, Genetics. 2018;**34**(4):797-810

[51] Emebiri L, Singh S, Tan M-K, Singh PK, Fuentes-Dávila G, Ogbonnaya F. Unravelling the complex genetics of Karnal bunt (Tilletia indica) resistance in common wheat (*Triticum aestivum*) by genetic linkage and genome-wide association analyses. G3: Genes, Genomes, Genetics. 2019;**9**(5):1437-1447. DOI: 10.1534/g3.119.400103

[52] Burow MD, Blake TK. Molecular Tools for the Study of Complex Traits: Molecular Dissection of Complex Traits. Boca Raton: CRC Press; 2019. pp. 13-30. DOI: 10.1201/9780429117770

[53] Ciucă M, Todorvska E, Kolev S, Nicolae R, Guinea I, Saulescu N. Marker-assisted selection (MAS) for drought tolerance in wheat using markers associated with membrane stability. FUNDULEA. 2009;**12**:7-12

[54] Gautam T, Saripalli G, Kumar A, Gahlaut V, Gadekar D, Oak M, et al.

Introgression of a drought insensitive grain yield QTL for improvement of four Indian bread wheat cultivars using marker assisted breeding without background selection. Journal of Plant Biochemistry and Biotechnology. 2021;**30**(1):172-183. DOI: 10.1007/s13562-020-00553-0

[55] Rai N, Bellundagi A, Kumar PK, Kalasapura Thimmappa R, Rani S, Sinha N, et al. Marker-assisted backcross breeding for improvement of drought tolerance in bread wheat (*Triticum aestivum* L. em Thell). Plant Breeding. 2018;**137**(4):514-526. DOI: 10.1111/pbr.12605

[56] Motomura Y, Kobayashi F, Iehisa JC, Takumi S. A major quantitative trait locus for cold-responsive gene expression is linked to frost-resistance gene Fr-A2 in common wheat. Breeding Science. 2013;**63**(1):58-67. DOI: 10.1270/jsbbs.63.58

[57] Ni F, Qi J, Hao Q, Lyu B, Luo M-C, Wang Y, et al. Wheat Ms2 encodes for an orphan protein that confers male sterility in grass species. Nature Communications. 2017;**8**(1):1-12. DOI: 10.1038/ncomms15121

[58] Xia C, Zhang L, Zou C, Gu Y, Duan J, Zhao G, et al. A TRIM insertion in the promoter of Ms2 causes male sterility in wheat. Nature Communications. 2017;**8**(1):1-9. DOI: 10.1038/ncomms15407

[59] Tucker EJ, Baumann U, Kouidri A, Suchecki R, Baes M, Garcia M, et al. Molecular identification of the wheat male fertility gene Ms1 and its prospects for hybrid breeding. Nature Communications. 2017;**8**(1):1-10. DOI: 10.1038/s41467-017-00945-2

[60] Wang Z, Li J, Chen S, Heng Y, Chen Z, Yang J, et al. Poaceae-specific MS1 encodes a phospholipid-binding protein for male fertility in bread wheat. Proceedings of the National Academy of Sciences. 2017;**114**(47):12614-12619. DOI: 10.1073/pnas.1715570114

[61] Kempe K, Rubtsova M, Gils M. Split-gene system for hybrid wheat seed production. Proceedings of the National Academy of Sciences. 2014;**111**(25):9097-9102. DOI: 10.1073/pnas.1402836111

[62] Longin CFH, Gowda M, Mühleisen J, Ebmeyer E, Kazman E, Schachschneider R, et al. Hybrid wheat: Quantitative genetic parameters and consequences for the design of breeding programs. Theoretical and Applied Genetics. 2013;**126**(11):2791-2801. DOI: 10.1007/s00122-013-2172-z

[63] Zhao Y, Li Z, Liu G, Jiang Y, Maurer HP, Würschum T, et al. Genome-based establishment of a high-yielding heterotic pattern for hybrid wheat breeding. Proceedings of the National Academy of Sciences. 2015;**112**(51):15624-15629. DOI: 10.1073/pnas.1514547112

[64] Würschum T, Leiser WL, Langer SM, Tucker MR, Longin CFH. Phenotypic and genetic analysis of spike and kernel characteristics in wheat reveals long-term genetic trends of grain yield components. Theoretical and Applied Genetics. 2018;**131**(10):2071-2084. DOI: 10.1007/s00122-018-3133-3

[65] Boeven PH, Longin CFH, Leiser WL, Kollers S, Ebmeyer E, Würschum T. Genetic architecture of male floral traits required for hybrid wheat breeding. Theoretical and Applied Genetics. 2016;**129**(12):2343-2357

[66] Lu Q, Lillemo M, Skinnes H, He X, Shi J, Ji F, et al. Anther extrusion and plant height are associated with Type

I resistance to Fusarium head blight in bread wheat line 'Shanghai-3/Catbird'. Theoretical and Applied Genetics. 2013;**126**(2):317-1334. DOI: 10.1007/s00122-012-1981-9

[67] Würschum T, Liu G, Boeven PH, Longin CFH, Mirdita V, Kazman E, et al. Exploiting the Rht portfolio for hybrid wheat breeding. Theoretical and Applied Genetics. 2018;**131**(7):1433-1442. DOI: 10.1007/s00122-018-3088-4

[68] Jiang Y, Schmidt RH, Zhao Y, Reif JC. A quantitative genetic framework highlights the role of epistatic effects for grain-yield heterosis in bread wheat. Nature Genetics. 2017;**49**(12):1741-1746. DOI: 10.1038/ng.3974

[69] Lassoued R, Macall DM, Smyth SJ, Phillips PW, Hesseln H. How should we regulate products of new breeding techniques? Opinion of surveyed experts in plant biotechnology. Biotechnology Reports. 2020;**26**:e00460. DOI: 10.1016/j.btre.2020.e00460

[70] Kumar K, Gambhir G, Dass A, Tripathi AK, Singh A, Jha AK, et al. Genetically modified crops: Current status and future prospects. Planta. 2020;**251**(4):1-27. DOI: 10.1007/s00425-020-03372-8

[71] Guo M, Ye J, Gao D, Xu N, Yang J. Agrobacterium-mediated horizontal gene transfer: Mechanism, biotechnological application, potential risk and forestalling strategy. Biotechnology Advances. 2019;**37**(1):259-270. DOI: 10.1016/j.biotechadv.2018.12.008

[72] Li S, Lin D, Zhang Y, Deng M, Chen Y, Lv B, et al. Genome-edited powdery mildew resistance in wheat without growth penalties. Nature. 2022;**602**:455-460. DOI: 10.1038/s41586-022-04395-9

[73] Ismagul A, Yang N, Maltseva E, Iskakova G, Mazonka I, Skiba Y, et al. A biolistic method for high-throughput production of transgenic wheat plants with single gene insertions. BMC Plant Biology. 2018;**18**(1):1-8. DOI: 10.1186/s12870-018-1326-1

[74] Pallotta MA, Warner P, Kouidri A, Tucker EJ, Baes M, Suchecki R, et al. Wheat ms5 male-sterility is induced by recessive homoeologous A and D genome non-specific lipid transfer proteins. The Plant Journal. 2019;**99**(4):673-685. DOI: 10.1111/tpj.14350

[75] Wang H, Sun S, Ge W, Zhao L, Hou B, Wang K, et al. Horizontal gene transfer of Fhb7 from fungus underlies Fusarium head blight resistance in wheat. Science. 2020;**368**(6493):5435

[76] Zhang L, Gao C, Chen C, Zhang W, Huang X-Y, Zhao F-J. Overexpression of rice OsHMA3 in wheat greatly decreases cadmium accumulation in wheat grains. Environmental Science & Technology. 2020;**54**(16):10100-10108. DOI: 10.1021/acs.est.0c02877

[77] Ohly H, Broadley M, Joy E, Khan M, McArdle H, Zaman M, et al. The BiZiFED project: Biofortified zinc flour to eliminate deficiency in Pakistan. Nutrition Bulletin. 2019;**44**(1):60-64. DOI: 10.1111/nbu.12362

[78] Jia M, Guan J, Zhai Z, Geng S, Zhang X, Mao L, et al. Wheat functional genomics in the era of next generation sequencing: An update. The Crop Journal. 2018;**6**(1):7-14. DOI: 10.1016/j.cj.2017.09.003

[79] Sharma S, Schulthess AW, Bassi FM, Badaeva ED, Neumann K, Graner A, et al. Introducing beneficial alleles from plant genetic resources into the wheat germplasm. Biology. 2021;**10**(10):982. DOI: 10.3390/biology10100982

[80] Sukumaran S, Krishna H, Singh K, Mottaleb KA, Reynolds M. Progress and prospects of developing climate resilient wheat in South Asia using modern pre-breeding methods. Current Genomics. 2021;**22**(6):440-449. DOI: 10.2174/1389202922666210705125006

[81] Rebetzke G, Jimenez-Berni J, Fischer R, Deery D, Smith D. High-throughput phenotyping to enhance the use of crop genetic resources. Plant Science. 2019;**282**:40-48. DOI: 10.1016/j.plantsci.2018.06.017

[82] Bhatta M, Sandro P, Smith MR, Delaney O, Voss-Fels KP, Gutierrez L, et al. Need for speed: Manipulating plant growth to accelerate breeding cycles. Current Opinion in Plant Biology. 2021;**60**:101986. DOI: 10.1016/j.pbi.2020.101986

[83] Reynolds MP, Lewis JM, Ammar K, Basnet BR, Crespo-Herrera L, Crossa J, et al. Harnessing translational research in wheat for climate resilience. Journal of Experimental Botany. 2021;**72**(14):5134-5157. DOI: 10.1093/jxb/erab256

[84] Hahn F, Sanjurjo Loures L, Sparks CA, Kanyuka K, Nekrasov V. Efficient CRISPR/Cas-mediated targeted mutagenesis in spring and winter wheat varieties. Plants. 2021;**10**(7):1481. DOI: 10.3390/plants10071481

[85] Schiml S, Puchta H. Revolutionizing plant biology: Multiple ways of genome engineering by CRISPR/Cas. Plant Methods. 2016;**12**(1):1-9. DOI: 10.1186/s13007-016-0103-0

[86] Jouanin A, Gilissen LJ, Schaart JG, Leigh FJ, Cockram J, Wallington EJ, et al. CRISPR/Cas9 gene editing of gluten in wheat to reduce gluten content and exposure—Reviewing methods to screen for coeliac safety. Frontiers in Nutrition. 2020;7:51. DOI: 10.3389/fnut.2020.00051

[87] Abe F, Haque E, Hisano H, Tanaka T, Kamiya Y, Mikami M, et al. Genome-edited triple-recessive mutation alters seed dormancy in wheat. Cell Reports. 2019;**28**(5):1362-1369. DOI: 10.1016/j.celrep.2019.06.090

[88] Sánchez-León S, Gil-Humanes J, Ozuna CV, Giménez MJ, Sousa C, Voytas DF, et al. Low-gluten, nontransgenic wheat engineered with CRISPR/Cas9. Plant Biotechnology Journal. 2018;**16**(4):902-910. DOI: 10.1111/pbi.12837

[89] Liang Z, Chen K, Li T, Zhang Y, Wang Y, Zhao Q, et al. Efficient DNA-free genome editing of bread wheat using CRISPR/Cas9 ribonucleoprotein complexes. Nature Communications. 2017;**8**(1):1-5. DOI: 10.1038/ncomms14261

[90] Rey M-D, Martín AC, Smedley M, Hayta S, Harwood W, Shaw P, et al. Magnesium increases homoeologous crossover frequency during meiosis in ZIP4 (Ph1 gene) mutant wheat-wild relative hybrids. Frontiers in Plant Science. 2018;**9**:509. DOI: 10.3389/fpls.2018.00509

[91] Kamiya Y, Abe F, Mikami M, Endo M, Kawaura K. A rapid method for detection of mutations induced by CRISPR/Cas9-based genome editing in common wheat. Plant Biotechnology. 2020;**37**(2):247-251. DOI: 10.5511/plantbiotechnology.20.0404b

[92] Lv J, Yu K, Wei J, Gui H, Liu C, Liang D, et al. Generation of paternal haploids in wheat by genome editing of the centromeric histone CENH3. Nature Biotechnology. 2020;**38**(12):1397-1401. DOI: 10.1038/s41587-020-0728-4

[93] Li CX, Xu WG, Guo R, Zhang JZ, Qi XL, Hu L, et al. Molecular marker assisted breeding and genome composition analysis of Zhengmai 7698, an elite winter wheat cultivar. Scientific Reports. 2018;**8**(1):1-8. DOI: 10.1038/s41598-017-18726-8

[94] Todkar L, Singh GP, Jain N, Singh PK, Prabhu KV. Introgression of drought tolerance QTLs through marker assisted backcross breeding in wheat (*Triticum aestivum* L.). Indian Journal of Genetics and Plant Breeding. 2020;**80**(2):209-212

[95] Zhang M, Fang T, Zhou X, Chen X, Li X, Feng J, et al. Combination of marker-assisted backcross selection of Yr59 and phenotypic selection to improve stripe rust resistance and agronomic performance in four elite wheat cultivars. Agronomy. 2022;**12**(2):497. DOI: 10.3390/agronomy12020497

[96] Bouguennec A, Lesage VS, Gateau I, Sourdile P, Jahier J, Lonnet P. Transfer of recessive skr crossability trait into well-adapted French wheat cultivar Barok through marker-assisted backcrossing method. Cereal Research Communications. 2018;**46**(4):604-615. DOI: 10.1556/0806.46.2018.043

[97] Ram H, Gupta N, Singh G, Kaur H, Kaur J, Srivastva P. Performance of marker-assisted backcross bread-wheat (*Triticum aestivum*) variety Unnat PBW 343 under diverse environments.

[98] Singh A, Jaiswal JP, Badoni S. Enhancing rust resistance in wheat through marker assisted backcross breeding. Indian Journal of Genetics and Plant Breeding. 2018;**78**(1):19-25

[99] Kumar TP, Ahlawat AK, Singh SK, Rathan ND, Rai A, Bajpai K, et al. Enhancing rust resistance in wheat through marker Quality evaluation of near isogenic lines of wheat developed through marker assisted backcross breeding for grain softness. Indian Journal of Genetics and Plant Breeding. 2022;**82**(1):56-64

[100] Rai A, Mahendru-Singh A, Raghunandan K, Kumar TP, Sharma P, Ahlawat AK, et al. Marker-assisted transfer of PinaD1a gene to develop soft grain wheat cultivars. 3 Biotech. 2019;**9**(5):1-10

[101] Rahman M, Davies P, Bansal U, Pasam R, Hayden M, Trethowan R. Marker-assisted recurrent selection improves the crown rot resistance of bread wheat. Molecular Breeding. Mar 2020;**40**(3):1-4. DOI: 10.1007/s11032-020-1105-1

[102] Guttieri MJ. Ms3 dominant genetic male sterility for wheat improvement with molecular breeding. Crop Science. 2020;**60**(3):1362-1372. DOI: 10.1002/csc2.20091

[103] Li H, Li S, Abdelkhalik S, Shahzad A, Gu J, Yang Z, et al. Development of thermo-photo sensitive genic male sterile lines in wheat using doubled haploid breeding. BMC Plant Biology. 2020;**20**(1):1

[104] Wang W, Pan Q, He F, Akhunova A, Chao S, Trick H, et al. Transgenerational CRISPR-Cas9 activity facilitates multiplex gene editing in allopolyploid wheat. The CRISPR Journal. 2018;**1**(1):65-74. DOI: 10.1089/crispr.2017.0010

[105] Jouanin A, Borm T, Boyd LA, Cockram J, Leigh F, Santos BA, et al. Development of the GlutEnSeq capture system for sequencing gluten gene families in hexaploid bread wheat with deletions or mutations induced by γ-irradiation or CRISPR/Cas9. Journal of

Cereal Science. 2019;**88**:157-166. DOI: 10.1016/j.jcs.2019.04.008

[106] Zhang S, Zhang R, Song G, Gao J, Li W, Han X, et al. Targeted mutagenesis using the Agrobacterium tumefaciens-mediated CRISPR-Cas9 system in common wheat. BMC Plant Biology. 2018;**18**(1):1-2. DOI: 10.1186/s12870-018-1496-x

Chapter 14

Gene Editing Improves the Agronomic Important Traits of Wheat – CRISPR-Cas9 and Cas12/Cpf1

Habtamu Kefale and Sewnet Getahun

Abstract

A hexaploid Wheat (*Triticum aestivum* L.) is the **3rd** most important staple food crop with 15% caloric intake next to maize and rice in the world. The global attention for wheat improvement are still encouraging. However, the population growth and demand for food at this time and in the next years could not be balanced. Due to this, huge investments have been established and performed to improve the most important agronomic traits of wheat. Among the new molecular tools and techniques that have given a big emphasis as it will have many concerns is gene editing. Many gene editing tools have been reported and being implemented including Zinc finger nuclease, transcription activator-like effector nuclease, and clustered regularly interspaced short palindromic repeats associated Cas9/12 system for targeted gene editing. Among these, clustered regularly interspaced short palindromic repeats associated Cas9/12 systems are very accurate and widely used for targeted gene editing. By using CRISPR-Cas mediated gene editing technique, important traits of wheat include disease and pest resistance, better grain and flour quality, gluten-free trait, better nutritional value, nitrogen use efficiency, threshability, and other yield components and has been edited and improved. Therefore, the use of gene editing technologies for wheat as well as other important crops improvement was irreversible.

Keywords: Cas12/Cpf1, CRISPR-Cas9, gene editing, genetic engineering, wheat

1. Introduction

The new approach and emerging technology of gene editing in crop plants and animals becomes a revolutionary science in the molecular era [1]. The conventional wheat crop genetic improvement or breeding progress has been described by the concept of genetic gain and measured by the difference between a selected population and its offspring population [2]. However, the global population growth is increasing at an increasing rate and is projected to reach 9.2billion in the coming 30 years [3, 4], and difficult to supply enough food and food products. Many studies suggested that improvement in genetic gain meet the growing population demand for agricultural products and food needs to utilize modern breeding techniques (tools and strategies) and platforms, implemented with improved agronomic practice, including improved field-based phenotyping with a better understanding of the genetic architecture of trait [2, 5].

Gene editing is among the new and growing technology of molecular science in crop improvement programs to improve the grain yield other agronomic important traits. The three most important gene-editing techniques widely used in the crop improvement program till this days are Zinc finger nuclease (ZFN), transcription activator-like effector nuclease (TALEN), and clustered regularly interspaced short palindromic repeats associated Cas9/12 (CRISPR-Cas9/12a) system for targeted gene editing [6, 7]. The development of CRISPR-based gene editing technologies recognizing distinct protospacer-adjacent motifs (PAMs), or having different spacer length/structure requirements broadens the range of possible genomic applications making them more preferred tools over ZFN and TALEN [8]. Clustered regularly interspaced short palindromic repeat (CRISPR)-CRISPR associated protein (CRISPR-Cas9/Cpf1) is a versatile, simple, and inexpensive system for precise sequence-specific modifications of DNA sequences including targeted mutagenesis for gene Knockout, single base substitution, and gene or allele replacement in vivo [6, 9]. CRISPR is a DNA fragment that contains non-contiguous short DNA repeats separated by spacers, which are snippets of varied sequences. CRISPR-associated (Cas) genes were anticipated to be related to CRISPR loci after they were found in the genome of Escherichia coli in 1987 [5].

Many genetic engineering activities have been done by different scientists across the world to get better traits related to grain yield, disease and pest resistance [3, 10], better grain and flour quality [11], gluten-free trait [12, 13], better nutritional value and nitrogen use efficiency [11] with the help of gene-editing tools. Similarly, in the USA Wang *et al.* [8] evaluated the natural and engineered variants of Cas12a (FnCas12a and LbCas12a) and Cas9 for their ability to induce mutations in endogenous genes controlling important agronomic traits in wheat. This review focuses on the application of the two CRISPR-Cas proteins (Cas9 and Cas12a) in the wheat crop improvement program. Therefore, this review highlights the current issues and advancements in wheat gene editing to improve the most important agronomic traits with aid of CRISPR-Cas proteins.

2. Gene editing for quality improvement

The higher demand for high-yielding varieties with premium grain quality in the continued economic development is a critical issue becomes increasing. Quality traits including grain yield, protein content, hectoliter weight, starch content are governed by many genes with a cumulative effect that is simultaneously affected by many factors, and it is more complicated in hexaploid wheat [11]. Improving this trait could not be a simple activity and not possible in conventional breeding or crop improvement program. Therefore, CRISPR Cas-mediated gene editing technologies have been used and created a great opportunity for allelic variations in a more faster and accurate manner [11]. Zhang *et al.* created and found allelic variations for grain hardiness, grain starch content, and dough color using genes of *pinb*, *waxy*, *ppo,* and *psy* in Fielder through *Agrobacterium* delivered CRISPR-Cas9 system. They had effectively obtained new wheat germplasms with better grain quality in hardiness, starch content, and dough color. Their improved grain quality wheat germplasms can be employed as donor parents in backcross breeding to improve the grain quality of premier wheat cultivars.

On the other hand, there is a need to have gluten-free wheat to overcome the risk of chronic disease (Coeliac disease (CD)). This disease is caused in genetically predisposed individuals by the ingestion of gluten proteins (gliadins and glutenins) from products of wheat, barley, and rye [12]. This human disease associated with wheat coeliac disease (CD) is an autoimmune reaction prevalent in 1–2% of the global population [12]. Even though gluten proteins are found in wheat, Jouanin *et al.* reported RNA interference (RNAi) silencing to down-regulate gliadin families which are capable of causing Coeliac disease. Therefore, an essential and strict lifelong treatment of CD is the consumption of gluten-free food sources and avoidance of gluten-containing products from wheat, rye, barley, and, in a rare case, oats [13].

3. Gene editing for disease resistance

Crop production is multi-task activity and is easily affected by many contributing factors for yield and quality reduction of the produce. Among the factors that greatly affect the quality and grain yield of wheat are biotic agents (pest, disease) and abiotic agents especially fertilization, soil acidity, drought, and cold stress besides its genetic potential. Although the conventional approach of crop protection (chemical and cultural) activities has been applied to protect the crop, the numbers, as well as the type of reported diseases and pests, become increasing [3]. Moreover, these classical breeding approaches to develop pest and disease-resistant varieties are laborious, cost-intensive, and not efficient. Therefore, the new technology and approach recently introduced gene editing via the CRISPR-Cas system has been utilized to protect the crops from pests and pathogens and to enhance disease and pest resistance among different crop plants i.e. Wheat, Rice, Cocoa, Tomato, and Grape *[3]*. In a similar study, fungal disease powdery mildew (*Blumeria graminis* f.sp. *tritici* (*Bgt*)) resistant gene (*TaEDR1*) is successfully introduced from *Arabidopsis thaliana* using gene editing via CRISPR-Cas9 technology to wheat [14]. In their study, they have cloned *Triticum aestivum* enhanced disease resistance-1 gene (*TaEDR1*) in hexaploid wheat and showed the knockdown *TaEDR1* mutants from VIGs or RNAi increased resistance to virulent *Bg*t isolates. Then they have generated wheat *edr1* plants by simultaneous modification of three homologs of TaEDR1 with the help of CRISPR Cas9 technology. Then, they have got wheat genotypes carrying TaEDR1 plants which did not exhibit mildew-induced cell death/disease symptoms. This candidate gene could be very important in the crop improvement program of wheat to overcome and reduce the problem of powdery mildew. Likewise, Brauer *et al.* [15] in Canada reported that gene editing of deoxynivalenol-induced transcription factor confers resistance to *Fusarium* head blight disease (*Fusarium graminearum)* in wheat. Therefore, gene editing has played a significant role in critically editing, improving, and developing candidate disease-resistant genes for wheat.

4. Gene editing for better nitrogen use efficiency

In the era of the green revolution, we remember what Norman Borlaug has contributed by developing a semi-dwarf wheat genotype that has a great efficiency for fertilizer response and then provides a higher grain yield. Although the semi-dwarf wheat type responded better than the older one, the crop has a great genetic potential to give more yield if its nitrogen use efficiency is edited using the molecular breeding technique CRISPR-Cas system. Moreover, scientists reported that the wheat crop has 40% nitrogen use efficiency whereas the remaining amount is released to the environment via leaching or volatilization [11]. To improve the NUE of wheat Zhang *et al.* [16] isolated and characterized three *TaARE1* homoeologs from the elite Chinese winter wheat cultivar ZM and tare1 transgene-free mutant lines with partial or triple-null alleles. Under hydroponic conditions, all transgene-free mutant lines demonstrated greater tolerance to N-deficit or starvation (**Figures 1** and **2**), as well as delayed senescence and higher grain yield in a field experiment under normal growth circumstances (**Figure 2**). When compared to the wild-type control, the AABBdd and aabbDD mutant lines had considerably improved nitrogen use efficiency, post-poned senescence, and increased grain production without showing any growth

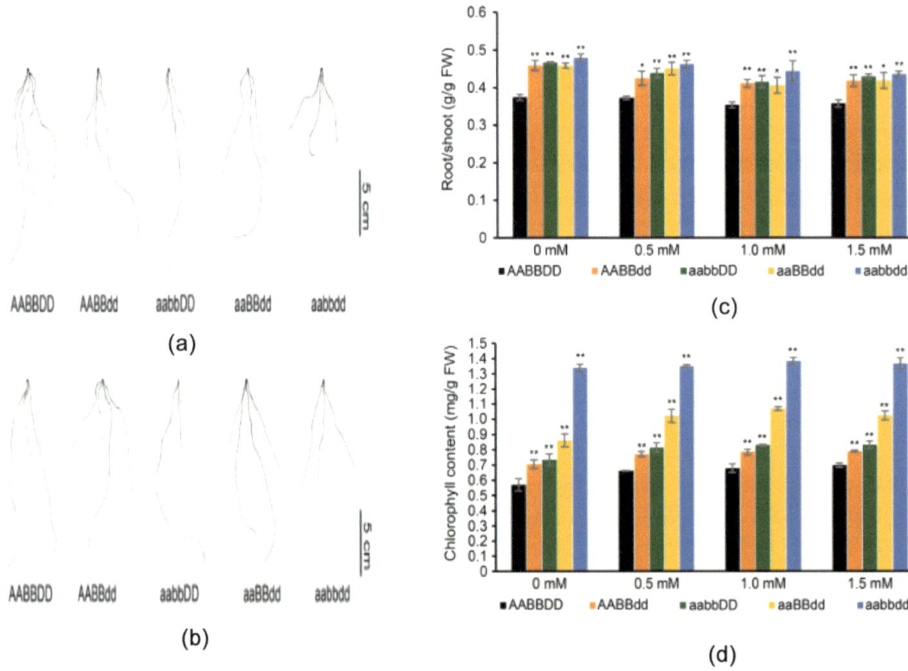

Figure 1.
Illustrates root morphology, root/shoot ratio, and chlorophyll content of taare1 mutant lines compared to the wild-type control. (a) Root morphology of wild-type and different taare1 mutant lines under N deficiency (0 mM NH4NO3) hydroponic condition (scale bars = 5 cm). (b) Root morphology of wild-type and different taare1 mutant lines under N supply (1.5 mM NH4NO3) hydroponic condition (scale bars = 5 cm). (c) Root/shoot ratio of wild-type and different taare1 mutant lines under different concentrations of N (0 mM NH4NO3,0.5 mM NH4NO3,1.0 mM NH4NO3, and 1.5 mM NH4NO3) hydroponic con- (d) Quantification of chlorophyll content in wild-type and different taare1 mutant lines under different concentrations of N (0 mM NH4NO3,0.5 mM NH4NO3,1.0 mM NH4NO3, and 1.5 mM NH4NO3) hydroponic conditions ([11], Journal of Integrative Plant Biology).

Figure 2.
Illustrates phenotypes of wild-type and different taare1 mutant lines in the field. (a) Plant phenotypes of wild-type and different taare1 mutant lines at the dough stage (scale bars = 10 cm). (b) Phenotypes of flag leaves of wild type and different taare1 mutant lines at the dough stage (scale bars = 5 cm). (c) Plant phenotypes of wild-type and different taare1 mutant lines at the dough stage in the field. (d) Plant phenotypes of wild-type and different taare1 mutant lines at the kernel ripe stage (scale bars = 10 cm). (e) Spike phenotypes of wild-type and different taare1 mutant lines at the kernel ripe stage (scale bars = 5 cm). (f) Grain size and appearance in wild-type and different taare1 mutant lines at the kernel ripe stage. The grains were aligned to illustrate grain length (a) and grain width (b) between wild-type and mutant lines (scale bars = 1 cm) ([11], Journal of Integrative Plant Biology).

abnormalities (**Figure 2**). For the first time, they were able to create novel wheat germplasm with better NUE and yield potential by modifying TaARE1 by genome editing tool.

5. Gene editing for yield component traits

Understanding the genetic basis of yield component traits in major crop plants holds a great promise to improve and utilize yield potential by allowing breeders to make informed decisions. Then, by assembling beneficial allelic combinations, it is possible to create new improved varieties [17]. Gene editing of the wheat homologs of *TONNEAU1*-recruiting motif encoding gene affects grain shape and weight in wheat by using CRISPR-Cas9 technology [18]. Likewise, Wang *et al.* showed that the

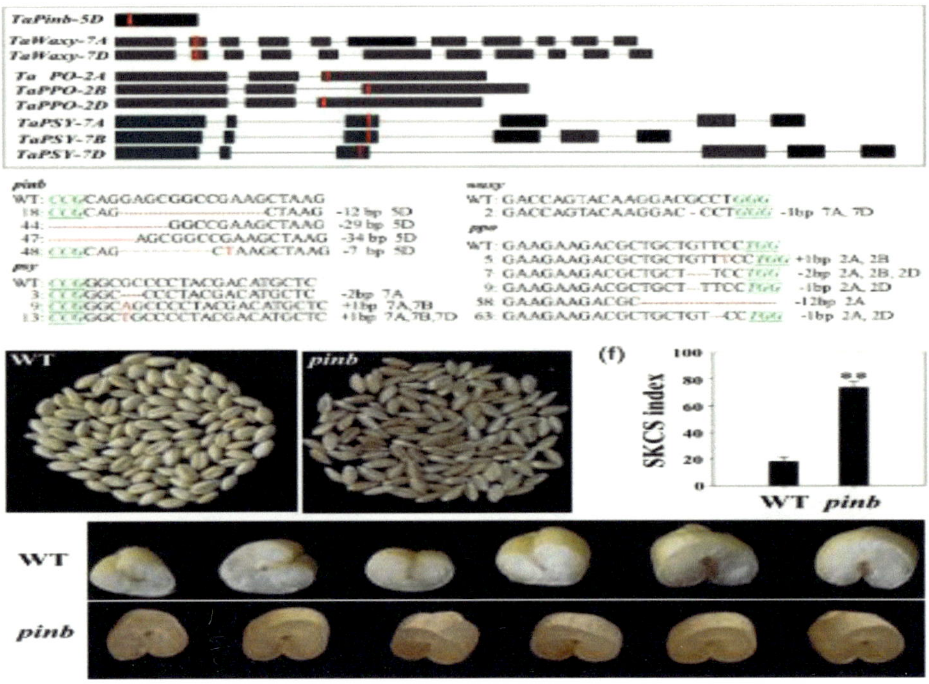

Figure 3.
The DNA sequence and phenotypic differences between edited and unedited lines. Source: Zhang et al. [11].

CRISPR-Cas9 gene editing of *TaGW7*, a homolog of *OsGW7* encoding a *TONNEAU1*-recruiting motif (TRM) protein affects grain shape and weight in allohexaploid wheat. Moreover, Wang *et al*. [8] also effectively used CRISPR LbCase12a-MGE gene-editing tool to generate heritable mutations in a wheat gene that controls grain size and weight. They found that utilizing altered Cas12a (LbCas12a-RVR) and Cas9 (Cas9-NG and xCas9) that can identify TATV and NG PAMs respectively, the range of editable loci in a wheat genome may be further broadened with Cas9 NG indicating greater editing efficiency on targets with a typical PAMs than xCas9 (**Figure 3**).

6. Gene editing for haploid wheat development

Wheat has two species cultivated in the world i.e. bread wheat (*Triticum aestivum* L.) and durum wheat (*Triticum durum* or *Triticum turgidum* subspe. *Durum*.) [19]. The bread wheat species is allohexaploid, (2n = 6x = 42) consists of three sub-genomes (A, B and G), and is cultivated for the purpose of bread because of lower gluten content. Whereas durum wheat is tetraploid, (2n = 4x = 28) has two sub-genomes (A and B) lacking D and is cultivated for pasta and macaroni. In the era of molecular breeding and omics crop improvement, scientists come up with the development of haploid paternal wheat. Conventional breeding of the most important cereal crops (maize, wheat, and rice) is based on the genetic mating of different parents with distinct traits to develop a single crop variety having desirable allelic combination may take also 8–10 years [20]. However, gene editing (GE) with haploid induction (HI) was successfully performed on (*T. aestivum* L., *Zea mays* L., *Hordeum vulgare* L., *Brassica napus* L) and possible to develop pure homozygous DH genotypes within two generations (**Figure 4**) [17].

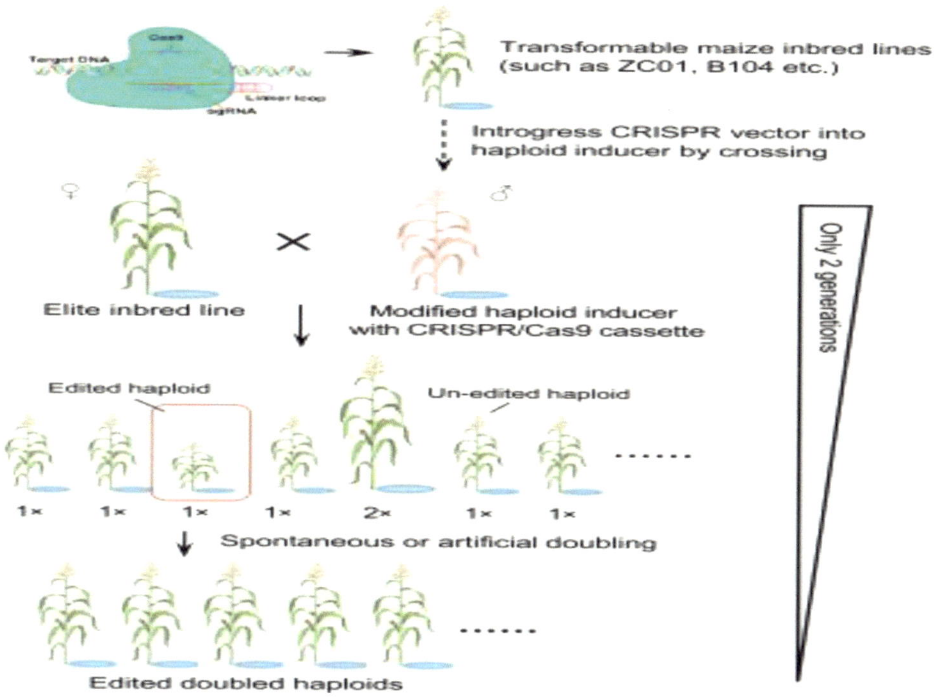

Figure 4.
Simple steps of Haploid inducer mediated editing (HIME) on maize crop.Source: Wang et al. [17].

Amin & Safwat [21], developed 120 Doubled haploid spring wheat genotypes from a cross of F1 of cross-pollination between wheat (*Triticum aestivum* L.) and millet (*Pennisetum glaucum*) with Indian cultivar (Kharchia) and Egyptian cultivar (Sakha 93). In this study, under normal conditions, agronomic traits of DH genotype traits (flowering time, number of spikelets per spike, plant height, spike length, and thousand seeds weight) revealed better performances. Therefore, to get the benefit of better performance of agronomic traits of wheat, doubled haploid (DH) genotype development is becoming a quick way (not more than 2 generations) of crop improvement [20].

In the conventional double haploid development, haploids may be induced either in vitro or in vivo methods [20]. Therefore, while performing the in vitro method of haploid production, isolated microspore culture, as well as an anther and ovary/ovule culture, needs colchicine/charcoal treatment and culturing the cell in the Petri-dish for the chromosome doubling but current in vivo methods are based on the modification of histone molecule H3 (CENH3). The conventional haploid development method takes a long time (6 years) than the haploid inducer mediated editing technology (**Figure 5**).

7. Other agronomic trait improvements of wheat

Free threshing in wheat is an advantage and leads to the selection of the domesticated Q allele, which is now present in almost all modern wheat varieties [22]. A study conducted by Liu *et al.* [23] to see the regulation mechanism of and improve

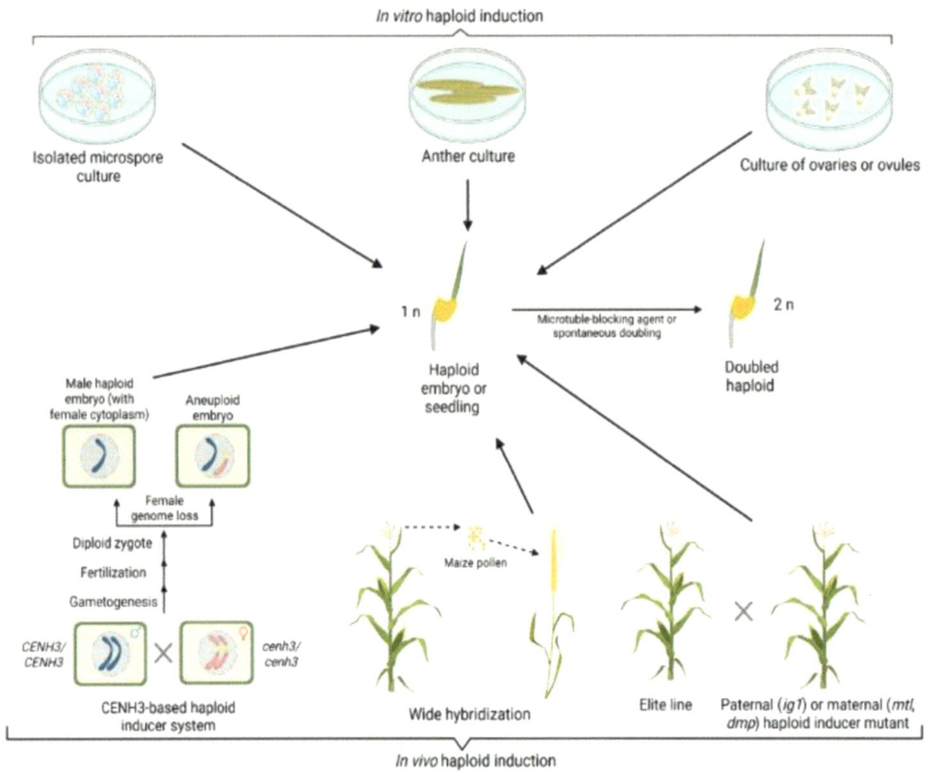

Figure 5.
The conventional haploid development procedure. Source: Bhowmik & Bilichak [20].

wheat spike and threshability using CRISPR-Cas9 obtained homozygous plants in the F1 generation with loss of function of only *TaAQ* or *TaDq* and simultaneous loss of function of *TaAQ* and *TaDq* to analyze the effect of these genes on wheat spikes and floret shapes. Then, two genes of *TaAQ* and *TaDq* were edited using CRISPR-Cas9 and resulted improved spike morphogenesis and grain threshability. This shows that the TaQ gene families are very important in the improvement of different traits of wheat. In wheat, the loss of function mutant of the (AP2) like transcription factor the Q gene changed the flowering time and spike architecture. Because of the benefit of free threshing in wheat, the domesticated wheat allele Q was chosen, and it is now found in nearly all current wheat cultivars. The domesticated allele Q confers a free threshing trait and a subcompact (i.e. partially compact) inflorescence (spike), while the pre-domesticated allele q, encodes an AP2 transcription factor [22].

8. Conclusion

The global demand for food is increasing at an alarming rate because of the population growth as expected to reach greater than 9 billion in 2050. It is difficult to supply enough food and feed this much population now and coming 30 years using the usual and conventional crop improvement technique and approach. Scientists and different organizations in the world performed different research activities and developed new and novel tools, procedures as well as protocols in wheat crop improvement. In this review manuscript, the gene-editing tools (CRISPR Cas systems) role and advancement were covered and highlighted. Based on this, many findings reported that CRISPR-Cas9 and 12 systems have been successfully used and implemented to improve agronomic important traits of wheat crop. Mainly, traits related to grain yield, disease and pest resistance, better grain, and flour quality, gluten-free trait, better nutritional value, and nitrogen use efficiency were improved with the help of gene editing tools especially CRISPR-Cas9 and 12 (Cpf1). Surprisingly, gene editing has been successfully implemented in the doubled haploid production and reduced the time required for fixing the trait by 6 years than the conventional method.

9. Future prospects

Although gene-editing techniques are used to improve the qualitative and quantitative traits of wheat, still the wheat genetic yield potential improvement is difficult because of its polygenic and polyploidy nature. Grain yield and most quality traits sometimes may not be improved simultaneously due to their indirect association. Moreover, the science of biotechnology is still growing and needs time, skill, and technology to explore the association of each trait with the grain yield of wheat since the ultimate goal of any crop improvement is an economic yield of the crop. Therefore, for the future scientists and organizations in the world shall create the technology used to detect multiple specific regions of DNA sequences at a time and improve by pyramiding the genes. Finally, the gene-editing technologies had the best features of no risk of chemicals like colchicine and short (2 years) doubled haploid wheat genotype development.

Authors' contributions

The first author drafted the manuscript, summarization of ideas, interpretation of the data, critical reviewing, synthesizing, and revision. The 2nd author's contribution is the critical evaluation of the manuscript and providing critical comments.

Funding

Not applicable.

Competing interests

The authors declare that they have no known competing financial interests or personal relationships that could have appeared to influence the work reported in this paper.

Ethics approval and consent to participate

Not applicable. Because I have not used the human research data for this paper.

Availability of data and material

This review is done by collecting original published research articles and I have cited the authors for each idea and data taken from their article and included in the reference list.

Abbreviations

UNDESA	United Nations Department of Economic and Social Affairs
TALEN	Transcription Factor Like Effector Nuclease
CRISPR	Clustered Regularly Interspaced Short Palindromic Repeats
ZFN	Zink Finger Nuclease
CD	Coeliac Disease
TaEDR1	*Triticum aestivum* enhanced Disease Resistance gene1
NUE	Nitrogen Use Efficiency
VIGs	Virus Induced Gene Silencing
Bgt	*Blumera Graminis* titici

Author details

Habtamu Kefale* and Sewnet Getahun
Department of Plant Science, College of Agriculture and Natural Resource, Debre Markos University, Ethiopia

*Address all correspondence to: habtie.k23@gmail.com; habtamu_kefale@dmu.edu.et

References

[1] Doudna JA, Charpentier E. The new frontier of genome engineering with CRISPR-Cas9. Journal of Science. 2014;**346**(6213):1258096. DOI: 10.1126/science.1258096

[2] Xu Y, Li P, Zou C, Lu Y, Xie C, Zhang X, et al. Enhancing genetic gain in the era of molecular breeding. Journal of Experimental Botany. 2017;**68**(11):2641-2666. DOI: 10.1093/jxb/erx135

[3] Paul NC, Park SW, Liu H, Choi S, Ma J, MacCready JS, et al. Plant and fungal genome editing to enhance plant disease resistance using the CRISPR/Cas9 system. Frontiers in Plant Science. 2021;**12**:1534. DOI: 10.3389/fpls.2021.700925

[4] United Nations Department of Economic and Social Affairs. Reconsidering Rural Development. New York City, United Nations: World Social Report; 2021

[5] Bao A, Burritt DJ, Chen H, Zhou X, Cao D, Tran LSP. The CRISPR/Cas9 system and its applications in crop genome editing. Critical Reviews in Biotechnology. 2019;**39**(3):321-336. DOI: 10.1080/07388551.2018.1554621

[6] Ahmar S, Saeed S, Khan MHU, Khan SU, Mora-Poblete F, Kamran M, et al. A revolution toward gene-editing technology and its application to crop improvement. International Journal of Molecular Sciences. 2020;**21**(16):1-28. DOI: 10.3390/ijms21165665

[7] Gupta D, Bhattacharjee O, Mandal D, Sen MK, Dey D, Dasgupta A, et al. CRISPR-Cas9 system: A new-fangled dawn in gene editing. Life Sciences. 2019;**232**(June):116636. DOI: 10.1016/j.lfs.2019.116636

[8] Wang W, Tian B, Pan Q, Chen Y, He F, Bai G, et al. Expanding the range of editable targets in the wheat genome using the variants of the Cas12a and Cas9 nucleases. Plant Biotechnology Journal. 2021:1-14. DOI: 10.1111/pbi.13669

[9] Li S, Zhang C, Li J, Yan L, Wang N, Xia L. Present and future prospects for wheat improvement through genome editing and advanced technologies. Plant Communications. 2021;**2**(4):100211

[10] Bisht DS, Bhatia V, Bhattacharya R. Improving plant-resistance to insect-pests and pathogens: The new opportunities through targeted genome editing. Seminars in Cell & Developmental Biology. 2019;**96**:65-76. DOI: 10.1016/J.SEMCDB.2019.04.008

[11] Zhang S, Zhang R, Gao J, Song G, Li J, Li W, et al. CRISPR/Cas9-mediated genome editing for wheat grain quality improvement. Plant Biotechnology Journal. 2021;**19**(9):1684-1686. DOI: 10.1111/pbi.13647

[12] Jouanin A, Gilissen LJWJ, Schaart JG, Leigh FJ, Cockram J, Wallington EJ, et al. CRISPR/Cas9 gene editing of gluten in wheat to reduce gluten content and exposure—Reviewing methods to screen for coeliac safety. Frontiers in Nutrition. 2020;**7**(April):51. DOI: 10.3389/fnut.2020.00051

[13] Scherf KA, Wieser H, Koehler P. Novel approaches for enzymatic gluten degradation to create high-quality gluten-free products. Food Research International. 2018;**110**:62-72. DOI: 10.1016/J.FOODRES.2016.11.021

[14] Zhang Y, Bai Y, Wu G, Zou S, Chen Y, Gao C, et al. Simultaneous modification of three homoeologs of TaEDR1 by

genome editing enhances powdery mildew resistance in wheat. Plant Journal. 2017;**91**(4):714-724. DOI: 10.1111/tpj.13599

[15] Brauer EK, Balcerzak M, Rocheleau H, Leug W, Schernthaner J, Subramanian R, et al. Enhanced reader. pdf. MPMI. 2020;**33**:553-560

[16] Zhang J, Zhang H, Li S, Li J, Yan L, Xia L. Increasing yield potential through manipulating of an ARE1 ortholog related to nitrogen use efficiency in wheat by CRISPR/Cas9. Journal of Integrative Plant Biology. 2021a;**63**(9):1649-1663. DOI: 10.1111/jipb.13151

[17] Wang B, Zhu L, Zhao B, Zhao Y, Xie Y, Zheng Z, et al. Development of a haploid-inducer mediated genome editing system for accelerating maize breeding. Molecular Plant. 2019a;**12**(4):597-602. DOI: 10.1016/j.molp.2019.03.006

[18] Wang W, Pan Q, Tian B, He F, Chen Y, Bai G, et al. Gene editing of the wheat homologs of TONNEAU1-recruiting motif encoding gene affects grain shape and weight in wheat. Plant Journal. 2019b;**100**(2):251-264. DOI: 10.1111/tpj.14440

[19] Smedley MA, Hayta S, Clarke M, Harwood WA. CRISPR-Cas9 based genome editing in wheat. Current Protocols. 2021;**1**(3):e65. DOI: 10.1002/cpz1.65

[20] Bhowmik P, Bilichak A. Advances in gene editing of haploid tissues in crops. Genes. 2021;**12**(9):1410. DOI: 10.3390/genes12091410

[21] Amin AY, Safwat G. Development of doubled haploid wheat genotypes using chromosome eliminating technique and assessment under salt stress. Journal of American Science. 2010;**6**(7):139-148

[22] Greenwood JR, Finnegan EJ, Watanabe N, Trevaskis B, Swain SM. New alleles of the wheat domestication gene Q reveal multiple roles in growth and reproductive development. Development (Cambridge). 2017; **144**(11):1959-1965. DOI: 10.1242/dev.146407

[23] Liu H, Wang K, Tang H, Gong Q, Du L, Pei X, et al. CRISPR/Cas9 editing of wheat TaQ genes alters spike morphogenesis and grain threshability. Journal of Genetics and Genomics. 2020;**47**(9):563-575. DOI: 10.1016/j.jgg.2020.08.004

Section 5

Wheat Quality

Chapter 15

Nutritional Quality of Wheat

Muhammad Javid Iqbal, Naureen Shams and Kalsoom Fatima

Abstract

The criteria of wheat quality are varied, which is suitable for one product may not have properties for another product. Wheat endosperm contains the proteins, carbohydrates, iron, and B-vitamins such as riboflavin and niacin. It also contains soluble fiber as well as trace minerals. Soluble fiber is considered to have health benefits that are not shared by insoluble part. It is the leading source of vegetal protein in human food, having a protein content of about 13%, relatively high as compared with other major cereals. Natural wheat has a number of medical properties, such as every component of the whole wheat grain contains elements that the person's body requires. Wheat comprises carbohydrates and gluten protein, which offer massive amounts of energy; inner bran coats, phosphates, and other mineral salts; and dietary fiber, which helps with bowel movements. Wheat protein and vitamins B and E aid to develop and rebuild muscle tissues. The wheat germs that are eliminated during the purification process are also high in important vitamin E, which could also lead to heart disease if not consumed. Constipation and other gastrointestinal disorders and nutritional diseases are common as a consequence of the lack of vitamins and minerals in refined wheat flour.

Keywords: wheat bran, nutritional quality, wheat varieties, gluten proteins, medicinal properties

1. Introduction

Wheat (*Triticum spp.*) is cereal crop that belong to the family *Poaceae* (order *Poales*). Wheat is a staple source of nutrients for around 40% of the world's population. Wheat has already been cultivated for millennia. Wheat was among the first cereals crop to be farmed, and that has been a staple diet throughout Europe, Western Asia, and Northern Africa for over 8000 years. This is most likely due to wheat's agricultural versatility, convenience of grain storage, and simplicity of flour conversion for a variety of cuisines. Wheat is probably the most frequently produced crop in the world, with over 218 million hectares under cultivation, and its global trade exceeds that of all other crops collectively. Wheat is a vital aspect of human diet, accounting for 20% of daily calories and protein. Wheat is indeed the second most powerful food crop in the undeveloped nations after rice in ensuring food security, with an estimated 80 million peasants depending on it for their survival [1].

1.1 Origin and evolution of wheat

About 10,000 years ago, the wheat was cultivated for first time, as part of the "Neolithic Revolution," which witnessed a shift from hunting and gathering to organized agricultural production. These early farmed wheat varieties were diploid (genome AA) (einkorn) and tetraploid (genome AABB) (emmer), and their hereditary linkages imply that they emerged in Egypt's south-eastern region. When hexaploidy bread wheat finally originated about 9000 years ago, cultivation had expanded to the Near East. Due to its high productivity and other quality parameters, landraces were selected by agriculturalists from wild inhabitants. It was considered as plainly nonscientific method of plant breeding. On the other hand, selective breeding or domestication was also responsible for genetic feature selection that distinguished them from their wild ancestors. Others have gone into further depth about the domestication disorder, but two characteristics stand out as particularly important. The first one is the failure of a spike to rupture as it reaches maturity, and this is main cause of seed loss during harvest. The non-shattering feature is regulated by mutations at the brittle rachis (Br) gene. It is a crucial trait for guaranteeing seed distribution in natural populations. The second important feature is a transition from hulled forms. In this condition, the glumes are securely bound to the grain, to free-threshing exposed forms. When a predominant mutation occurred at the gene Q, it improved the impact of recessive mutations at tenacious glume locus or Tg locus. All cultivated forms of diploid, tetraploid, and hexaploid wheat have a strong rachis, aside from the spelt bread wheat strain. Moreover, einkorn, emmer, and spelt really are hulled early cultivated varieties. On the other hand, modern tetraploid and hexaploid wheat varieties are free-threshing. Einkorn and emmer had been domesticated from natural populations. Bread wheat, despite einkorn and emmer, has been grown in agriculture for a long time. Bread wheat has been developed created by crossing cultivated emmer with an unrelated wild grass named *Triticum tauschii* (also called *Aegilops tauschii* and *Ae. squarrosa*). A novel hexaploid genome AABBDD was selected by the researchers due to their superior qualities, and this hybridization probably occurred numerous times separately. As compared with batter adapted species, the modern wheat species are unable to live in wild because during the adaption, genetic changes occurred. In the 1880s, John Bennet Lawes beautifully proved this by enabling a portion of Rothamsted's famed long-term Broadbalk experiment to revert to its original state. As a result, in 1882, he left a portion of the wheat crop unharvested and tracked its progress throughout the years. Weeds took over after a successful crop in 1883, and the few remnant wheat plants (spindly with little ears) were removed and photographed in 1885. The genomes of tetraploid and hexaploid wheat are closely related to wild and cultivated einkorn A genomes. It was investigated that the D genome of hexaploid wheat is obtained from T genome tauschii. The B genome of tetraploid and hexaploid wheat, on the other hand, is most likely derived from the S genome found in Aegilops' Sitopsis portion, with Aegilops speltoides being the closest existing species. Ae. speltoides' S genome is also the most similar to T. timopheevi's G genome, a tetraploid species possessing both the A and G genotypes [2].

1.2 Cultivated wheat today

Hexaploid bread wheat currently accounts for 95% of global wheat output, while for the leftover, 5% tetraploid durum wheat was taken into consideration. Durum wheat is batter adapted in dry Mediterranean climate than bread wheat. It is commonly known as pasta wheat. It is utilized to bake bread and also used in South Africa to prepare regional cuisines. These regional cuisines include couscous and bulgar. Faro is the Italian name for wheat varieties such as einkorn, emmer, and spelt. These wheat verities are still cultivated in small quantities in some areas, including as Spain, Turkey, the Balkans, and the Indian subcontinent [3].

1.3 Why has wheat been so successful?

Despite its recent origins, bread wheat has enough genetic variability to allow the formation of approximately 25,000 varieties that are adapted to a variety of temperate climates [4]. If enough water and mineral fertilizers are available, as well as good pest and pathogen control, yields can surpass 10 tonnes per hectare. This is why it became more favorable in temperate climates as compared with other crops. Wheat is harvested by two methods such as conventional method and mechanical combine harvesters. If pests and water content were controlled, then wheat can be stored properly. For wheat storage, moisture content should be less than 15% [2].

2. Types of wheat

Wheat is divided into three categories: species, commercial types, and growth habits. There are 16 species based on these, two commercial forms are available: one is bread (*Triticum aestivum*), and second one is macaroni (*Triticum durum*), and three growing patterns (winter habit wheat, spring wheat, and facultative wheat). Winter wheat lies dormant during a winter freeze.

2.1 Major cultivated species of wheat

*2.1.1 Bread wheat (*T. aestivum*)*

T. aestivum is known as bread wheat that produces about 95% of total wheat. *T. aestivum* (bread wheat L. 2n = 42, hexaploid, AABBDD genomes) is classified into hard wheat and soft wheat depending upon the grain hardness. Bread wheat is used as flour that is used in many types of food diversity and baked products.

*2.1.2 Pasta wheat or durum wheat (*T. durum*)*

T. turgidum *L. var. durum Desf.* (2n = 28, tetraploid, AABB genomes) accounts for roughly 35–40 million metric tonnes of total global production. This variety is adapted in hot and dry conditions. Mediterranean Sea includes the wheat-cultivated countries. This tetraploid variety of wheat is used to synthesize pasta so that it's called as pasta wheat. In the Middle Eastern countries, durum wheat is ground into the flour and used as feed, and grain grits are also used in the Saudi Arabia [5]. Other species of wheat is less important; these species are cultivated according to the demand of market.

These are the following:

Einkorn is diploid species.

Emmer is tetraploid variety.

Spelt is hexaploidy variety.

Spelt, emmer are different from other varieties because their grains are not obtained by threshing. Harvesting of grains is classified as properties of wheat according to marketing point of view. Purchasers' classification of wheat is different according to uses and cultivation and profit [1, 6]. For the sake of the marketplace, cultivated wheat grain that undergoes trade is categorized based on grain qualities. Wheat buyers have been using categorizations to help them decide which wheat to buy because each category has its own set of applications. With this technique, wheat growers may identify which wheat varieties are perhaps the most lucrative to cultivate.

2.2 Classes used in the United States are

2.2.1 Durum

Semolina flour is made from quite a rigid, transparent, luminous grain.

2.2.2 Hard red spring

Bread and firm caramelized items are made with a hard, brownish, protein rich wheat. Hard red springtime wheat is often used to make baking powder and high-gluten flours. The Minneapolis Grain Exchange is where it's mostly exchanged.

2.2.3 Hard red winter

Rigid brown and highly protein wheat grains are used for bread and bakery products. This flour is used as protein agent in the pie crusts. Hard red wheat is used as unbleached form in the market and traded by Kansas City board of trade.

2.2.4 Soft red winter

Cakes, biscuits, muffins, and pastries are made using a low-protein wheat. With bakery products, you can use cake flour, pastry flour, self-rising flour, and pastry flour. The Chicago Board of Trade is the market for it.

2.2.5 Hard white

Hard and light in color, not transparent, white, and less protein. This variety is dry and present in temperate grassland that is used for the bread and brewing.

2.2.6 Soft white

This variety is of white color and very soft containing less protein and grown in the moisture places. Soft white wheat is used for pie and pastry. This variety is very expensive in the market and has a great demand [7].

3. Uses of wheat

Wheat is a valuable source of carbs in most developed nations and globally protein source used for human food, animals used it as dietary fiber. It contains minerals, fat, and vitamins, which are source of micronutrients and dietary fiber. Meat-based diet is less important than the wheat-based diet [6, 8, 9]. It is fruitful for the heath, which is approved by EFSC; the wheat fiber is very important for the diarrhea glucose response and cholesterol control. The significance of wheat is shown by the EU framework 6 in the Healthgrain Program [1]. Dough is the sticky flour after water mixing that has viscoelastic characteristics [10]. Due to fermentation, it has some swelling problem used in the bun and bread and releases the carbon dioxide [11]. Gluten polymer is related to the coeliac disease, a chronic inflammation that affects the European countries. Respiratory system is also infected, and food allergies are related to wheat [12]. Other wheat-related disorders, such as asthmatic and food sensitivities, have been reported, prompting a lot of wheat investigation in the human healthcare profession. Wheat demand has been increasing over the world, particularly in countries where agricultural production is difficult. Consumption is increased worldwide, and production is affected by climate change. Wheat is also used for animal's feed. Some low-quality wheat is used in the industry to make glue, paper adhesive, and several other products such as alcohols [1].

4. Wheat gluten proteins and processing properties

Dough formation of wheat from wheat grains has been used in the formation of bread, biscuits, cakes, pasta, noodles, and pastry that make the wheat superior to other crops of template areas. The wheat dough has the ability to store the protein, which forms gluten.

Gluten is the best protein fraction that was discovered by chemist Baccari in 1728. This was prepared by washing the wheat and preparing the dough and adding the salt solution that contains cohesive mass, which has 75% protein and starch. The gluten protein is prepared in an essential pure form by simple method depending upon the characteristics. First, they are insoluble in water and soluble in alcohol that was called as prolamin. Second, the individual gluten protein has strong covalent bond force, which permits the gluten protein to separate as a cohesive mass. Gluten protein has high biological and chemical importance that can be discussed in the literature [2].

4.1 What is the origin of gluten?

In the endoplasmic reticulum, gluten releases the protein during the protein synthesis that was transferred to lumen of endoplasmic reticulum and stored. Storage protein follows the two routes:

Golgi-dependent route.

Golgi-independent route [13].

4.2 Nutritional contents

Wheat is essential for the health of people due to having a large number of diet contents and nutritional value. Its important can be guessed to see the developed countries that can use only bread, noodle, cakes, pastry, and lactogen. Carbohydrates 55% and 20% of food calories are present in the wheat grains. Carbohydrates 78%, protein 14%, fat 2%, minerals 2.5%, and vitamins such as thiamine and vitamin B, as well as minerals such as zinc and iron, selenium, and magnesium make up a small percentage of the diet [14–16]. Wheat has pericarp that is classified as true seed.

Protein is stored in the endosperm; the protein contents are about 72%. Wheat grains are also rich in pantothenic acid, riboflavin and some minerals, sugars, etc. The barn, which consists of pericarp Testa and aleurone, is also a dietary source for fiber, potassium, phosphorus, magnesium, calcium, and niacin in small quantities.

Wheat kernels are a treasure trove of nutrients important to human nutrition. Endosperm accounts for around 83% of the weight of the kernel and is the origin of white flour. The endosperm comprises the majority of the protein, carbohydrates, iron, and numerous B-complex vitamins such as riboflavin, niacin, and thiamine in the total kernel. Bran makes up roughly 14.5% of the weight of the kernel [17–20]. Bran can be present in the whole wheat flour and can also be purchased individually. Protein is present in small amounts in the bran and significant amounts of the B-complex vitamins described above, trace minerals, and indigestible cellulose fiber termed dietary flour, among the nutrients in the whole wheat. Wheat germ is the wheat kernel's embryo. High amount of protein, lipids, and numerous B vitamins is present in wheat germ and embryo [21]. Wheat germ is high in minerals and low in salt and cholesterol. It is high in vitamin E, magnesium (Mg), thiamin, pantothenic acid, niacin phosphorus, and zinc (Zn) as well as small amount of ubiquinone (ubiquinone) and PABA (para-aminobenzoic acid) are found in it [15, 19, 22].

Wheat germ is abundant in fiber, with 1 g per tablespoon. A high-fiber diet can help regulate immune function (i.e., reduce constipation) and may be suggested for individuals who are at risk for colon disease, heart disease, or diabetes.

5. Types of wheat flours and its uses

5.1 All-purpose flour

Endosperm of the wheat kernel is used for all-purpose flour, and it is separated from barn and germ during grinding process. Hard wheat is used for the manufacturing of all-purpose flour. It can be prepared by using a blend of soft and hard wheat. So, it can be used for varieties of baked food such as noodles, cookies, cakes, pastries, and yeast breads.

Enriched all-purpose flour is fortified with iron and B vitamins in proportions that are equivalent to or greater than whole wheat flour. Enriched bleached all-purpose chlorine is added to flour to help it develop, shape the gluten, and enhance the cooking condition. Despite the fact that chlorine somehow does not kill the nutrients, it does reduce the risk of degradation or contamination. Enriched unbleached all-purpose flour is off-white in appearance after being discolored by oxygen in the atmosphere during the aging. Unbleached and bleached flour are similar according to the maturational value.

5.2 Bread flour

From the endosperm of the wheat kernel, bread flour is obtained. It is mostly used by commercial bakers and major retailers. It has a higher protein concentration than all-purpose flour. It is typically used for breadmaking process.

5.3 Self-rising flour

When salt and baking powder were added in all-purpose flour then, this flour is known as self-rising flour. It's one cup containing 11/2 teaspoons baking powder and

1/2 teaspoon salt. If you modify the salt and baking powder quantities, you can use self-rising flour instead of all-purpose flour in a recipe.

5.4 Whole wheat flour

Whole wheat flour has a rough appearance and contains endosperm of the wheat kernel, germ, and barn. Gluten development is slowed when bran is present. Whole wheat flour–based baked goods are often heavier and thicker than white flour–based goods.

6. Other flours

6.1 Cake flour

Soft wheat flour was used in the milling process. Cakes, cookies, crackers, and pastries are all good candidates. Gluten-free and low in nutritional value.

6.2 Pastry flour

Wheat flour that is gluten-free and silky smooth and used in the milling process. Protein content is comparable to cake flour; however, carbohydrate content is lower.

6.3 Gluten flour

Bakers use it in conjunction with low-protein flours to improve baking efficiency and generate massive gluten bread.

6.4 Durum flour

Produced as a by-product of semolina production. Used to create commercial noodles in the United States.

6.5 Farina

Hard wheat endosperm that has been coarsely mashed. Many breakfast cereals in the United States contain this component. It's also used to make low-cost pasta with a low saturated fat content. There is no cholesterol, low sodium, and sugar-free, but it contained high Mn, P, and dietary fibers [23].

7. Medicinal properties of wheat

Wheat has various therapeutic properties when grown naturally. The nutrients that are required by human body all are present in the whole wheat grain. Wheat provides heat and energy through starch and gluten. Phosphates and other mineral salts are found in the inner bran coatings. The outer bran provides much-needed supplements and the nondigestible component that facilitates digestion. Wheat germ, vitamins B and E, and protein assist the development and muscle recovery. Wheat germ, which is extracted during the purification process, is high in important vitamin

E, which can cause heart disease if not consumed. Constipation and other gastro-intestinal problems and nutritious diseases have become more common as a result of the lack of vitamins and minerals in refined wheat flour. Whole wheat protects against various illnesses such as constipation, heart disease, diverticulum, obesity, appendicitis, ischemia, and diabetes [24]. There have been several claims linking wheat, specifically wheat gluten, to a variety of medical disorders, extending from unlikely tales in the mainstream media to science research [25]. Autoimmune illnesses such as rheumatoid arthritis, which could be more common among celiac patients and family, are among them [26]. It may be easier to imagine pathways for links between disorders with a comparable immune background than it is to explain the well-known link across wheat, celiac disease, and schizophrenia. Others with sporadic idiopathic ataxia (gluten ataxia), migraines, acute psychoses, and a variety of neurological disorders have been documented [23].

The phenolic acid cross-linking may limit the health benefits of soluble fiber, which are not shared by insoluble fiber. Insoluble fiber, on the other hand, may help transfer phenolic antioxidants to the intestine, potentially lowering the risk of colorectal cancer [17, 27, 28]. Some doctors prescribe a gluten-free, casein-free diet because of the link with autism. Some of these effects are immune-mediated, while non-immune-mediated effects are extremely impossible to articulate and evaluate. They could be caused by the release of bioactive peptides. These peptides are produced from gluten protein. Gluten is an important source of large number of peptides such as opioid peptides [28] as well as an angiotensin-converting enzyme inhibitor [29].

7.1 Wheat bran

Use of wheat barn as dietary fibers helps in the prevention of various gastric and digestive ailments. Some of these are cancer of colon, intestinal cancer, irritable bowel syndrome (IBS). In addition to these, wheat barn also aids in diminishing the risk of hemorrhoid and hiatal hernia, hypertension, breast cancer, hypercholesterolemia, gallbladder diseases, and type 2 diabetes [24, 30]. Being rich in iron and phosphorus, it helps in easing the consumption by increasing stool output and bowel frequency. It has a lot of fiber on the exterior, which helps to balance nutrient uptake and excretion.

7.2 Wheat germ

Wheat germ is a rich source of a number of vitamins and minerals, which has increased its employment in both skin care lines as well as for persons of all ages as a source of nourishment. Its antibilious, antihidrotic, antivenous, vitamin E and minerals such as Zn, Fe, Cu and Mg. Its oil extract is also a huge source of vitamins such as E, D, and A, protein contents, and lecithin. In the vast field of skin care, germ oil finds its employment as anti-skin irritant and alleviates skin dryness and cracking. Due to its antioxidant properties, it finds wide application as carrier oil in number of products. Its external application to the skin increases and improves the circulation of blood, which in turn helps repair the skin damaged by harmful rays of the sun. Germ oil also aids in the prevention of dermatitis. The oil extracted from wheat germ has a shelf life of near about 6–8 months. Wheat germ also holds immense nutritional values as it contains fatty acids vital for healthy growth of the body. They form 3% by weight of the grain but contain only 26% of vitamins, proteins, and minerals.

Being rich in various vitamins and minerals, it gives very good results even with lesser amount and is generally added as carrier oil [23].

7.3 Wheat stem, fruit, and seed

Different parts of the wheat have different nutritional and medicinal properties. Employment of young stems of wheat in curing of biliousness and intoxication is widely known. Removal of skin blemishes is done by using ash. Wheat fruits have antipyretic and sedative properties. The light grain has antihidrotic properties. It treats the night sweats and spontaneous sweating.

Sex harmones in the seed are used in China and are believed to increase the fertility of the female. The seed sprouts have antibilious and vinous and constructive characteristics. Diseases such as malaise, sore throat, abdominal coldness and spasms, constipation, and cough can be treated using the seed sprouts of wheat. The plant of wheat is also believed to contain anticancer properties [18].

8. Ways to treat some common aliments

8.1 Internal rejuvenation

The 8% protein content of wheat comprising eight essential amino acids has phenomenal rejuvenation properties. The essential amino acids are present in a perfect and delicate percentage and offer amazing healing effects ranging from skin to muscular and all other organ systems. Firstly, the protein content of the wheat is metabolized into amino acids. This nourishes the heart and lungs, healthy skin and hair, tendons and ligaments, brain, central nervous system, and glandular network, as well as forming durable muscles and clearer vision. The energy and nutritive benefit that come with the wheat are because of B-complex vitamins, especially thiamin, riboflavin, and niacin, rejuvenating the skin and circulatory system. In addition to all these nutritional benefits of wheat, it also nourishes the hormonal system, which in turn helps in healing the wounds and regulating the blood pressure. Wheat contains nutrients that are important for maintaining internal water equilibrium, such as Fe, P, and K. Wheat thus aids in the restoration of internal balance [21, 27].

8.2 Tooth disorders

As chewing wheat consumes some time, it helps in exercising the teeth gems, which acts as a facial exercise too. When eaten along with other food items, it promotes the chewing of other food too, which eases up the process of digestion. This juice of wheat acts to relieve sore throat and pyorrhea. Chewing wheat grass cleanses and draws the toxins and bacterial growth in the gums, thus preventing tooth decay and tooth aches.

8.3 Constipation

In milling process of flour, bran of wheat is considered as a by-product and is complete in nutrient values as compared with flour. It has qualities of being best agent to treat constipation. Fruits and vegetables are also taken for this purpose but are less effective than bran. Bran is concentrated in cellulose, which exists in massive form in the intestine and has a function to inhibit as well as to cure constipation. It has efficiency in removal of constipation by continuous elastic contraction and relaxation of the intestine.

8.4 Skin diseases

Chlorophyll inhibits bacterial growth, which causes diseases in organisms. Normal activity and growth of cells are important for a healthy skin. If growth became abnormal or affected because of bacterial cells, then wheat grass therapy is used to stop the bacterial cells nourishment. In this therapy, intake of wheat grass juice is recommended. This flour is used to cure wounds and ulcer. A paste of wheat flour, vinegar by boiling these together is used if a surface burns or itching occurs. It also acts as sterilizer.

8.5 Digestive system disorders

There are many disorders in digestive system and tracts. Wheat grass juice is used to cure these disorders. Enema is given to get rid of constipation. For this purpose, first of all, neem water is given, then 90–120 ml wheat grass juice is given after 20 minutes. This will treat sickness in colon, ulcerative colitis, and mucous.

8.6 Circulatory disorders

For treatment of circulatory diseases, wheat grass juice is taken. Wheat is rich in chlorophyll and improves working of the lungs and heart as chlorophyll is rich in hemoglobin (iron). It will lower down the effect of CO_2. Because of these advantages, wheat grass juice is of immense importance [31].

8.7 Wheat for treating boils

For the treatment of boils, which have pus in them, formerly, this was done by surgeon's knife. But now wheat flour is used. Alse (from shop) is grounded in powder form and added in fried wheat flour. Add one table spoon of water in it. Allow this mixture to thick with continuous stirring. Allow it to cool down. Bandage was made by this paste and bound on the boil. This will prove to be beneficial and provide relief. Warm water is used to wash the boil. Ointment is used on daily basis. Clean it on daily basis.

8.8 Wheat for treating scars

Wheat flour for treating scars is commonly in use. Grounded paste of roasted wheat flour is made. Oil is extracted from this paste by pressing paste between a thin cloth. This oil treats itching.

8.9 Wheat for curing chest pain

Wheat flour is used for curing chest pain. Paste is made by heating wheat, barn, and coarse salt. Paste is placed on a bandage and rubbed on the chest. This will provide relief.

8.10 Wheat for tonsil pain

Wheat flour is used for treating tonsil pain. This is treated by making a halwa by heating wheat flour and water. This paste is placed in a bandage and placed on a tonsil.

8.11 Wheat for treating acne or pimples

Using whole wheat flour, make a fine paste. On pimples, apply this paste after being kept 1 h in the refrigerator. After that, simply wash it away. Make a habit of it.

A meta-analysis of data from much more than 30 well-designed animal studies examining the anticancer properties of wheat bran, the section of the grain with the largest amount of the insoluble dietary fiber's cellulose and lignin, was published in 1998 by scientists at Wayne State University in Detroit. They concluded that animals-provided wheat bran had a 32% lower risk of colon cancer, and they want to do a meta-analysis of human research to confirm their findings. Wheat bran (WB) proved to be more important than oat or maize bran at controlling colon tumorigenesis.

9. Other uses of wheat

Straw can be used for a variety of purposes, including biofuel, thatched roofs, and garden mulch. Paper is made from a fiber collected from the stems. After the seed has been gathered, the stems are split into usable pieces and steeped for 24 h in clear water. The fibers produce a tan-green paper. Laundry, resizing textiles, and other uses for the seed's starch. Chappatis are a popular wheat-based dish in India, Pakistan, and Iran. Dalia would be a whole flour that is used to make them. Wheat in the crushed form known as Dalia is particularly nutritious. In the past, it was a highly popular Indian meal. It's made by immersing two tablespoons of crushed or shredded wheat in water for 30 minutes and then slowly boiling it until the water almost evaporates. After that, to taste, add milk and honey. It's a healthy breakfast option. However, fresh data from the long-running Nurses' Health Study at Brigham Women's Hospital/Harvard University School of Public Health discovered that women who ate a high-fiber diet had the same risk of colon cancer as those who ate a low-fiber diet in early 1999. Investigators are expecting confirmation proof before modifying dietary requirements because this study conflicts hundreds of others performed over the previous 40 years [23].

A meta-analysis of findings from more than 30 well-designed animal experiments examining the anticancer impacts of wheat bran, the component of the wheat with the greatest percentage of the insoluble dietary fiber's cellulose and lignin, was published in 1998 by scientists at Wayne State University in Chicago. They discovered a 32% significantly lower risk of colorectal cancer in animals fed wheat bran, and they want to do a meta-analysis of research involving human subjects to confirm their findings. Wheat bran is abundant in the whole flours. Wheat bran (WB) proved to be more effective than oatmeal or maize bran at inhibiting colorectal cancer. According to Liu et al. (2021), overconsumption of whole grains and dietary fiber has been linked to a lower incidence of liver cancer and chronic liver disease mortality. According to the researchers, primary liver cancer is the most common malignancy and the third largest cause of cancer-related death globally, and its number of fatalities has been on the rise in the United States. Liver cancer 5-year survival rates increased from 11.7 to 21.3% between 2000 and 2011, illustrating the importance of prevention and treatment for this lethal illness [31].

10. Chemical composition and nutritional quality of wheat

Wheat is one of the most extensively grown staple foods on the planet. Wheat grains may be treated into semolina, flours, and other products, which makes it very essential component for employing its nutrients in numerous of meal items including bread, pasta, and other bakery products. The granules of different wheat kinds come in many different shapes. They are long, cylindrical, and narrowly flattened, although they have always been thought to be oval in shape. The grain is 5–9 mm long, weighs 35–50 mg, and also has a wrinkle down one side where it has been formerly linked to the wheat flower. It contains 13–17% bran, 2–3% germ, and 80–85% floury endosperm. Barn covers the grain's inner half and is made up of multiple layers of cells rich in minerals and vitamin B. Since many bran fibers are insoluble in water, it can preserve the interior region. Pentosans, cellulose, and polymers based on arabinose and xylose, all of which have been firmly bound to proteins, constitute this fiber's chemical content. Bran dry matter contains minerals, proteins, and carbohydrates 72% and 16%, respectively. Certain amino acids in the outer layer of endosperm and flour are in remarkably different proportions. Glutamine and proline concentrations are about half; however, arginine concentrations are tripled, and alanine, asparagine, glycine, histidine, and lysine levels are all doubled of what they have been in wheat flour. Aleurone has the largest concentration of proteins and enzymes; these enzymes and proteins are required for germination. It is also high in dietary fibers and contains fats, proteins along with ash 1.5%, 13% and 0.5%, respectively. It also has a high concentration of vitamin E. The valuable nutritional contents of the edible wheat have been listed in **Table 1**.

10.1 Protein composition of wheat

Wheat contains between 10% and 18% protein by dry weight basis. The distinct protein fractions that can be isolated from powdered wheat include albumins, glutenins, globulins, and gliadins. Albumin is water-soluble, glutenins are soluble in dilute acid or NaOH solutions, globulins are insoluble in water but soluble in a dilute NaCl solution, and gliadins are soluble in 70% ethyl alcohol. Albumins and globulins are prominent in seed coats, aleurone cells, and germs. Gliadins and glutenins are storage proteins make up for 75%, while globulins and albumins only comprise 25% of overall amino acid composition. The storage proteins such as gliadins and glutenins are absent in seed coat layers and germ but occur abundantly in the mealy endosperm.

10.2 Carbohydrate composition of wheat

In most cases, the polymerization of glucose monomers culminates in starch production, and this starch is stored energy form of cereals. Amylose and amylopectin were discovered to be two different forms of polymers based on their chemical composition. Amylose has a mostly linear structure that results from the 1,4-linkage of monomer units comprising 1000–5000 glucose units. Amylopectin is branched polymer, which contained 20–25 glucose monomers in each chain. Under normal

S#	Nutritional value of wheat	Units	White wheat flour	Whole grain flour
1	Energy	kcal	364	340
2	Protein	gm	10.3	13.2
3	Total fat	gm	1.0	2.5
4	Carbohydrates	gm	73.6	61.3
5	Fiber	gm	2.7	10.7
6	Calcium	mg	15	34
7	Iron	mg	1.2	3.6
8	Magnesium	mg	22	137
9	Phosphorus	mg	108	357
10	Potassium	mg	107	363
11	Sodium	mg	2.0	2.0
12	Zinc	mg	0.7	2.6
13	Thiamin (B1)	mg	0.1	0.5
14	Riboflavin (B2)	mg	0.04	0.2
15	Niacin (B3)	mg	1.3	5
16	Vit. B6	mg	0.04	0.4
17	Folate	DFE	26	44

Table 1.
*Nutritional facts of wheat (*USDA datasheet).*

circumstances, wheat grain comprises approximately 20–30% amylose and 70–80% amylopectin. Starch makes between 60 and 75% of the dry weight of wheat. Wheat seeds come in two sizes: one is small (5–8 m), which is spherical in shape; and second one is large (25–40 m), which is lenticular in shape; and they develop 15 and 10–30 days after pollination, respectively. The smaller one takes up approximately 80% of the casing.

10.3 Fatty acid composition of wheat

Fatty acid (FA) synthesis rates in ripe wheat seeds fluctuate. Wheat requires the chemical component for lipid synthesis known as acetyl coenzyme A. Glycerides, phospholipids sphingosine, waxes, and the isoprenoid series are examples of synthetic compounds. Underdevelopment of malonyl-CoA, NADPH, dehydration, and condensation produce palmitic acid, which is then converted to stearic acid via a different mechanism. The germ contains the highest proportion of lipids 11%, while the endosperm's bran, proteins, and starch also have significant amounts. Majority of the binding lipids are phosphatidyl choline, phosphatidyl ethanolamine, phosphatidyl serine, and lysophosphatidic derivatives because they contain free -OH group on glycerol monomers-sitosterol, campestral, and saturated sterols C28 and C29 are the most common sterols. According to studies, the three fractions have a high level of linoleate (C18:2) and lesser amounts of palmitate (C16:0) in total lipids as well as triglycerides. It's composed of indigestible lignin and plant polysaccharide components in the human gastrointestinal tract.

10.4 Soluble dietary fibers

Pectic compounds and hydrocolloids are components that are soluble in water.

10.5 Insoluble dietary fibers

Components including cellulose, hemicellulose, and lignin are water-insoluble, and wheat grains are rich in insoluble fiber. Arabinoxylan is a type of insoluble fiber, which is regarded to be a suitable site for the fermenting of short-chain fatty acids (SCFAs), especially butyrate. Because butyrate fermentation took place in the colon, we might conclude that there is an abundance of butyrate in the colon. Butyrate is hypothesized to promote bowel health and lower cancer risk through a range of techniques when found in excessive amount in the colon. Dietary fiber consumption has numerous benefits, including prevention against cardiovascular disorders, serum cholesterol stabilization, glucose uptake and increased insulin modulation, and prevention of constipation and diverticular disease. The increased awareness of the possible health advantages of high-fiber diets has inspired a spike in interest in whole-grain and bran bread consumption. Phytates are potentially hazardous compounds present in high quantity in whole wheat flours. The bulk of inorganic phosphorus (Pi) is stored as phytate in mature grain seeds, which produces complexes with minerals such as Ca^{2+}, Fe^{3+}, Zn^{2+}, and Mg^{2+}, lowering their bioavailability. Soluble fibers such as (13,14)-D-glucan (also known as -glucan) have been shown to have immunostimulatory characteristics and to affect glycemic, insulin, and lipid reactions to meals.

11. Micronutrient malnutrition

Micronutrient deficiency raises morbidity and mortality rates, lowers productivity levels, stymies national development efforts, leads to persistently high rates of population growth, and lowers the livelihood and standard of living for all those affected.

11.1 Zn deficiency

Micronutrient deficiency in Zn impacts both agriculture and individuals. Zinc deficiency is now widely recognized as a severe health risk factor and a major cause of mortality.

Zinc insufficiency is consistently ranked among the 20 most important aspects in the world, and fifth among the 10 most important concerns in developing nations, according to a WHO evaluation of the risk factors for the development of illnesses and disorders. Zinc deficiency is linked to a number of major health issues, involving delays in muscle hypertrophy, immune function, and level of academic achievement, as well as an increased risk of illness, DNA damage, and cancer development.

11.2 Fe deficiency

Fe deficiency has been associated to an increased risk of tissue hypoxia and heart failure in young newborns and pregnant women, both of which can result in death. The bulk of maternal deaths during childbirth is assumed to be caused by maternal anemia, which is aggravated by blood loss during labor, and 20% of all maternal deaths are caused by maternal anemia. Babies born to iron-deficient mothers are frequently dwarfed and ill, and attention problems, poor fine motor skills, and loss of memory in children are all causes of Fe deficiency. Pregnancy-related iron insufficiency has been connected to permanent brain damage in the fetus as well as irreversible cognitive growth in their children. Iron deficiency in pregnant women has been associated to premature and low birth weight, which can lead to significant issues such as immunologic malfunction and development failure.

11.3 Fe and Zn contents in wheat

In whole grain wheat germplasm, the Fe and Zn percentages are substantial, as well as the effects of the climate on these values were studied. According to reports and surveys, Zn concentrations in various nations range between 20 and 35 mg kg^1. The majority of the seed-Zn is found in the embryonic and aleurone layer such as 150 mg kg^1, although the endosperm contains just a small amount such as 15 mg kg^1. Fe levels ranged from 28.8 to 56.5 mg kg^1 in wheat grain from plants grown in Mexico in 1994. Clearly, wheat germplasm has considerable genetic variability to considerably raise Fe and Zn contents in wheat grain.

11.4 Se contents in wheat

Because of its antioxidant, anticancer, and antiviral characteristics, selenium (Se) is a necessary component for humans and other species. According to one study, soils are typically deficient in bioavailable Se, resulting in Se deficiency in many countries' food production systems. Despite the fact that Se content in wheat grain fluctuates significantly, spanning from 0.02 to 0.60 mg kg1 for most of the world's wheat, an Australian study found that wheat accounted for almost half of most people's Se intake. Selenite is a mineral that is totally soluble and easily absorbed by plants, makes up the majority of Se in alkaline soils while in acidic soils, Se is usually found as insoluble selenides and elemental Se.

Selenites, which are completely miscible and quickly absorbed by plants, make up the majority of Se in alkaline soils. In acidic, poorly oxygenated soils, Se is usually found as insoluble selenides and elemental Se [32, 33].

12. Conclusion

Wheat is enjoyed by consumers around the globe in a wide variety of most suitable products such as breads, biscuits, cookies, cakes, noodles, breakfast cereals, etc. Many different types of bakery formulations have been developed in different regions of the world based on the traditional food habits of the people. The behavior of wheat flour dough under mechanical manipulation and the quality of the completed product are both influenced by the rheological qualities of wheat. As the baking industry gets increasingly mechanized, understanding rheological behavior and dough qualities becomes very critical. Water, sugar, yeast, oxidizing and reducing agents, and emulsifiers are some of the key ingredients used in bakery products to improve dough handling, taste, and life span. Nutritional quality of wheat is much important as it is basic staple food for the masses. Nutritional attributes play a vital role toward the health status of the consuming population, which ultimately affect the economy of the nation.

Author details

Muhammad Javid Iqbal[1*], Naureen Shams[2] and Kalsoom Fatima[1]

1 Department of Chemistry, COMSATS University Islamabad, Lahore, Pakistan

2 Department of Education, University of Agriculture, Faisalabad, Pakistan

*Address all correspondence to: javidiqbal@cuilahore.edu.pk; imjavid@gmail.com

References

[1] Giraldo P et al. Worldwide research trends on wheat and barley: A bibliometric comparative analysis. Agronomy. 2019;**9**(7):352

[2] Shewry PR. Wheat. Journal of Experimental Botany. 2009;**60**(6): 1537-1553

[3] Padulosi S. Hulled Wheats: Proceedings of the First International Workshop on Hulled Wheats, 21-22 July 1995, Castelvecchio Pascoli, Tuscany, Italy. Vol. 4. Roma RM, Italy: Bioversity International; 1996

[4] Feldman M. Wheats. Evolution of crop plants. Journal of Experimental Botany. 1995;**60**(6):1537-1553

[5] Peña R. Wheat for bread and other foods. In: Bread Wheat Improvement and Production. Rome: Food and Agriculture Organization of the United Nations; 2002. pp. 483-542

[6] Shewry PR, Hey SJ. The contribution of wheat to human diet and health. Food and Energy Security. 2015;**4**(3):178-202

[7] Gooding MJ. The Wheat Crop. Wheat: Chemistry and Technology. 4th ed. Amsterdam, Netherlands: Elsevier; 2009. pp. 19-49

[8] Sarwar MH et al. The importance of cereals (Poaceae: Gramineae) nutrition in human health: A review. Journal of Cereals and Oilseeds. 2013;**4**(3):32-35

[9] Lafiandra D, Riccardi G, Shewry PR. Improving cereal grain carbohydrates for diet and health. Journal of Cereal Science. 2014;**59**(3):312-326

[10] Orth R, Shellenberger J. Origin, production, and utilization of wheat. Wheat Chemistry and Technology. 1988;**3**:1-14

[11] Hoseney RC, Rogers DE. The formation and properties of wheat flour doughs. Critical Reviews in Food Science & Nutrition. 1990;**29**(2):73-93

[12] Tye-Din JA, Galipeau HJ, Agardh D. Celiac disease: A review of current concepts in pathogenesis, prevention, and novel therapies. Frontiers in Pediatrics. 2018;**6**:350

[13] Kumamaru, T., et al., Protein body biogenesis in cereal endosperms, in Endosperm. 2007, Springer. Berlin/ Heidelberg, Germany p. 141-158.

[14] Abdel-Aal E-S, Sosulski F, Hucl P. Origins, characteristics, and potentials of ancient wheats. Cereal Foods World. 1998;**43**(9):708-715

[15] Shewry PR et al. Comparative field performance over 3 years and two sites of transgenic wheat lines expressing HMW subunit transgenes. Theoretical and Applied Genetics. 2006;**113**(1):128-136

[16] Topping D. Cereal complex carbohydrates and their contribution to human health. Journal of Cereal Science. 2007;**46**(3):220-229

[17] Blechl A et al. Transgenic wheats with elevated levels of Dx5 and/or Dy10 high-molecular-weight glutenin subunits yield doughs with increased mixing strength and tolerance. Journal of Cereal Science. 2007;**45**(2):172-183

[18] Drankhan K et al. Antitumor activity of wheats with high orthophenolic content. Nutrition and Cancer. 2003;**47**(2):188-194

[19] Shewry Jones P. HD, transgenic wheat: Where do we stand after the first

12 years? Annals of Applied Biology. 2005;**147**(1):1-14

[20] Uauy C et al. A NAC gene regulating senescence improves grain protein, zinc, and iron content in wheat. Science. 2006;**314**(5803):1298-1301

[21] Adams ML et al. Evidence of low selenium concentrations in UK bread-making wheat grain. Journal of the Science of Food and Agriculture. 2002;**82**(10):1160-1165

[22] Shewry P. The HEALTHGRAIN Programme Opens New Opportunities for Improving Wheat for Nutrition and Health. Hoboken, New Jersey, United States: Wiley Online Library; 2009

[23] Kumar P et al. Nutritional contents and medicinal properties of wheat: A review. Life Sciences and Medicine Research. 2011;**22**(1):1-10

[24] Hadjivassiliou M et al. Gluten ataxia in perspective: Epidemiology, genetic susceptibility and clinical characteristics. Brain. 2003;**126**(3):685-691

[25] Carter JW, Madl R, Padula F. Wheat antioxidants suppress intestinal tumor activity in min mice. Nutrition Research. 2006;**26**(1):33-38

[26] Neuhausen SL et al. Co-occurrence of celiac disease and other autoimmune diseases in celiacs and their first-degree relatives. Journal of Autoimmunity. 2008;**31**(2):160-165

[27] Alvarez M et al. Silencing of HMW glutenins in transgenic wheat expressing extra HMW subunits. Theoretical and Applied Genetics. 2000;**100**(2):319-327

[28] Yoshikawa M, Takahashi M, Yang S. Delta opioid peptides derived from plant proteins. Current Pharmaceutical Design. 2003;**9**(16):1325-1330

[29] Motoi H, Kodama T. Isolation and characterization of angiotensin I-converting enzyme inhibitory peptides from wheat gliadin hydrolysate. Food/Nahrung. 2003;**47**(5):354-358

[30] Garvin DF, Welch RM, Finley JW. Historical shifts in the seed mineral micronutrient concentration of US hard red winter wheat germplasm. Journal of the Science of Food and Agriculture. 2006;**86**(13):2213-2220

[31] Wang L et al. Two new species and one new record of Hyalopsora (Pucciniastraceae) on ferns in China. Phytotaxa. 2021;**527**(1):41-50

[32] Irge DD. Chemical Composition and Nutritional Quality of Wheat, Teff (Eragrostis tef (Zucc) Trotter), Barley (Hordeum vulgare L.) and Rice (Oryza sativa)-A Review. Food Science and Quality Management. 2017;**59**:6-12

[33] Šramková Z, Gregová E, Šturdík E. Chemical composition and nutritional quality of wheat grain. Acta Chimica Slovaca. 2009;**2**(1):115-138

Chapter 16

Beyond Bread and Beer: Value-Added Products from Wheat

Timothy J. Tse, Farley Chicilo, Daniel J. Wiens and Martin J.T. Reaney

Abstract

Although wheat (*Triticum aestivum*) and related cereals [Barley (*Hordeum vulgare*), Rye (*Secale cereale*) are primarily used for producing baked goods and beverages, cereal crops can be used to create many value-added goods beyond these traditional products. Fractionation of cereal grains and extraction of valuable phytochemicals allows greater access to materials for use in food additives and nutritional supplements. Fermentation for beverage and fuel bioethanol production results in not only renewable fuel, but also a range of other coproducts, including nootropics. In addition to traditional grain fermentation, straw fermentation is also discussed, which further utilizes the whole plant. The main by-product of cereal grain fermentation, wheat stillage, can undergo a range of processes to enhance its value as a animal feeds, as well as extraction of useful compounds. These methods provide a glimpse of the many sequential and divergent processes that may bring us closer to realizing the full potential of wheat and related cereal grains.

Keywords: added-value products, wheat, fermentation, bioactive compounds, phytochemicals bioethanol, fractionation, protein

1. Introduction

Wheat is the world's second most produced grain [1], with global production forecasted to reach a new record of 780 million tonnes for 2021–2022 [2]. Wheat end-uses include food products (e.g., bread, pasta) [1], other consumer goods (e.g., hair products, skin care, cleaning agents) [1], and industrial applications (e.g., renewable fuels) [3]. In Canada, wheat varieties are often grouped by their functional properties and are categorized as Western Canadian or Eastern Canadian varieties, depending on the regions where they are grown [4]. These varieties are often used for the food and feed industry, although those with high starch and lower protein content are typically used as animal feed or for biofuel production [4, 5].

Opportunities to increase the value of wheat start while it is still growing, as limited grazing can allow the wheat to serve as both fresh feed for cattle and produce a harvestable crop at the end of the season [6]. Under suitable climatic conditions (i.e., low early-season rainfall) grazing by domestic animals can even improve crop yields by reducing lodging and improving the crops' response to late-season rain [7]. After harvest, the most common processes for adding value to wheat are food related.

In areas with high wheat production, flour mills can use the largest portion of available wheat by far, with most of the remaining wheat going towards production of breakfast food, pet food, and feed for livestock [8, 9].

In addition to traditional uses of wheat, fractionation of whole wheat grains can add considerable value. It is predicted that by 2024 the wheat starch market will exceed $4 Billion (USD) owing to significant use of wheat in agriculture-based industries [10]. The extraction of starch from wheat is an involved process requiring steeping and degermination [10]. After fractionation, bioactive compounds can be extracted and concentrated from individual wheat fractions, such as vitamin E from the germ and bran layers [11].

In North America, wheat is one of the predominant feedstocks for starch-based bioethanol production, especially where this cereal crop is locally available and abundant [12]. Wheat bioethanol is a renewable source of fuel, and its use can reduce greenhouse gas emissions when used to displace petroleum-based gasoline. Several valuable co-products can also arise from ethanol fermentation, including nootropic compounds, organic acids, glycerol, and a variety of fusel alcohols [13].

The recovery of grain by-products also holds crucial opportunities for wheat valorization, as waste products such as wheat straw also have value. Historically, wheat straw is incorporated into soil after harvest to improve soil quality, or removed and used as a component of animal feed or building material [14, 15]. Straw has also been investigated as a resource for liquid biofuel production (e.g., ethanol, butanol) [16, 17].

After alcoholic fermentation, additional processing can increase the total added value by utilizing the leftover stillage. Although fermentation consumes most available starches and sugars, proteins from both the wheat and yeast are left in the stillage after the ethanol is evaporated from the mash [18]. This protein-rich mixture can be integrated in animal feeds, as well as have valuable components removed to produce industrial chemicals [19].

In this chapter, we review value-added processes and products that go beyond the use of wheat for flours and beverages. These processes include fractionating whole wheat grain, extracting bioactive phytochemicals, fermenting both milled wheat and wheat straw for biofuel production, as well as the use of the resulting byproducts from wheat fermentation. Together, these processes provide a multitude of paths towards enhancing the total value of wheat.

2. Wheat fractionation

Mature cereal grain like those from wheat, rye, and barley are primarily composed of starch, protein, and cell wall polysaccharide (**Figure 1**). Typically, these materials account for 90% of grain dry weight [20]. Many bioactive components of wheat grains (e.g., polyphenols, phytic acid, phenolics, and minerals) are concentrated in the bran [21–23], specifically the aleurone layer [24]. The aleurone layer contains the highest amounts of bioactive compounds with antioxidant activity [25–28]. Additionally, wheat germ is also a source of value-added compounds including vitamin E (e.g., tocopherols) and oil.

Wheat bran obtained by milling constitutes about 15% of the mass of milled grain, and is composed primarily of the outer pericarp, inner pericarp, testa, hyaline layer, embryo, aleurone layer, and residual starchy endosperm (**Figure 1**) [20–22].

Starch accounts for ~60–70% of the mass of wheat grain [29] and is primarily composed of amylose and amylopectin [30], making it suitable for a wide variety of industrial applications, food products, and other consumer goods.

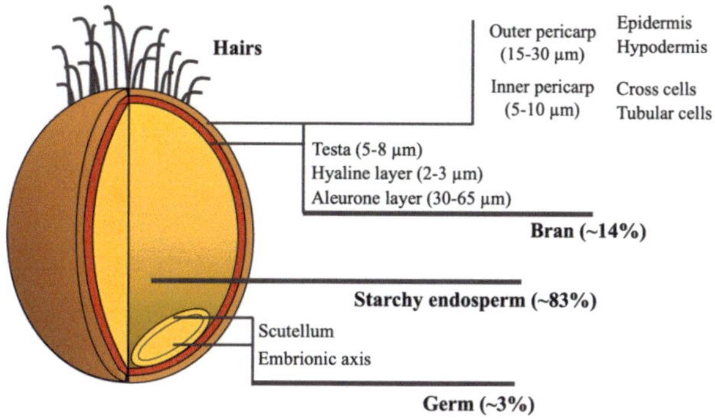

Figure 1.
Anatomy of wheat grain.

After milling, the endosperm is the primary product that makes up white flour (refined flour), while whole wheat flour (whole-grain flour) consists of the bran, germ, and endosperm [31]. Compared to white flour, whole wheat flour is richer in vitamins, phenolic acids, and minerals, as the whole kernel is included [31]. Meanwhile, due to the absence of bran, germ, and the aleurone components, white flour is more suited for use in baked food products that require leavening [31]. Further fractionation and milling processes can be used to isolate valuable bioactive phytochemicals.

Protein fractionation and extraction is another method of wheat valorization. Protein content can be influenced by the application of nitrogen fertilizer, which can result in an increased proportion of gliadin proteins, thus affording wheat that produces dough with increased extensibility [32]. Proteins make up to 7–22% of wheat grain dry weight [33], although it is unevenly distributed in the grain. For example, 5.1% of protein has been reported for the pericarp, 5.7% for the testa, 22.8% for the aleurone, and 34.1% for the germ [34].

Solvent fractionation can produce several protein concentrates with unique properties. Wheat protein fractions, like all proteins, are classified based on differences in their solubility: albumins are water soluble, globulins are soluble in dilute salt solutions, prolamins are soluble in 70% aqueous ethanol, and glutenin is soluble in dilute acids/bases [11]. Each protein fraction imparts different functionality to wheat products and when isolated can be used in specific applications. For example, wheat albumin can be used as a nutraceutical that controls blood sugar [35]. Wheat globulin has been found to increase dough stiffness and can be used in noodle products to improve both extensibility and hardness [36].

The other two fractions of wheat protein, gliadin and glutenin, are together know as gluten. Gluten is commonly marketed in two forms; vital wheat gluten, which can be hydrated using water to recover its elastic properties, and nonvital wheat gluten, which is irreversibly denatured [37]. Although nonvital wheat gluten is typically used as an ingredient for protein enrichment, vital wheat gluten can also be used for its structural properties [38]. Vital wheat gluten can be used to fortify flour and increase the elastic properties required for bread-making or can be used to produce textured protein products used to imitate or extend meat. Wheat gluten can also be used as a stabilizing agent for foods, particularly, commercially produced sauces. Vital wheat gluten has recently been gaining attention for its use as a biodegradable polymer suited for the manufacture of packaging materials.

Natural antimicrobials are more readily incorporated into wheat bioplastics than traditional plastics. Protein-based plastics are prepared at lower temperatures using conditions that are less likely to volatilize and/or degrade antimicrobials [39]. Additionally, wheat gluten has been proposed as a sensing material to monitor carbon dioxide accumulation within food packaging [40]. In the future, gluten-based foods might be available in biodegradable gluten-based packaging that acts as a matrix for antimicrobials such as essential oils, and provides feedback on whether the food inside is spoiling!

3. Bioactive phytochemicals in wheat

Whole wheat is a rich source of bioactive phytochemicals including, flavonoids, phytic acid, phenolic acids, carotenoids, tocopherols, alkylresorcinols, benzoxazinoids, phytosterols, γ-oryzanol, β-glucan, and lignans (**Table 1**) [75, 76]. Methods for extraction of these phytochemicals (**Table 1**) are further detailed in Luthria et al., [76]. Most of the phenolic acids (e.g., ferulic acid) exist in the form of bound insoluble complexes and are primarily aggregated in the cell wall matrix of whole grain [77]. These compounds have anti-inflammatory properties that promote gastrointestinal health [23]. They can also act as antioxidants to prevent heart disease and lower the incidence of colon cancer [23]. Consequently, wheat phenolic acids can have great value when formulating functional food products. Destruction of the cell wall matrix is necessary to increase the accessibility of these bioactive compounds. This can be accomplished using a variety of processes including the use of enzymes, fermentation [78], steam explosion, and ultra-fine grinding [79].

Other antioxidant compounds include pigments (e.g., carotenoids). These are typically found in the germ, aleurone, and endosperm fractions [80], although the distribution of these compounds can vary depending on the type of wheat (e.g., einkorn, durum, and common wheat) [81–83]. Carotenoid content in wheat grain can range between 1.8 and 5.8 mg/g [84]. These pigments (e.g., lutein, zeaxanthin, β-carotene) can have provitamin A activity and provide protection against cardiovascular disease and UV-inducing skin damage, as well as yielding products that impart improved antioxidant capacity and can mitigate oxidative stress [11, 80].

Wheat is also a moderate source for vitamin E (e.g., α-tocopherol) [11], providing approximately 5-17 mg of α-tocopherol equivalent per 100 g [85, 86]. The sum of all tocopherols and tocotrienols (a.k.a. tocols) in wheat is in the range of 49–58 mg/g [87–89]. Wheat germ tocols are primarily α- and β-tocopherols, whereas tocols of the pericarp, testa, and aleurone are enriched in tocotrienols [11]. α-Tocopherol is a fat-soluble antioxidant that protects cell membranes with high contents of polyunsaturated fatty acids against oxidative damage [90, 91]. When consumed, these tocols mitigate the production of reactive oxygen/nitrogen species and modulating signal transduction [90] thereby boosting immune response [91].

Wheat β-glucans, lignans, and phytosterols have been investigated as treatments for hypercholesterolemia and cardiovascular disease [92–94]. Importantly, a major lignan in wheat bran was identified to be secoisolariciresinol diglucoside (SDG), which is known to be converted into the mammalian lignans enterodiol and enterolactone by intestinal microflora [95, 96]. Wheat also contains lariciresinol diglucoside. These lignan metabolites function as antioxidants and free radical scavengers, leading to decreased risk of cancer development [95, 97] and hypocholesterolemic properties [98]. These compounds are predominantly concentrated in the outer layers of the grain (e.g., seed coat and pericarp), and in the aleurone layer; however, small concentrations can be found in the inner endosperm [99]. Compared to other cereal grains, wheat

Compound	Location in grain	Use	Extraction method	Market size (US dollars)
Alkylresorcinols	Outer membrane of wheat grain	Biochemical markers of whole grain diet	CSE, SFE, UAE [41–46]	—
Benzoxazinoids	Roots and leaves	Antiallergenic, anti-inflammatory, anticancer, and appetite-suppressing effects	ASE [47]	—
β-Glucan	Aleurone cell walls of wheat bran	Hypocholesterolemic properties	A/BH [48]	$0.73B by 2025 [49]
Carotenoids	Germ, aleurone, and endosperm fractions	Antioxidant	CSE, SFE [50–53]	$1.74B by 2025 [54]
Flavonoids	Germ	Anti-inflammatory	UAE, MAE, PLE, SFE [55]	$1.06B by 2025 [56]
γ-Oryzanol	Wheat bran	Hypocholesterolemic properties	UAE, CSE [57]	$2.0B by 2022 [58]
Lignans	Wheat bran	Antioxidant	A/BH, ED [59]	$0.59B by 2027 [60]
Phenolic acids	Outer membrane of wheat grain	Anti-inflammatory	CSE, MAE, PLE, SPE, ED [61–66]	$2.1B by 2025 [67]
Phytic acids	Outer membrane of wheat grain	Antioxidant	A/BH [68]	$0.83B by 2028 [69]*
Phytosterols	Wheat bran and germ	Production of therapeutic steroids, nutrition, and cosmetics	ASE, A/BH [70, 71]	$1.3B by 2027 [72]
Tocopherols/ toocotrienols	Germ, pericarp, testa, and aleurone	Antioxidant	CSE, SFE [51–53, 73]	$11.94B by 2025 [74]

Abbreviations: Conventional solvent extraction (CSE); microwave-assisted extraction (MAE); pressurized liquid extraction (PLE); solid-phase extraction (SPE); supercritical fluid extraction (SFE); ultrasonic-assisted extraction (UAE); enzymatic digest (ED); acid/base hydrolysis (A/BH).
**Denotes market size for nutraceuticals.*

Table 1.
Use, market size, and extraction methods for bioactive phytochemicals derived from wheat.

is not rich in lignans [100], although they can be a source of SDG. Approximately, 2 mg/g of secoisolariciresinol can be found in the common wheat germ, refined flour, and whole grain flour [99]. Overall, total lignan content in common wheat can range from 2 to 52 mg/g in the germ, refined flour, or whole grain flour [99].

Other value-added components include oil content in wheat bran (3–4%) and germ (7–9%) [101]. Due to the presence of antioxidant bioactive phytochemicals the health benefits of these oils have been investigated [101]. For example, wheat germ oil and wheat bran oil were found to contain high amounts of polyunsaturated fatty acids, as well as bioactive compounds tocopherols, carotenoids, and oryzanol like compounds (e.g., steryl ferulates) [101]. Wheat germ oil can be obtained via mechanical pressing of separated wheat germ, whereas supercritical CO_2 extraction [102, 103], or solvent extraction [104] can be used to recover oil from both wheat germ and wheat bran. The content of bioactive compounds in these oils could be of sufficient quantity to mitigate cardiovascular disease, diabetes, cancer, and other diseases [101]. They have also been utilized in a range of medicinal (e.g., fish oil production), cosmetic (e.g., shampoo), insect control, vitamin, feed, and food products [104]. More recently, the utilization of wheat germ oil and wheat bran fiber have successfully been applied as a fat replacer in developing low-fat beef patties, resulting in better quality, stability, and reduced cholesterol content [105]. Since wheat germ is a by-product of wheat milling and contains extractable oil, isolation and purification of these oils can add significant value as a source of bioactive phytochemicals.

In general, the stability of wheat bioactive phytochemicals can be influenced by processing (milling, fermentative proofing, baking, enzymatic hydrolysis, extrusion, cooking, steaming, malting, etc.) [106] and storage conditions (temperature, light, pressure, time, etc.) [107]. For example, cooking of wheat grain can result in a 55% loss of tocopherol content [108], and increased temperatures and pressures can result the degradation of antioxidant pigments [109]. Therefore, depending on the wheat variety, food processing and milling methods can greatly affect the concentration and activity of wheat bioactive compounds [76, 110].

4. Wheat grain in biofuel production

Bioethanol can be produced using any sugar or starch-rich crop, and is an increasingly attractive fuel type as it reduces reliance on the limited supply of fossil fuels. Biofuels generated from renewable feedstocks can contribute to the reduction of greenhouse gases when compared to crude oil. Globally, the total production of ethanol is over 26 billion gallons in 2020 [111]. In Canada, wheat is the second most common feedstock used for bioethanol, and wheat fermentation is of interest for developing local and renewable energy supplies across many parts of Asia, Europe, and North America.

Ethanol production for biofuels is typically accomplished with a simplified process. Wheat entering the process is typically coarsely milled then rapidly heated with steam to destroy any microbial contamination and produce a thick mash. Sugars are released from the starch by an enzymatic process called saccharification. Initially, starch is treated with heat tolerant alpha-amylase that can function at temperatures as high as 95°C. Subsequently, glucoamylase is added to release more sugars. Other enzymes are optionally added to decompose pectins and hemicellulose. After saccharification, nutrients are added to accelerate the fermentation [112]. Fermentation using yeast (typically *Saccharomyces cerevisiae*) converts the sugar to ethanol, with many factors affecting the final ethanol yield [113]. After fermentation, the mixture, called beer, is transferred to a distillation system where ethanol is distilled from the beer [13]. Finally, the ethanol mixture is further purified through rectification and dehydration, resulting in a final ethanol concentration of 95% or higher [114].

A relatively recent improvement to the fermentation process, referred to as very high gravity (VHG) fermentation, involves the use of the highest possible concentration of sugar in the mash. Select yeasts strains have been identified that can tolerate both the high initial sugar concentration and high ethanol concentrations (>15%) [114, 115]. By fermenting high concentration solutions, considerable amounts of water can be saved, allowing plants to operate at higher capacity without the need for additional space and equipment [116]. The yield of ethanol from wheat can be improved through selection of high starch and low protein cultivars [117].

During fermentation of wheat grain, additional value-added compounds are also produced including glycerol, succinic acid, acetic acid, lactic acid, and α-glycerylphosphorylcholine (e.g., a nootropic compound) [13, 118]. The nootropic compound α-GPC has been investigated as a treatment for Alzheimer's disease and strokes [119]. Production of α-GPC varies significantly with cultivar, and several cultivars have produced promising amounts of this substance when fermented [118, 120].

5. Wheat straw/waste in biofuel production

Despite the potential positives of bioethanol as a replacement for fossil fuels, there have been many criticisms of the use of food sources for bioethanol production. A major concern is that using crops in large-scale production of bioethanol will divert food to the energy sector. It is feared that this competition will increase food prices and contribute to the scarcity of available products as the world population grows [121, 122]. As production of biofuel expands, the materials used must be based on non-food sources such as byproducts, waste, or agricultural losses to remain economically viable and sustainable. Most wheat produced is used for human consumption; it is grown in over 120 countries, and accounts for approximately 1/5 the world's caloric needs [123]. To avoid competing with food crops for agricultural land, bioethanol can instead be produced using less nutrient-rich parts of the crop, such as the straw.

Wheat straw is one of the most abundant agricultural byproducts and is of low commercial value. Straw is primarily used for cattle feed, disposed of, or even burned as waste. On average, 1 kg of straw is collected for every 1.3 kg of grain [124], and this straw can be used as a feedstock for producing bioethanol. Pre-treatment of lignocellulosic products such as wheat straw is required prior to fermentation and is performed using hydrolysis to make cellulose more conducive to enzyme action [125]. Commonly, steam explosion is used as a pre-treatment method and combined with an acid catalyst in wheat straw bioethanol production [124, 126–128]. Other studies have demonstrated success in using H_2SO_4 prior to steam treatment to improve sugar, and therefore ethanol, yields [129]. Production of biofuels from cellulose-rich materials is generally more complex and requires new technologies, but the prices of raw materials such as wheat straw are significantly less and act as an incentive for the biofuel industry to pursue lignocellulosic resources. Currently, the overall production of bioethanol from grain is less costly than from wheat straw [130], however, as the technology develops and government restrictions on greenhouse emissions increase, the use of cellulose-rich materials in bioethanol production is likely to grow significantly.

Biofuel production also provides an opportunity to reclaim damaged and spoiled crops [131]. Damaged grains (e.g., discoloration, breakage, cracking, fungi infection, insect damage, chalky grain) used for ethanol production can reduce feedstock costs by a factor of 10 when compared to grains of higher quality [131]. Alcoholic fermentations of wheat damaged from some of these materials, such as *Fusarium* fungal infections, can produce stillage/wet grains that are unsuitable to use as feed for cattle due to the presence of mycotoxins [132], however, these byproducts need not go to waste. Fermented *Fusarium* infected wheat can be fed to black soldier fly larvae which are able to degrade the toxins, allowing the wheat's nutrients to be recovered [133, 134]. The larvae can then be dried and sold as a highly nutritional and protein-rich feed ingredient.

6. Adding value to fermented wheat byproducts

Alcoholic fermentation of wheat depletes available simple sugars and starches, as these are used for the yeast to grow and produce ethanol. However, after the ethanol is distilled from the beer, the remaining stillage contains protein, oils, fiber, and non-starch carbohydrates. These residual nutrients can also add significant value to fermented wheat through a variety of different processing options. Most commonly, whole stillage is separated into thin stillage (liquid containing suspended solids) and distillers' wet grains (wet solid portions) using physical processing techniques such as screening and centrifugation [135]. The thin stillage can then be dewatered and heated, resulting in a condensed syrup known as distillers' soluble. By mixing the syrup with wet grains as they dry, a nutritional cattle feed referred to as distillers' dry grains with soluble (DDGS) is produced [136].

Compared to unprocessed stillage/wet grains, DDGS has a much greater shelf life and is more readily transported [137]. The DDGS also contains a higher concentration of protein (~38%) than unfermented wheat [19], and the sale of cattle feed typically provides 10–20% of an ethanol producer's revenue [137]. Fermented wheat can also be further enhanced as a feed product through various protein concentration methods [18], including secondary fermentation using lactic acid bacteria. Secondary fermentation can result in an increase in higher value compounds such as 1,3-propandiol, and a feed with up to 60% protein [138] and greater probiotic content [139].

Alternatively, protein from DDGS can be solubilized and extracted to allow amino acids to be removed individually [19]. By leaving essential amino acids in the DDGS, the product can retain value as an animal feed, while allowing valuable non-essential amino acids to be extracted. The extracted amino acids can include aspartic acid, glutamine, glycine, L-arginine, L-lysine, L-phenylalanine, proline, and serine [19]. These non-essential amino acids can have further value added through various chemical transformations, such as aspartic acid into acrylamide [140].

Protein can also be concentrated by the removal of other substances, such as fiber, from the fermented wheat products. Fiber can be removed via aspiration of DDGS coming from dry-grind ethanol facilities [141]. This can allow the fiber-rich fraction to be valorized through the extraction of phytosterols, which are concentrated in the fraction. The reduced fiber content of the remaining fraction results in higher protein and fat content, improving its value as a feed [142]. DDGS can also be fractionated into high protein and high fiber fractions through sieving [143]. By using a combination of both air classification (winnowing) and sieving, a fiber fraction of around 50% reduced protein, and a fraction with an additional 30% protein can be produced, compared to the whole DDGS [117]. Oil is another component of wheat that is still present after fermentation. Although there are patented techniques for extracting oil from corn DDGS and thin stillage [144, 145], the lower oil content of wheat and other grains has resulted in oil extraction techniques remaining largely undeveloped for fermented wheat products.

7. Conclusion

Wheat is one of the largest grain crops produced in the world, second only to corn, with United States and Canada being the two largest exporters of wheat globally.

Typically, the end products of wheat are used as food sources for either human or animal consumption, however, the production of biofuel using feedstocks such as wheat is steadily expanding. The fractionation and appropriate selection of harvested wheat crops is crucial to increase economic value; wheat varieties can be tailored for specific applications, such as biofuel production, or the extraction of bio-active phytochemicals. Furthermore, the stability, quality, and concentration of bioactive compounds can be affected by the processing and storage conditions of wheat, which contribute to the value of these products in cosmetics or antioxidant and anti-inflammatory supplements. Typically, biofuel production uses wheat grain with high starch content to create efficient and productive fermentations. This type of biofuel production is under increasing criticism for the diversion valuable food sources and agriculturally productive land. By using grain production by-products such as wheat straw, or damaged grained crops, ethanol production can be accomplished with reduced greenhouse gas emissions and less expensive, renewable feedstocks. Lignocellulosic by-products can add significant value to wheat through fermentation and will be of growing interest with increasing societal and economic pressures to reduce dependency on petroleum. Even after fermentation, the remaining stillage can be used as high protein feed, and have valuable products diverted to industrial streams. Overall, these processes can substantially enhance the value of wheat and reduce agricultural waste. The future will likely lead to new techniques and further improvements in wheat processing.

Acknowledgements

This work was supported by the Saskatchewan Agricultural Development Fund (20190155, 20190154, 20180281, 20180248, 20180255, 20170133); National Sciences and Engineering Research Council of Canada Discovery Grant (RGPIN-2018-06631); and Mitacs (IT19122, IT16156).

Conflict of interest

Dr. Martin J.T. Reaney is the founder of, and has an equity interest in, Prairie Tide Diversified Inc. (PTD, Saskatoon, SK, Canada: previous company name is Prairie Tide Chemicals Inc.).

Author details

Timothy J. Tse*, Farley Chicilo, Daniel J. Wiens and Martin J.T. Reaney*
University of Saskatchewan, Saskatoon, Canada

*Address all correspondence to: timothy.tse@usask.ca and martin.reaney@usask.ca

References

[1] Statista. Worldwide Production of Grain in 2020/21, by Type [Internet]. 2021. Available from: https://www.statista.com/statistics/263977/world-grain-production-by-type/ [Accessed: December 21, 2021]

[2] Sosland Publishing. USDA Forecasts Records for World Wheat Production, Use and Trade [Internet]. 2021. Available from: https://www.world-grain.com/articles/15837-usda-forecasts-records-for-world-wheat-production-use-and-trade [Accessed: October 15, 2021]

[3] Talebnia F, Karakashev D, Angelidaki T. Production of bioethanol from wheat straw: An overview on pretreatment, hydrolysis and fermentation. Bioresource Technology. 2010;**101**:4744-4753. DOI: 10.1016/j.biortech.2009.11.080

[4] Canada Grain Commission. Canadian Wheat Classes [Internet]. 2019. Available from: https://grainscanada.gc.ca/en/grain-quality/grain-grading/wheat-classes.html [Accessed: October 17, 2021]

[5] Bradford H, Hayes P. Canada Biofuels Annual 2019: GAIN Report Number: CA19017. 2019. USDA Foreign Agricultural Service. 2019

[6] Kilcher MR. Effect of cattle grazing on subsequent grain yield of fall rye (*Secale cereale* L.) in southwestern Saskatchewan. Canadian Journal of Plant Science. 1982;**62**:795-796

[7] Virgona JM, Gummer FAJ, Angus JF. Effects of grazing on wheat growth, yield, development, water use, and nitrogen use. Australian Journal of Agricultural Research. 2006;**57**(12). DOI: 10.1071/AR06085

[8] Bureau of the Census. Census of Manufacturers: Grain Mill Products. Washington, DC: U.S. Department of Commerce; 1992

[9] Lou J, Wilson W. Value-Added Wheat Products: Analysis of Markets and Competition. Agricultural Economics Report No. 386. Fargo, ND: North Dakota State University; 1998

[10] Wood L. Global Wheat Starch Market Report 2018-2024: Key Growth Drivers, Market Trends and Key Players—Market to Exceed $4 Billion—ResearchAndMarkets.com [Internet]. 2018. Available from: https://www.businesswire.com/news/home/20180321005489/en/Global-Wheat-Starch-Market-Report-2018-2024-Key-Growth-Drivers-Market-Trends-and-Key-Players---Market-to-Exceed-4-Billion---ResearchAndMarkets.com [Accessed: December 02, 2021]

[11] Uthayakumaran S, Wrigly C. Chapter 5 - Wheat: Grain-quality characteristics and management of quality requirements. In: Wrigley C, Batey I, Miskelly D, editors. Cereal Grains. 2nd ed. Cambridge, MA, USA: Woodhead Publishing; 2017. pp. 91-134. DOI: 10.1016/b978-0-08-100719-8.00005-x

[12] Saunders J, Izydorczyk M, Levin DB. Limitations and challenges for wheat-based bioethanol production. In: Economic Effects of Biofuel Production. 1st ed. London, UK: IntechOpen; 2011. pp. 429-452. DOI: 10.5772/20258

[13] Tse TJ, Wiens DJ, Chicilo F, Purdy SK, Reaney MJT. Value-added compounds from ethanol fermentations—A review. Fermentation. 2021;7:267. DOI: 10.3390/fermentation7040267

[14] Elbashiry EMA, Chen J, Tuo W, Ren Y, Guo Z. Review of the pretreatment methods for wheat straw building materials. Journal of Reinforced Plastics & Composites. 2018;**37**(1):35-48. DOI: 10.1177/0731684417730442

[15] Moir J, Letts, J. Thatch: Thatching in England 1790-1940. English Heritage Research Transactions 5; Earthscan Publications Ltd.; 1999. p. 228

[16] Glithero NJ, Wilson P, Ramsden SJ. Straw use and availability for second generation biofuels in England. Biomass Bioenergy. 2013;**55**:311-321. DOI: 10.1016/j.biombioe.2013.02.033

[17] Qureshi N, Saha BC, Cotta MA, Singh V. An economic evaluation of biological conversion of wheat straw to butanol: A biofuel. Energy Conversion and Management. 2013;**65**:456-462. DOI: 10.1016/j.enconman.2012.0.015

[18] Ratanapariyanuch K, Shim YY, Wiens DJ, Reaney MJT. Grain thin stillage protein utilization: A review. Journal of the American Oil Chemists' Society. 2018;**95**:933-942. DOI: 10.1002/aocs.12056

[19] Villegas-Torres MF, Ward JM, Lye GJ. The protein fraction from wheat-based dried distiller's grain with solubles (DDGS): Extraction and valorization. New Biotechnology. 2015;**32**(6):606-611

[20] Shewry PR, Hawkesford MJ, Piironen V, Lampi A-M, Gebruers K, Boros D, et al. Natural variation in grain composition of wheat and related cereals. Journal of Agricultural and Food Chemistry. 2013;**61**:8295-8303. DOI: 10.1021/jf3054092

[21] Health Canada. Novel food Information—Wheat Bran Extract as a Novel Food Ingredient. [Internet]. 2015. Available from: https://www.canada.ca/en/health-canada/services/food-nutrition/genetically-modified-foods-other-novel-foods/approved-products/wheat-bran-extract-food-ingredient.html [Accessed: November 24, 2021]

[22] Hemery Y, Chaurand M, Holopainen U, Lampi A-M, Lehtinen P, Piironen V, et al. Potential of dry fractionation of wheat bran for the development of food ingredients, part I: Influence of ultra-fine grinding. Journal of Cereal Science. 2011;**53**:1-8

[23] Laddomada B, Caretto S, Mita G. Wheat bran phenolic acids: Bioavailability and stability in whole wheat-based foods. Molecules. 2015;**20**:15666-15685. DOI: 10.3390/molecules200915666

[24] Ciudad-Mulero M, Barros L, Fernandes Â, Ferreira ICFR, Callejo MJ, Matallana-González MC, et al. Potential health claims of durum and bread wheat flours as functional ingredients. Nutrients. 2020;**12**:504. DOI: 10.3390/nu12020504

[25] Delcour JA, Rouau X, Courtin CM, Poutanen K, Ranieri R. Technologies for enhanced exploitation of the health-promoting potential of cereals. Trends in Food Science & Technology. 2012;**25**:78-86. DOI: 10.1016/j.tifs.2012.01.007

[26] Esposito F, Arlotti G, Bonifati AM, Napolitano A, Vitale D, Fogliano V. Antioxidant activity and dietary fibre in durum wheat bran by-products. Food Research International. 2005;**38**:1167-1173. DOI: 10.1016/j.foodres.2005.05.002

[27] Mareo Anson NM, van der Berg R, Havenaar R, Bast A, Haenen G. Ferulic acid from aleurone determines the antioxidant potency of wheat grain (*Triticum aestivum* L.). Journal of Agricultural and Food Chemistry. 2008;**56**:5589-5594. DOI: 10.1021/jf800445k

[28] Zhou K, Su L, Yu L. Phytochemicals and antioxidant properties in wheat bran. Journal of Agricultural and Food Chemistry. 2004;**52**:6108-6114. DOI: 10.1021/jf049214g

[29] Toepfer EW, Polansky MM, Eheart JF, Slover HT, Hepburn FN, Quackenbush FW. Nutrient composition of selected wheats and wheat products XI Summary. Cereal Chemistry. 1972;**49**:173-186

[30] Shewry PR, Hey SJ. The contribution of wheat to human diet and health. Food and Energy Security. 2015;**4**:178-202. DOI: 10.1002/fes3.64

[31] Taccari M, Aquilanti L, Polverigiani S, Osimani A, Garofalo C, Milanovic V, et al. Microbial diversity of type I sourdoughs prepared and Back-slopped with Wholemeal and refined soft (*Triticum aestivum*) wheat flours. Journal of Food Science. 2016;**81**:1996-2005. DOI: 10.1111/1750-3841.13372

[32] Godfrey D, Hawkesford MJ, Powers SJ, Millar S, Shewry PR. Effects of crop nutrition on wheat grain composition and end use quality. Journal of Agricultural and Food Chemistry. 2010;**58**:3012-3021. DOI: 10.1021/jf9040645

[33] Vogel KP, Johnson VA, Mattern PJ. Protein and lysine content of grain, endosperm, and brand of wheats from USDA world wheat collection. Crop Science. 1976;**16**:655-660. DOI: 10.2135/cropsci1976.0011183X001600050014x

[34] Jensen SA, Martens H. The botanical constituents of wheat and wheat milling fractions. II. Quantification by amino acids. Cereal Chemistry. 1983;**60**:170-177

[35] Saito S, Oishi S, Shudo A, Sugiura Y, Yasunaga K. Glucose response during the night is suppressed by wheat albumin in healthy participants: A randomized controlled trial. Nutrients. 2019;**11**(187):1-9. DOI: 10.3390/nu11010187

[36] Zhang LL, Guan EQ, Yang YL, Liu YX, Zhang TJ, Bian K. Impact of wheat globulin addition on dough rheological properties and quality of cooked noodles. Food Chemistry 2021;**362**:130-170. DOI: 1016/j.foodchem.2021.130170

[37] Tedrus G de AS, Ormenese R de CSC, Speranza SM, Chang YK, Bustos FM. Estudo da adição de vital gluten à farinha de arroz, farinha de aveia e amido de trigo na qualidade de pães. Ciência e Tecnologia de Alimentos. 2001;**21**:20-25. DOI: 10.1590/s0101-20612001000100006

[38] Ortolan F, Steel CJ. Protein characteristics that affect the quality of vital wheat gluten to be used in baking: A review. Comprehensive Reviews in Food Science and Food Safety. 2017;**16**(3):369-381. DOI: 10.1111/1541-4337.12259

[39] Gómez-Heincke D, Martínez I, Partal P, Guerrero A, Gallegos C. Development of antimicrobial active packaging materials based on gluten proteins. Journal of the Science of Food and Agriculture. 2016;**96**(10):3432-3438. DOI: 10.1002/jsfa.7525

[40] Bibi F, Guillaume C, Gontard N, Sorli B. Wheat gluten, a bio-polymer to monitor carbon dioxide in food packaging: Electric and dielectric characterization. Sensors & Actuators, B: Chemical. 2017;**250**:76-84. DOI: 10.1016/j.snb.2017.03.164

[41] Christian HG, Anna EM, Reinhold C, Ralf MS. Development and validation of an HPLC method for the determination of alk(en)ylresorcinols using rapid ultrasound-assisted extraction of mango peels and rye grains. Food Chemistry.

2015;**169**:261-269. DOI: 10.1002/9781119207917.ch21

[42] Gunenc A, Hadinezhad M, Tamburic-Ilincic L, Mayer PM, Hosseninian F. Effects of region and cultivar on alkylresorcinols content and composition in wheat bran and their antioxidant activity. Journal of Cereal Science. 2013;**57**:405-410. DOI: 10.1016/j.jcs.2013.01.003

[43] Holt MD, Moreau RA, DerMarderosian A, McKeown N, Jacques PF. Accelerated solvent extraction of alkylresorcinols in food products containing uncooked and cooked wheat. Journal of Agricultural and Food Chemistry. 2012;**60**:4799-4802. DOI: 10.1021/jf3001422

[44] Landberg R, Dey ES, Francisco JDC, Aman P, Kamal-Eldin A. Comparison of supercritical carbon dioxide and ethyl acetate extraction of alkylresorcinols from wheat and rye. Journal of Food Composition and Analysis. 2007;**20**:534-538. DOI: 10.1016/j.jfca.2007.02.009

[45] Rebolleda S, Beltrán S, Sanz MT, González-Sanjosé ML, Solaesa ÁG. Extraction of alkylresorcinols from wheat bran with supercritical CO_2. Journal of Food Engineering. 2013;**119**:814-821. DOI: 10.1016/j.jfoodeng.2013.07.008

[46] Sampietro DA, Jimenez CM, Belizán MM, Vattuone MA, Catalán CA. Development and validation of a micromethod for fast quantification of 5-n-alkylresorcinols in grains and whole grain products. Food Chemistry. 2013;**141**:3546-3551. DOI: 10.1016/j.foodchem.2013.06.069

[47] Pedersen H, Laursen B, Mortensen A, Fomsgaard I. Bread from common cereal cultivars contains an important array of neglected bioactive benzoxazinoids. Food Chemistry. 2011;**127**:1814-1820. DOI: 10.1016/j.foodchem.2011.02.070

[48] Maheshwari G, Sowrirajan S, Joseph B. Extraction and isolation of β-glucan from grain sources—A review. Journal of Food Science. 2017;**82**:1535-1545. DOI: 10.1111/1750-3841.13765

[49] Grand View Research. Beta-glucan Market Size, Share & Trends Analysis Report by Source (Cereal, Mushroom, Yeasts, Seaweed), by Application (F&B, Personal Care, Pharmaceuticals, Animal Feed), By Type, By Region, and Segment Forecasts, 2018-2025. San Francisco, CA, USA: Grand View Research; 2017

[50] Durante M, Lenucci MS, Laddomada B, Mita G, Caretto S. Effects of sodium alginate bead encapsulation on the storage of durum wheat (*Triticum durum* Desf.) bran oil extracted by supercritical CO_2. Journal of Agricultural and Food Chemistry. 2012;**60**:10689-10695. DOI: 10.1021/jf303162m

[51] Ge Y, Ni Y, Yan H, Chen Y, Cai T. Optimization of the supercritical fluid extraction of natural vitamin E from wheat germ using response surface methodology. Journal of Food Science. 2002;**67**:239-243. DOI: 10.1111/j.1365-2621.2002.tb11391.x

[52] Gelmez N, Kincal NS, Yener ME. Optimization of supercritical carbon dioxide extraction of antioxidants from roasted wheat germ based on yield, phenolic and tocopherol contents, and antioxidant activities of the extracts. Journal of Supercritical Fluids. 2009;**48**:217-224. DOI: 10.1016/j.supflu.2008.11.002

[53] Hidalgo A, Brandolini C, Pompei C, Piscozzi R. Carotenoids and tocols of einkorn wheat (*Triticum monococcum ssp. monococcum* L.). Journal of Cereal

Science. 2006;**44**:182-193. DOI: 10.1016/j.jcs.2006.06.002

[54] Grand View Research. Carotenoids Market Size, Share & Trends Analysis Report by Source (Natural, Synthetic), by Product (Beta-Carotene, Lutein, Lycopene, Astaxanthin, Zeaxanthin, Canthaxanthin), by Application, and Segment Forecasts, 2018-2025. San Francisco, CA, USA: Grand View Research; 2016

[55] Chaves JA, de Souza MC, da Silva LC, Lachos-Perez D, Torres-Mayanga PC, da Fonseca Machado AP, et al. Extraction of flavonoids from natural sources using modern techniques. Frontiers in Chemistry. 2020;**8**:507887. DOI: 10.3389/fchem.2020.507887

[56] Grand View Research. Flavonoids Market Size, Share & Trends Analysis Report by Product (Anthocyanins, Flavones, Anthoxanthins), by Application, By Region, and Segment Forecasts, 2018-2025. San Francisco, CA, USA: Grand View Research; 2016

[57] Loypimai P, Moongngarm A, Sittisuanjik A, Khamanan S. Optimization of tocols and γ-oryzanol extraction from rice bran using ultrasound and soybean oil as a green solvent. Food Research. 2020;**4**:2322-2332. DOI: 10.26656/fr.2017.4(6).360

[58] Grand View Research. Gamma Oryzanol Market Size, Share & Trends Analysis Report by Application (Sports Supplement, Cosmetics, Pharmaceuticals, Animal Feed), and Segment Forecasts to 2022. San Francisco, CA, USA: Grand View Research; 2016

[59] Cukelj N, Jakasa I, Sarajlija H, Novotni D, Curic D. Identification and quantification of lignans in wheat bran by gas chromatography-electron capture detection. Talanta. 2011;**84**:127-132. DOI: 10.1016/j.talanta.2010.12.025

[60] Grand View Research. Lignans Market Size, Share & Trends Analysis Report by Source (Plant Resins, Cereals & Grains), by Application (Food & Beverages, Animal Nutrition), By Region (APAC, North America) and Segment Forecasts, 2020-2027. San Francisco, CA, USA: Grand View Research; 2020

[61] Angioloni A, Collar C. Effects of pressure treatment of hydrated oat finger millet and sorghum flours on the quality and nutritional properties of composite wheat breads. Journal of Cereal Science. 2012;**56**:713-719. DOI: 10.1016/j.jcs.2012.08.001

[62] Buranov A, Mazza G. Extraction and purification of ferulic acid from flax shives, wheat and corn bran by alkaline hydrolysis and pressurized solvents. Food Chemistry. 2009;**115**:1542-1548. DOI: 10.1016/j.foodchem.2009.01.059

[63] Hromadkova Z, Kostalova Z, Ebringerova A. Comparison of conventional and ultrasound-assisted extraction of phenolics-rich heteroxylans from wheat bran. Ultrasonics Sonochemistry. 2008;**15**:1062-1068. DOI: 10.1016/j.ultsonch.2008.04.008

[64] Inglett GE, Rose D, Stevenson DG, Chen D, Biswas A. Total phenolics and antioxidant activity of water and ethanolic extracts from distillers dried grains with soluble with or without microwave irradiation. Cereal Chemistry. 2009;**86**:661-664. DOI: 10.1094/CCHEM-86-6-0661

[65] Irakli MN, Samanidou VF, Biliaderis CG, Papadoyannis IN. Development and validation of an HPLC-method for determination of free and bound phenolic acids in cereals after solid-phase extraction. Food Chemistry. 2012;**134**:1624-1632. DOI: 10.1016/j.foodchem.2012.03.046

[66] Meneses N, Martins S, Teixeira JA, Mussatto SI. Influence of extraction solvents on the recovery of antioxidant phenolic compounds from brewer's spent grains. Separation and Purification Technology. 2013;**108**:152-158. DOI: 10.1016/j.seppur.2013.02.015

[67] Grand View Research. Polyphenols Market Size, Share & Trends Analysis Report by Product (Grape Seed, Green Tea, Cocoa), by Application (Beverages, Food, Feed, Dietary Supplements, Cosmetics), and Segment Forecasts, 2019-2025. San Francisco, CA, USA: Grand View Research; 2019

[68] Ebrahimian M, Motamedi H. Utilization of wheat bran as a source for phytic acid production. International Food Research Journal. 2016; **23**:2436-2439

[69] Grand View Research. Nutraceuticals Market Size, Share & Trends Analysis Report by Product (Dietary Supplements, Functional Food, Functional Beverages), by Region, and Segment Forecasts, 2020-2028. San Francisco, CA, USA: Grand View Research; 2021

[70] Dunford NT, Irmak S, Jonnala R. Effect of the solvent type and temperature on phytosterol contents and compositions of wheat straw, bran, and germ extracts. Journal of Agricultural and Food Chemistry. 2009;**57**:10608-10611. DOI: 10.1021/jf9022383

[71] Nurmi T, Nyström L, Edelmann M, Lampi AM, Pirronen V. Phytosterols in wheat genotypes in the HEALTHGRAIN diversity screen. Journal of Agricultural and Food Chemistry. 2008;**56**:9710-9715. DOI: 10.1021/jf8010678

[72] Grand View Research. Phytosterols Market Size, Share & Trends Analysis Report by Product (Beta-sitosterol, Campesterol, Stigmasterol), by Application (Pharmaceuticals, Cosmetics, Food Ingredients), By Region, and Segment Forecasts, 2020-2027. San Francisco, CA, USA: Grand View Research; 2020

[73] Ge Y, Yan H, Hui B, Ni Y, Wang S, Cai T. Extraction of natural vitamin E from wheat germ by supercritical carbon dioxide. Journal of Agricultural and Food Chemistry. 2002;**50**:685-689. DOI: 10.1021/jf010615v

[74] Grand View Research. Mixed Tocopherols Market Size, Share & Trends Analysis Report by Product (Alpha Rich, Gamma Rich, Gamma-Delta Rich), by Application (Dietary Supplements, Cosmetics, Pharmaceuticals), and Segment Forecasts, 2018-2025. San Francisco, CA, USA: Grand View Research; 2017

[75] Gani A, Wani SM, Masoodi FA, Hameed G. Whole-grain cereal bioactive compounds and their health benefits: A review. Journal of Food Processing & Technology. 2012;**3**:146. DOI: 10.4172/2157-7110.1000146

[76] Luthria DL, Lu Y, Maria John KM. Bioactive phytochemicals in wheat: Extraction, analysis, processing, and functional properties. Journal of Functional Foods. 2015;**18**:910-925. DOI: 10.1016/j.jff.2015.01.001

[77] Lu Y, Fuerst PE, Lv J, Morris CF, Yu L, Fletcher A, et al. Phytochemical profile and antiproliferative activity of dough and bread fractions made from refined and whole wheat flours. Cereal Chemistry. 2015;**92**:271-277. DOI: 10.1094/cchem-10-14-0214-R

[78] Yin ZN, Wu WJ, Sun CZ, Liu HF, Chen WB, Zhan QP, et al. Antioxidant and anti-inflammatory capacity of ferulic acid released from wheat bran by

solid-state fermentation of *Aspergillus niger*. Biomedical and Environmental Sciences. 2019;**32**:11-21. DOI: 10.3967/bes2019.002

[79] Rosa NN, Barron C, Gaiani C, DuFour C, Micard V. Ultra-fine grinding increases the antioxidant capacity of wheat bran. Journal of Cereal Science. 2013;**57**:84-90. DOI: 10.1016/j.jcs.2012.10.002

[80] Masisi K, Diehl-Jones WL, Gordon J, Chapman D, Moghadasian MH, Beta T. Carotenoids of aleurone, germ, and endosperm fractions of barley, corn and wheat differentially inhibit oxidative stress. Journal of Agricultural and Food Chemistry. 2015;**63**:2715-2724. DOI: 10.1021/jf5058606

[81] Abdel-Aal E-SM, Young JC, Rabalski I, Hucl P, Fregeau-Reid J. Identification and quantification of seed carotenoids in selected wheat species. Journal of Agricultural and Food Chemistry. 2007;**55**:787-794. DOI: 10.1021/jf062764p

[82] Giordano D, Locatelli M, Travaglia F, Bordiga M, Reyneri A, Coïsson JD, et al. Bioactive compound and antioxidant activity distribution in roller-milled and pearled fractions of conventional and pigmented wheat varieties. Food Chemistry. 2017;**233**:483-491. DOI: 10.1016/j.foodchem.2017.04.065

[83] Siebenhandl S, Grausgruber H, Pellegrini N, Del Rio D, Fogliano V, Pernice R, et al. Phytochemical profile of main antioxidants in different fractions of purple and blue wheat, and black barley. Journal of Agricultural and Food Chemistry. 2007;**55**:8541-8547. DOI: 10.1021/jf072021j

[84] Kruger JE, Reed G. Enzymes and color. In: Pomeranz Y editor. Wheat: Chemistry and Technology.3rd ed. St. Paul, MN: AACC International; 1988. pp. 411-500

[85] Beta T, Duodu K. Bioactives: Antioxidants. In: Wrigley C, Corke H, Seetharaman K, Faubion J. editors. Encyclopedia of Food Grains. Vol. 2. Oxford, UK: Elsevier; 2016. pp. 277-282

[86] Bramley PM, Elmadfa I, Kafatos A, Kelly FJ, Manios Y, Roxborough HE, et al. Vitamin E. Journal of the Science of Food and Agriculture. 2000;**80**:913-938. DOI: 10.1002/(SICI)1097-0010(20000515)80:7<913::AID-JSFA600>3.0.CO;2-3

[87] Bock MA. Minor constituents of cereals. In: Kulp K, Ponte JGJ. editors. Handbook of Cereal Science and Technology. 2nd ed. New York: Marcel Dekker; 2000. pp. 479-504

[88] Chung OK, Ohm J. Cereal lipids. In: Kulp K, Ponte JGJ. editors. Handbook of Cereal Science and Technology. 2nd ed. New York: Marcel Dekker; 2000. pp. 417-477

[89] Chung OK, Ohm J, Ram MS, Park S-H, Howitt C. Wheat Lipids. In: Khan K, Shewry PR, editors. Wheat: Chemistry and Technology. 4th Ed. Saint Paul, MN, USA: American Association of Cereal Chemists Inc.; 2009. pp. 363-399

[90] Lee GA, Han SN. The role of vitamin E in immunity. Nutrients. 2018;**11**:1614. DOI: 10.3390/nu10111614

[91] Lewis ED, Meydani SN, Wu D. Regulatory role of vitamin E in the immune system and inflammation. IUBMB Life. 2019;**71**:487-494. DOI: 10.1002/iub.1976

[92] Higuchi M. Antioxidant properties of wheat bran against oxidative stress. In: Preedy VR, Zibadi S, editors. Wheat and Rice in Disease Prevention and Health. San Diego, USA: Academic Press; 2014. pp. 181-199

[93] Quílez J, García-Lorda P, Salas-Salvadó J. Potential uses and benefits of phytosterols in diet: Present situation and future directions. Clinical Nutrition. 2003;**22**:343-351. DOI: 10.1016/s0261-5614(03)00060-8

[94] Sima P, Vannucci L, Vetvicka V. β-Glucans and cholesterol (review). International Journal of Molecular Medicine. 2018;**4**:1799-1808. DOI: 10.3892/ijmm.2018.3411

[95] Qu H, Madl RL, Takemoto DJ, Baybutt RC, Wang W. Lignans are involved in the antitumor activity of wheat bran in colon cancer SW480 cells. The Journal of Nutrition. 2005;**135**:598-602. DOI: 10.1093/jn/135.3.598

[96] Yang X, Guo Y, Tse TJ, Purdy SK, Mustafa R, Shen J, et al. Oral pharmacokinetics of enriched secoisolariciresinol diglucoside and its polymer in rats. Journal of Natural Products. 2021;**84**:1816-1822. DOI: 10.1021/acs.jnatprod.1c00335

[97] Davies MJ, Bowey EA, Adlercreutz H, Rowland IR, Rumsby PC. Effects of soy or rye supplementation of high-fat diets on colon tumor development in azoxymethane-treated rats. Carcinogenesis. 1999;**20**:927-931. DOI: 10.1093/carcin/20.6.927

[98] Guo Y, Yang X, Tse TJ, Purdy SK, Mustafa R, Shen J, et al. Hypocholesterolemic efficacy of secoisolariciresinol diglucoside and its polymer in rat. Journal of Natural Products. 2021;**84**:2845-2850. DOI: 10.1021/acs.jnatprod.1c00549

[99] Rodríguez-García C, Sánchez-Quesada C, Toledo E, Delgado-RodríguezM, GaforioJJ. Naturally lignan-rich foods: A dietary tool for health promotion. Molecules. 2019;**24**:917. DOI: 10.3390/molecules24050917

[100] Landette JM. Mammalian lignans: A review of source, intake, metabolism, intestinal bacteria, health. Food Research International. 2012;**46**:410-424. DOI: 10.1016/j.foodres.2011.12.023

[101] Kumar GS, Krishna AGG. Studies on the nutraceuticals composition of wheat derived oils wheat bran oil and wheat germ oil. Journal of Food Science and Technology. 2015;**52**:1145-1151. DOI: 10.1007/s13197-013-1119-3

[102] Piras A, Rosa A, Falconieri D, Procedda S, Dessi MA, Marongju B. Extraction of oil from wheat germ by supercritical CO_2. Molecules. 2009;**14**:2573-2581. DOI: 10.3390/molecules14072573

[103] Zacchi P, Daghero J, Jaeger P, Eggers R. Extraction/fractionation and deacidification of wheat germ oil using supercritical carbon dioxide. Brazilian Journal of Chemical Engineering. 2006;**23**:105-110

[104] Ghafoor K, Özcan MM, AL-Juhaimi F, Babiker EE, Sarker ZI, Ahmed IAM, Ahmed MA. Nutritional composition, extraction, and utilization of wheat germ oil: A review. European Journal of Lipid Science and Technology. 2017;**119**:1600160. DOI: 10.1002/ejlt.201600160

[105] Khalid A, Sohaib M, Nadeem MT, Saeed F, Imran A, Imran M, et al. Utilization of wheat germ oil and wheat bran fiber as fat replacer for the development of low-fat beef patties. Food Science & Nutrition. 2021;**9**:1271-1281. DOI: 10.1002/fsn3.1988

[106] Batifoulier F, Verny M-A, Chanliaud E, Remesy C, Demigne C. Effect of different breadmaking methods on thiamine, riboflavin and pyridoxine contents of wheat bread. Journal of Cereal Science. 2005;**42**:101-108

[107] Reddy MB, Love M. The impact of food processing on the nutritional quality of vitamins and minerals. Advances in Experimental Medicine and Biology. 1999;**459**:99-106

[108] Wyatt CJ, Perez Carballido S, Mendez RO. α- and γ- tocopherol content of selected foods in the Mexican diet: Effect of cooking losses. Journal of Agricultural and Food Chemistry. 1998;**46**:4657-4661. DOI: 10.1021/jf9800716

[109] Hashemi SMB, Pourmohammadi K, Gholamhosseinpour A, Es I, Ferreira DS, Mousavi Khaneghah A. Fat-soluble vitamins. In: Barba FJ, Saraiva JMA, Cravotto G, Lorenzo JM, editors. Innovative Thermal and Non-Thermal Processing, Bioaccessibility and Bioavailability of Nutrients and Bioactive Compounds. Cambridge, MA, USA: Woodhead Publishing; 2019. pp. 267-289. DOI: 10.1016/b978-0-12-814174-8.00009-3

[110] Li Y, Ma D, Sun D, Wang C, Zhang J, Xie Y, et al. Total phenolic, flavonoid content, and antioxidant activity of flour, noodles, and steamed bread made from different colored wheat grains by three milling methods. Crop Journal. 2015;**3**:328-334. DOI: 10.1016/j.cj.2015.04.004

[111] Annual Ethanol Production: U.S. and World Ethanol Production. Renewable Fuels Association. Available from: https://ethanolrfa.org/markets-and-statistics/annual-ethanol-production [Accessed: November, 2021]

[112] Patni N, Pillai S, Dwivedi A. Wheat as a promising substitute of corn for bioethanol production. Procedia Engineering. 2013;**51**:355-362

[113] Tse TJ, Wiens DJ, Reaney MJT. Biofuel production—A review of factor affecting ethanol yield. Fermentation. 2021;**7**:268. DOI: 10.3390/fermentation7040268

[114] Swain M, Mohanty S. Bioethanol production from corn and wheat: food, fuel, and future. In: Ramesh CR, Ramachandran S, editors. Bioethanol Production from Food Crops. San Diego, USA: Academic Press; 2019. pp. 45-49. DOI: 10.1016/C2017-0-00234-3

[115] Puligundla P, Smogrovicova D, Obulam V, Ko S. Very high gravity (VHG) ethanolic brewing and fermentation: A research update. Journal of Industrial Microbiology & Biotechnology. 2011;**38**(9):1133-1144

[116] Thomas K, Hynes S, Ingledew W. Practical and theoretical considerations in the production of high concentrations of alcohol by fermentation. Process Biochemistry. 1996;**31**:321-331

[117] Saunders J, Izydorczyk M, Levin D. Limitations and challenges for wheat-based bioethanol production. In: Dos Santos-Bernades M, editors. Economics Effects of Biofuel Production. London, UK: InTech; 2011. pp. 429-452. DOI: 10.5772/20258

[118] Oyeneye A, Shen J, Shim YY, Tse TJ, Reaney MJT. Production of α-glycerylphosphorylcholine and other compounds from wheat fermentation. ACS Omega. 2020;**5**:12486-12494. DOI: 10.1021/acsomea.0c01352

[119] Moreno MDJM. Cognitive improvement in mild to moderate Alzheimer's dementia after treatment with the acetylcholine precursor choline alfoscerate: A multicenter, double-blind, randomized, placebo-controlled trial. Clinical Therapeutics. 2003;**25**:178-193. DOI: 10.1016/s0149-2918(03)90023-3

[120] Tse TJ, Wiens DJ, Shen J, Beattie AD, Reaney MJT. *Saccharomyces cerevisiae* fermentation of 28 barley and 12 oat cultivars. Fermentation. 2021;7:59. DOI: 10.3390/fermentation7020059

[121] Machado K, Seleme R, Maceno M, Zattar I. Carbon footprint in the ethanol feedstocks cultivation—Agricultural CO_2 emission assessment. Agricultural Systems. 2017;**157**:140-145

[122] Malaca J, Freire F. Addressing land use change and uncertainty in the life-cycle assessment of wheat-based bioethanol. Energy (Oxford). 2012;**45**(1):519-527

[123] Enghiad A, Ufer D, Countryman A, Thilmany D. An overview of global wheat market fundamentals in an era of climate concerns. International Journal of Agronomy. 2017;**2017**:1-15

[124] Ruiz H, Silva D, Ruzene D, Lima L, Vicente A, Teixeira J. Bioethanol production from hydrothermal pretreated wheat straw by a flocculating Saccharomyces cerevisiae strain—Effect of process conditions. Fuel (Guildford). 2012;**95**:528-536

[125] Erdei B, Barta Z, Sipos B, Réczey K, Galbe M, Zacchi G. Ethanol production from mixtures of wheat straw and wheat meal. Biotechnology for Biofuels. 2010;**3**(1):16

[126] Ballesteros I, Negro MJ, Oliva JM, Cabanas A, Manzanares P, Ballesteros M. Ethanol production from steam-explosion pretreated wheat straw. Applied Biochemistry and Biotechnology. 2006;**12**:496-508

[127] Hu G, Heitmann JA, Rojas OJ. Feedstock pretreatment strategies for producing ethanol from wood, bark, and forest residues. BioResources. 2008;**1**:270-294

[128] Ingrao C, Matarazzo A, Gorjian S, Adamczyk J, Failla S, Primerano P, et al. Wheat-straw derived bioethanol production: A review of life cycle assessments. The Science of the Total Environment. 2021;**781**:146751

[129] Linde M, Jakobsson EL, Galbe M, Zacchi G. Steam pretreatment of dilute H_2SO_4-impregnated wheat straw and SSF with low yeast and enzyme loadings for bioethanol production. Biomass and Bioenergy. 2008;**4**:326-332

[130] Lantz M, Prade T, Ahlgren S, Björnsson L. Biogas and ethanol from wheat grain or straw: Is there a trade-off between climate impact, avoidance of iLUC and production cost? Energies (Basel). 2018;**11**(10):2633

[131] Razdan N, Kocher GS. Utilization of damaged and spoiled wheat grains for bioethanol production. Bioscience Biotechnology Research Communications. 2018;**11**:658-673. DOI: 10.21786/bbrc/11.4/17

[132] Canadian Food Inspection Agency. RG-8 Regulatory Guidance: Contaminants in Feed (formerly RG-1, Chapter 7) [Internet]. 2017. Available from: https://inspection.canada.ca/animal-health/livestock-feeds/regulatory-guidance/rg-8/eng/1347383943203/1347384015909?chap=0 [Accessed: December 05, 2021]

[133] Gulsunoglu Z, Aravind S, Bai Y, Wang L, Kutcher HR, Tanaka T. Deoxynivalenol (DON) accumulation and nutrient recovery in black soldier fly larvae (*Hermetia illucens*) fed wheat infected with fusarium spp. Fermentation. 2019;**5**:83. DOI: 10.3390/fermentation503008

[134] Wiens DJ, Tse TJ, Reaney MJT. Using black soldier fly larvae to reclaim and concentrate nutrients

from Fusarium head blight infected wheat. Conference Presentation. Entomological Society of Canada/ Entomological Society of Ontario 2021 Joint Annual Meeting, Niagara Falls, Ontario, Canada

[135] Wu YV, Sexson KR, Lagoda AA. Protein-rich residue from wheat alcohol distillation: Fractionation and characterization. Cereal Chemistry. 1984;**61**:423-427

[136] Bhadra R, Muthukumarappan K, Rosentrater KA, Kannadhason S. Drying kinetics of distillers wet grains (DWG) under varying condensed distillers Solubles (CDS) and temperature levels. Cereal Chemistry. 2011;**88**(5):451-458

[137] Rosentrater KA, Ileleji K, Johnston DB. Manufacturing of fuel ethanol and distillers grains—Current and evolving processes. In: Liu KS, Rosentrater KA, editors. Distillers Grains—Production, Properties, and Utilization. 1st ed. Boca Raton: AOCS Publishing; 2012. pp. 73-102

[138] Ratanapariyanuch K, Shim YY, Emami S, Reaney MJT. Production of protein concentrate and 1,3-propanediol by wheat-based thin stillage fermentation. Journal of Agricultural and Food Chemistry. 2017;**65**:3858-3867

[139] Pedersen C, Jonsson H, Lindberg J, Roos S. Microbiological characterization of wet wheat distillers' grain, with focus on isolation of lactobacilli with potential as probiotics. Applied and Environmental Microbiology. 2004;**70**:1522-1527

[140] Scott E, Peter F, Sanders J. Biomass in the manufacture of industrial products—The use of proteins and amino acids. Applied Microbiology and Biotechnology. 2007;**75**:751-762. DOI: 10.1007/s00253-007-0932-x

[141] Singh V, Moreau RA, Hickes KB, Belyea RL, Staff CH. Removal of fiber from distillers dried grains with solubles (DDGS) to increase values. Transactions of ASAE. 2002;**45**:389-392

[142] Thacker P, Deep A, Beltranena E. Use of a post-production fractionation process improves the nutritional value of wheat distillers grains with solubles for young broiler chicks. Journal of Animal Science and Biotechnology. 2013;**4**(18):1-6. DOI: 10.1186/2049-1891-4-18

[143] Liu KS. Fractionation of distillers dried grains with solubles (DDGS) by sieving and winnowing. Bioresource Technology. 2009;**100**:6559-6569. DOI: 10.1016/j.biortech.2009.07.053

[144] Cantrell DF, Winsness DJ. Method of recovering oil from thin stillage. United States Patent # US 8,008,517 B2. 2011

[145] Haas M. Extraction and use of DDGS lipids for biodiesel production. In: Liu KS, Rosentrater KA, editors. Distillers Grains—Production, Properties, and Utilization. 1st ed. Boca Raton: AOCS Publishing; 2012. pp. 487-502